Metallurgy and Civilisation: Eurasia and Beyond

Metallurgy and Civilisation: Eurasia and Beyond

Proceedings of the 6th International Conference on the Beginnings of the Use of Metals and Alloys (BUMA VI)

Edited by Jianjun Mei and Thilo Rehren

Archetype Publications

in association with the University of Science and Technology Beijing and the Institute for Archaeo-Metallurgical Studies, London

First published 2009 by Archetype Publications Ltd. in association with the University of Science and Technology Beijing and the Institute for Archaeo-Metallurgical Studies, London.

Archetype Publications Ltd.
6 Fitzroy Square
London W1T 5HJ
www.archetype.co.uk

The views and practices expressed by individual authors are not necessarily those of the editors or the publisher.

ISBN: 978-1-904982-49-4

British Library Cataloguing in Publication Data
A catalogue record for this book is available from the British Library.

Cover illustrations

Main image (front and back): Furnace after one week of continuous operation. (Photo: B. Asmus; see contribution by B. Asmus, pp. 155–62.)
Front cover (right top to bottom):
The Gulcho River not far from Sary Tash and the important crossing leading to the east (Xinjiang), to the south (Murghab), to the west (Sogdiana) and to the north (Osh in Fergana). (Photo: J. Barnes.)
The iron ox (Tang dynasty) excavated at the Pujin port, Yongji county, Shanxi province, China. (Photo: Yongsheng Liu.)
Mould fragments for a *ding* vessel (H66:95) from the Lijia site. (Photo: W. Zhou; see contribution by W. Zhou *et al.*, pp. 62–72.)
Black-colour mottle on an ancient bronze sword found in south-western China (see contribution by Yao and Sun, pp. 85–9).
Back cover: Sample ZGM40 microstructure, etched, two metals (see contribution by W. Zhou *et al.*, pp. 62–72).

Printed on acid-free paper

Typeset by Caz Jones Design, Norwich
Printed and bound in Great Britain by Latimer Trend Ltd., Plymouth

Contents

Foreword by Weidong Luo, Chancellor, University of Science and Technology Beijing vii
Foreword by Robert Maddin, Chairman of the BUMA Standing Committee ix
Preface xi
Acknowledgements xiii
List of contributors xvii
Introduction xxi

Early metallurgy across Eurasia

Ancient metallurgy in the Eurasian steppes and China: problems of interactions 3
Evgenij Chernykh

Early metallurgy in China: some challenging issues in current studies 9
Jianjun Mei

Metal trade in Bronze Age Central Eurasia 17
Liangren Zhang

Documentary and archaeological evidence for an antique copper-nickel alloy (*baitong*) production in southern China and its exportation to India 26
François Widemann

Metal trade between Europe and Asia in classical antiquity 35
Alessandra Giumlia-Mair, Michel Jeandin and Ken'ichi Ota

The black bronzes of Asia 44
Paul Craddock, Maickel van Bellegem, Philip Fletcher, Richard Blurton and Susan La Niece

Bronze casting technologies in ancient China

Origins and evolution of the casting technology of Anyang bronze ritual vessels: an exploratory survey 55
Yu Liu

Three Western Zhou bronze foundry sites in the Zhouyuan area, Shaanxi province, China 62
Wenli Zhou, Jianli Chen, Xingshan Lei, Tianjin Xu, Jianrong Chong and Zhankui Wang

New research on lost-wax casting in ancient China 73
Weirong Zhou, Yawei Dong, Quanwen Wan and Changsui Wang

Incipient metallurgy in Yunnan: new data for old debates 79
Tzehuey Chiou-Peng

A study of the surface craft of weapons from the Ba-Shu region of ancient China 85
Zhihui Yao and Shuyun Sun

Production of signature artifacts for the nomad market in the state of Qin during the late Warring States period in China (4th–3rd century BCE) 90
Katheryn M. Linduff

Ancient iron and steel technologies in Asia

An early iron-using centre in the ancient Jin state region (8th–3rd century BC) 99
Rubin Han and Hongmei Duan

From western Asia to the Tianshan Mountains: on the early iron artefacts found in Xinjiang 107
Wu Guo

South Indian Iron Age iron and high carbon steel: with reference to Kadebakele and comparative insights from Mel-siruvalur 116
Sharada Srinivasan, Carla M. Sinopoli, Kathleen D. Morrison, Rangaiah Gopal and Srinivasa Ranganathan

Survival of traditional Indian ironworking 122
Vibha Tripathi and Prabhakar Upadhyay

Fine structures: mechanical properties and origin of iron of an ancient steel sword excavated from an old mound in Japan 129
Masahiro Kitada

Specialisation in iron- and steel-making in the early Middle East and Central Asia: myths, assumptions and a reassessment of early manuscript evidence 134
Brian Gilmour

Ancient metallurgical and manufacturing processes

The early history of lost-wax casting 147
Christopher J. Davey

A natural draught furnace for bronze casting 155
Bastian Asmus

The liquation process utilised in silver production from copper ore: the transfer to and development in Japan 163
Eiji Izawa

A technical study of silver samples from Xi'an, Shaanxi province, China, dating from the Warring States period to the Tang dynasty 170
Junchang Yang, Paul Jett, Lynn Brostoff and Michelle Taube

Scientific analysis of lead-silver smelting slag from two sites in China 177
Pengfei Xie and Thilo Rehren

Foreword

On 15–20 September 2006, the 6th International Conference on the Beginnings of the Use of Metals and Alloys (BUMA VI) was held with great success at the University of Science and Technology Beijing (USTB). Over 100 delegates from around the world attended the conference. Among them, more than 60 came from foreign countries such as Britain, USA, Russia, Germany, India, Sweden, France, Japan, Italy, South Korea, Denmark and Canada. At the opening of the conference, as the chancellor of the USTB, I expressed my warm welcome to all the delegates for their participation in the conference and for visiting our university. It was my great pleasure to see that BUMA had once again returned to its 'birthplace', since it was in this university that the first BUMA conference was held in 1981, with about 20 delegates.

I would once again like to thank the China Numismatic Museum, the Chinese Society for the History of Science and Technology and the Institute for Archaeo-Metallurgical Studies in London for providing financial support to the BUMA VI conference. Their generosity contributed greatly to the success of BUMA VI.

On the occasion of the publication of the BUMA VI proceedings, I wish to express my profound respect and gratitude to Professor Tsun Ko, one of the two founders of the BUMA conference series. Without his far-reaching vision and vigorous efforts, this university would not have carried out research into the history of metallurgy as a characteristic field of studies, and the Institute of Historical Metallurgy and Materials (IHMM) would not have developed into an internationally well-known research centre for the history of metallurgy.

Born in June 1917, and now a Senior Academician of the Chinese Academy of Sciences, Professor Ko is a world-renowned metallurgist, historian of science and educator, and a pioneering researcher of the history of China's metallurgy and materials science. In 1953 he returned to China from Britain with his wife and children. The following year he was assigned to a professorship at Beijing University of Iron and Steel Technology, as our university was formerly known, and has remained here ever since. His rigorous attitude and tireless efforts in nurturing talent have inspired the profound admiration and affection of his students.

Professor Ko began to lead research into the history of metallurgy in China in 1974, and his research achievements over the past 30 years have been outstanding. In 1994, he received major awards for his contributions from the national and local governments. Even in his 10th decade, he has continued to play an active and innovative role in the reform of engineering education in China, and in guiding major research projects of our university. His energy and dedication never fail to astonish and inspire us. With his enduring thirst for knowledge, he is the embodiment of the saying 'continuing to study while living to a ripe old age'.

In recent years, under the guidance of Professor Ko, adopting an approach involving the use of modern techniques to investigate ancient metallurgical remains, the IHMM has published a series of outstanding research results, which have received recognition around the world. In 2004, on the basis of the IHMM, the Research Centre for Science, Technology and Civilisation was established, further expanding the scope of research in the history of science and technology in our university. We hope that this Centre will make more and more contributions to revealing Chinese culture to the world. It is in this light that I am particularly pleased to see the publication of the BUMA VI proceedings, *Metallurgy and Civilisation*, and hope it will further advance academic exchanges and understandings among scholars across the whole world.

Professor Weidong Luo
Chancellor
University of Science and Technology Beijing

Foreword

The BUMA conferences

On a visit to Beijing in 1978, I was made aware of exciting studies on early metallurgy that were going on in China. But it was not until my return visit the following year that I finally made contact with Professor Ko. I say finally because over the previous 35 years, there were occasions when we almost met. In 1944, Professor Ko was in Kunming on his way to Birmingham, and I was in Kunming with the US Air Force. In 1953, Professor Ko left Birmingham just before I arrived there. During my year in Birmingham I learned of Professor Ko's work, so I knew of him before I actually met him in 1979. It was Professor Ko who made me aware of the truly exciting discoveries in early metallurgy made by him and his group of scientists and scholars. I learned, for example, of the vast scale of cast iron production that went on so early, the astounding copper mines at Tonglushan (Mount Verdigris), among so many other discoveries. It seemed to me that it was a great pity that most of the publications of these exciting finds were not known outside China. Certainly, some information was filtering out (with considerable delay) via the publication *Kaogu Xuebao*, but the large majority of studies were only published in provincial journals which seldom reached the worldwide scientific community except for those articles and journals carried out by travelers. Furthermore, I learned about the existence of the Ancient Metals Group formed by Professor Ko, perhaps the largest interdisciplinary gathering of scientists, engineers, archaeologists and historians anywhere in the world, carrying out studies on a large array of questions raised by historians and archaeologists. During our meeting, Professor Ko and I recognized the need for an international gathering so that scientists and scholars could share information and learn of each other. We were in total agreement that the group should include members from many disciplines, to interact with archaeologists to help answer the many questions raised by their excavation findings. This group should therefore consist of physical scientists, zoologists, botanists, biological scientists, geologists, conservators and scholars such as archaeologists, historians and art historians. We aimed at that time to hold the first conference in Beijing about two years later in 1981 or 1982.

When I returned to the States, I approached the National Science Foundation with the idea that it should fund a group from the US to attend such a conference. The Foundation was receptive to the idea and I subsequently submitted a proposal for financial support for travel and subsistence of up to 10 scientists/scholars to participate in the conference. I submitted a projected budget and a list of 10 colleagues from the US. The proposal was approved. Among the 10 was the late Cyril Stanley Smith. The conference, to be known later as BUMA I, was exciting to say the least. Quoting from a passage by Cyril Stanley Smith in the Preface to the proceedings of BUMA II:

> I was privileged to be at the one (BUMA I) in Beijing in 1981, and I remember vividly the excitement of learning what the collaboration between Chinese archaeologists and metallurgists had achieved, then little known in the West: the casting techniques used in making ceremonial bronze vessels of exactly known provenance, putting on sounder base much that had previously only been inferred; the construction of ancient furnaces; molds of both clay and metal used for mass production; the unsuspectedly vast scale of cast iron production and use; the copper mines at Tonglushan (Mount Verdigris); and perhaps the most striking, the great diversity of techniques used to generate beautiful detail in metallic and nonmetallic objects recovered from the fifth century B.C. tomb of the Marquis Yi of Zeng in the Hubei Provincial Museum.

Professor Smith captured the excitement felt by many of us at that first conference. BUMA I was very well attended, with many scientists and scholars from a variety of disciplines from throughout China. Although I personally wrote or telephoned an invitation to a number of colleagues from various European countries, few were able to make the trip. I recall that the late Professor Jobe from Britain, who happened to be in Beijing at the time, attended.

It was not until five years later in 1986 that the next BUMA conference (BUMA II) was held: this one was in Zhengzhou, Henan province, China, in mid-October. Sixty invitations were sent out to individual scholars/scientists, and, initially, all but two accepted, but disappointingly only a few actually attended. Henan has been called 'the cradle of Chinese civilization' and Zhengzhou was an excellent venue for such a gathering. During the conference we were able to visit many places of historic interest.

After a hiatus of six years, BUMA III convened in Sanmenxia, Henan province, in 1994, again hosted by Professor Ko. A number of Japanese scientists and scholars were also present to observe the conference since the venue for BUMA IV was announced to be in Japan. The Japanese scholars, particularly the late Professor Yunoshin Imai, had earlier expressed an interest in hosting a BUMA conference and at Sanmenxia it was decided to hold BUMA IV in Japan. We also decided that we should attempt to have regular meetings every four years and that the meetings should be held in Pacific Rim countries. The chief reasoning behind these thoughts was that although there were many conferences on early technology scheduled each year in other parts of the world, there were few in this region. Setting the frequency at four years seemed at the time to be about the optimum in order to hear about new studies instead of recycling old studies. Of course, we realized that these guidelines were not written indelibly, and, as research studies have accelerated in this region, there should be no reason why a four-year frequency had to be the rule.

Four years after BUMA III, Japanese scholars and scientists sponsored a 'forum' in Matsue, Shimane province, organized by the late Professor Yunoshin Imai as a prelude to their BUMA IV conference, which was held also in Matsue in May of 1998. The program contained sessions on the history of iron and steel technology in the Far East, the history of iron and steel in the West, and sessions on archaeology, copper and bronze, silver and gold. The conference was well attended by scholars from the Far East but rather sparsely by our colleagues from the West. At this conference, we learned that our South Korean colleagues had expressed an interest in hosting the BUMA V conference scheduled for 2002. As had been done in the past, a number of scientists/scholars from South Korea, the next host country, had been invited to attend BUMA IV. Hence, many were present in Matsue.

BUMA V was convened in Gyeongju, South Korea, in April 2002, organized by a committee chaired by Professor Hyung-Yong Ra. Gyeongju was the capital of the Shilla Kingdom that flourished from 57 BCE to 935 CE; it was for all those years the center of the golden age of Korean culture. In Gyeongju, royal tombs, pagodas and statues dot the city and surrounding hills. Here, as with the other conferences, there were sessions on metals and civilizations, iron and copper metallurgy, as well as on the noble metals. As at Matsue four years earlier, attendance from colleagues in the Far East was good while attendance from the West was sparse.

We come now to BUMA VI, organized by Professor Jianjun Mei. He has put together an excellent program with perhaps a better international gathering than any of the previous BUMA conferences. The oral and poster presentations have been exciting and stimulating. He and the group of scientists he assembled have produced an excellent program. On behalf of all of us, I thank Professor Mei and his associates for their effort and their results.

Finally, some thoughts about future conferences: my overall impressions from past conferences are that future BUMA conferences must make a concerted effort to induce scholars and scientists from a greater number of disciplines and from a greater number of countries, particularly from the West, to attend. I view our mission as that of representatives of a large array of people interested in addressing questions asked by the archaeologist. In my view, the archaeologist is like the conductor of an orchestra and needs the input of each of his orchestra members in order to present a clear rendition of the entire score. For example, when the archaeologist excavates seeds, he calls in a botanist, if metals, a metallurgist, and if bones then a zoologist, etc. In this way the conductor/archaeologist is able to create a harmonious symphony. If he attempts to answer those questions without the help of the experts, he risks producing dissonance, a cacophonous jumble of sounds.

Another serious impression from the past is the absolute necessity of the publication of the proceedings of the conference. They cannot be left to the memory of those who may have attended. The published proceedings constitute a history of what was the state of our understanding of particular subjects at the time. This is a *must*; the proceedings must be published otherwise the conferences fade away with time.

Lastly, our early thought to hold these conferences in a Pacific Rim country appears to be justifiably flexible: BUMA VII will be held in India in 2009.

This, then, is a brief history of the conferences we know as the BUMA conferences. I have greatly valued my participation in these conferences and hope also to be present for BUMA VII.

Thank you all for your many contributions.

Robert Maddin
University Professor, Emeritus, University of Pennsylvania
Chairman of the BUMA Standing Committee

Preface

A brief reflection on BUMA VI

In the preceding pages, Professor Maddin has summarised some of the developments that generated and shaped the series of conferences organised under the title of the Beginnings of the Use of Metals and Alloys (BUMA) from its inception in the late 1970s to its most recent incarnation in Beijing in 2006. As a relative newcomer to Asian archaeometallurgy, this was my first BUMA participation, and I can only comment on it in comparison to other conferences I have attended.

A quick ad hoc analysis of the presentations shows the impressive range of topics covered; there were papers on gold, silver, copper, arsenical copper, bronze, tin, lead, iron, zinc, aluminium, and on ceramic, moulds and furnaces. Not only were metal artefacts covered, but also various production processes. However, there were relatively fewer papers on slags, crucibles, and other process waste materials.

In contrast to this broad range of materials, there is greater focus in the geographical range. Forty-two papers (about 85%) concerned Asia in the relatively narrow sense that is the region east of the Urals and Mesopotamia, but not including Arabia, the Middle East and western Asia. That is remarkable in that these latter areas play a major and active role in archaeometallurgical research; indeed, the term *archaeo-metallurgy* was coined by Beno Rothenberg while working in Timna in southern Israel. A further seven papers discussed European material, and two papers were based on African material. Historically, BUMA was conceived as a circum-Pacific event; in view of this, it is remarkable that no paper discussed the rich archaeometallurgy of the Americas. But on second thoughts, this is understandable and justified. The impetus for BUMA was a meeting between Professor Ko and Professor Maddin, triggered by the latter's desire to learn about the impressive work done in China. Thus, the organisational drive to establish BUMA was circum-Pacific, but the academic nucleus was clearly formed by Chinese archaeometallurgy.

The rich contribution by archaeological papers was remarkable, and in refreshing contrast to other, sometimes more materials science-heavy presentations in similar meetings elsewhere. The close and fruitful interaction between archaeologists and scientists, and the primacy of archaeological questions, are part of the important legacy of the founders of BUMA, and we should treasure and support it for the future. In my view, for archaeometallurgy to be relevant beyond being an interesting pastime of materials scientists, it has to be an integral part of archaeology, embedded in wider archaeological enquiries and agendas. One material bias though does stand out, and this has the potential to become a hallmark of BUMA: the emphasis on bronze casting and related topics. The crucial role of bronze in early Chinese society and the truly impressive level of craftsmanship and metallurgical skill demonstrated in so many facets of bronze work are more than enough to sustain many more international gatherings of scholars. Similarly, a significant number of papers concerned early iron, particularly cast iron; and again, there is a sound reason and justification for this in the important role which this diverse and fascinating group of alloys plays in early Asian metallurgy.

In summary, it appears to me – and I stress that this is my personal opinion rather than an agreed policy of the BUMA Advisory Board – that the strength of the 'brand' BUMA lies in its clearly defined focus on Asian metallurgy, the region east of the Urals and Mesopotamia; in a focus on smelting and manufacturing techniques, particularly but by no means exclusively on bronze and iron; and in the archaeological relevance of its papers, with emphasis on cultural interactions and evolutions over time and space. These foci have emerged by natural selection, by the individual and independent decisions of all participants (and those who decided not to offer their work for presentation), rather than by a top-down policy. This strong trend alone should ensure that this is a solid and sustainable profile, making BUMA a well-defined and recognisable set of conferences. I am sure that no-one would want to exclude participants with other interests and topics within archaeometallurgy from attending and presenting their work. Comparative studies and case studies from other regions can only increase the richness of the overall conference experience.

It is satisfying that the proceedings of BUMA VI will be published in time for BUMA VII. I look forward to seeing the bouquet of academic offerings in Bangalore, and how they continue to develop BUMA into the future.

Thilo Rehren
Professor of Archaeological Materials and Technologies
University College London, Institute of Archaeology

Acknowledgements

It goes without saying that the publication of a book such as *Metallurgy and Civilisation* would be impossible without the help, support and cooperation from many individuals and institutions. First of all, we would like to thank all the authors for their valuable contributions to this volume. Their patient cooperation and support were a great driving force behind the production of this book. We are most grateful to Professor Tsun Ko, Professor Robert Maddin, Dr Alessandra Giumlia-Mair, Dr Paul Craddock, Professor Katheryn M. Linduff, Dr Donald Wagner, Dr Tzehuey Chiou-Peng, Dr Bryan Hanks, Dr Liangren Zhang, Dr Nigel Meeks, Dr Nick Boroffka, Professor Vibha Tripathi, Dr Sharada Srinivasan, Professor Eiji Izawa, Dr Robert Murowchick, and Susan Bennett, for their encouragement, support and help in various ways during a long process of discussion, review, revision and editing over the past two and a half years.

The preparation of this book has received much appreciated support and assistance from many colleagues at the Institute of Historical Metallurgy and Materials, University of Science and Technology Beijing. We would like to thank Professor Rubin Han, Professor Shuyun Sun, Professor Yanxiang Li, Professor Xiaocen Li, Associate Professor Xiuhui Li, Associate Professor Wei Qian and Associate Professor Meifang Zhang, not just for their support of this book, but also for their immense contribution to the success of hosting BUMA VI (6th International Conference on the Beginnings of the Use of Metals and Alloys) in Beijing, 15–20 September 2006.

Raul Carstocea and John Moffett deserve our special heartfelt thanks for their substantial contribution to the editing of all the manuscripts. We would also like to thank James Black, our publisher at Archetype, for his patience and cooperation.

Dr Mei would like to express his gratitude to the Needham Research Institute for providing him with office space and other facilities when he worked in Cambridge in the first half of 2009. He is also grateful to the State Scholarship Fund for awarding him a scholarship that enabled him to work on the final submission of all manuscripts in Britain.

The publication of this book has received a grant in association with the '211 Project for Supporting Higher Education in China' from the University of Science and Technology Beijing. Further financial support has been provided by the Institute for Archaeo-Metallurgical Studies, London, through the generosity of its benefactors. These substantial contributions are gratefully acknowledged.

Jianjun Mei and Thilo Rehren
May 2009

鸣谢

本书为“211工程”三期建设项目资助出版。

List of contributors

Bastian ASMUS
Institute of Archaeology, University College London, 31–34 Gordon Square, London WC1H 0PY, UK
(b.asmus@ucl.ac.uk)

Maickel van BELLEGEM
Department of Conservation, Documentation and Scientific Research, The British Museum, London WC1B 3DG, UK
(mbellegem@thebritishmuseum.ac.uk)

Richard BLURTON
Department of Asia, The British Museum, London WC1B 3DG, UK
(rblurton@thebritishmuseum.ac.uk)

Lynn BROSTOFF
Research Chemist, Preservation Research and Testing Division, Library of Congress, 101 Independence Avenue, SE, Washington, DC 20540-4560, USA
(lbrostoff@loc.gov)

Jianli CHEN
School of Archaeology and Museology, Peking University, Beijing 100871, China
(jianli_chen@pku.edu.cn)

Evgenij CHERNYKH
Institute of Archaeology of RAS, Ul. Dm. Ulianova 19, 117036 Moscow, Russia
(archmetal@rambler.ru)

Tzehuey CHIOU-PENG
230 International Building MC-483, 910 South 5th Street Champaign, Illinois 61820, USA
(tzehuey@illinois.edu)

Jianrong CHONG
Provincial Institute of Archaeology, Xi'an, Shaanxi 710054, China
(chjrong2003@yanhoo.com.cn).

Paul CRADDOCK
Department of Conservation, Documentation and Scientific Research, The British Museum, London WC1B 3DG, UK
(pcraddock@thebritishmuseum.ac.uk)

Christopher J. DAVEY
Australian Institute of Archaeology and Archaeology Program, La Trobe University, Victoria, 3086, Australia
(director@aiarch.org.au)

Yawei DONG
Ezhou Museum, Hubei 436000, China
(bwgdyw@163.com)

Hongmei DUAN
Intellectual Property Publishing House, No.1 Madiannancun, Madianqiao Haidian District, Beijing 100088, China
(duanhongmei@cnipr.com)

Philip FLETCHER
Department of Conservation, Documentation and Scientific Research, The British Museum, London WC1B 3DG, UK
(pfletcher@thebritishmuseum.ac.uk)

Brian GILMOUR
Research Laboratory for Archaeology and the History of Art, University of Oxford, Dyson Perrins Building, South Parks Road, Oxford OX1 3QY, UK
(brian.gilmour@rlaha.ox.ac.uk)

Alessandra GIUMLIA-MAIR
AGM Archeoanalisi, via E.Toti 8, I - 39012 Merano, Italy
(giumlia@yahoo.it)

Rangaiah GOPAL
Department of Archaeology, Museums and Heritage, Palace Complex, Mysore 570 001, India

Wu GUO
Institute of Archaeology, Chinese Academy of Social Sciences, 27 Wangfujing Street, Dongcheng District, Beijing 100710, China
(gwbluesea@gmail.com)

Rubin HAN
Institute of Historical Metallurgy and Materials, University of Science and Technology Beijing, Beijing 100083, China
(hanrubin@metall.ustb.edu.cn)

Eiji IZAWA
Professor Emeritus, Kyushu University (Economic Geology and Mining Engineering), 2-15-4 Sakurayamate, Shingu-machi, Fukuoka-ken 811-0113, Japan
(izawa@mine.kyushu-u.ac.jp)

Michel JEANDIN
Ecole des Mines de Paris, F- 91003 Evry, France
(michel.jeandin@ensmp.fr)

Paul JETT
Head, Department of Conservation and Scientific Research, Freer Gallery of Art/Arthur M. Sackler Gallery, Smithsonian Institution, PO Box 37012 (MRC 707) Washington, DC 20013-7012, USA
(jettpa@si.edu)

Masahiro KITADA
National Institute of Materials Science, Tsukuba Ibaraki, 305-0047, Japan
(kitada.masahiro@palette.plala.or.jp)

Susan LA NIECE
Department of Conservation, Documentation and Scientific Research, The British Museum, London WC1B 3DG, UK
(slaniece@thebritishmuseum.ac.uk)

Xingshan LEI
School of Archaeology and Museology, Peking University, Beijing 100871, China
(leixingshan@263.net)

Katheryn M. LINDUFF
128 Frick Fine Arts Building, Department of Art History, University of Pittsburgh, Pittsburgh, PA 15260, USA
(Linduff@pitt.edu)

Yu LIU
Institute of Archaeology, Chinese Academy of Social Sciences, Beijing 100710, China
(liuyu88@126.com)

Jianjun MEI
Director, Institute of Historical Metallurgy and Materials, University of Science and Technology Beijing, Beijing 100083, China
(meijianjun12@yahoo.com.cn)

Kathleen D. MORRISON
Director, Center for International Studies, University of Chicago, 1126 E. 59th St., Chicago, IL 60637, USA
(morrison@uchicago.edu)

Ken'ichi OTA
Q inverse (LLP), Arubo 102B, 2-14 Mino 7, Chome Mino Osaka 562-0001 Japan
(otakenichi@aol.com)

Srinivasa RANGANATHAN
National Institute of Advanced Studies, Bangalore, India
(rangu2001@yahoo.com)

Thilo REHREN
UCL Institute of Archaeology, University College London, 31–34 Gordon Square, London WC1H 0PY, UK
(th.rehren@ucl.ac.uk)

Carla M. SINOPOLI
Museum of Anthropology, University of Michigan, 1109 Geddes Avenue, Ann Arbor, MI 48109-1079, USA
(sinopoli@umich.edu)

Sharada SRINIVASAN
National Institute of Advanced Studies, Bangalore, India
(sharadasrinivasa@yahoo.com)

Shuyun SUN
Institute of Historical Metallurgy and Materials, University of Science and Technology Beijing, 30 Xueyuanlu, Beijing 100083, China
(sunshuyun@metall.ustb.edu.cn)

Michelle TAUBE
National Museum of Denmark, Department of Conservation, I. C. Modewegsvej, Brede, DK-2800 Kgs. Lyngby, Denmark
(michelle.taube@natmus.dk)

Vibha TRIPATHI
Department of AIHC and Archaeology, Banaras Hindu University, Varanasi 221005, India
(vibhahist@gmail.com)

Prabhakar UPADHYAY
Department of AIHC and Archaeology, Banaras Hindu University, Varanasi 221005, India
(prabhakarupadhyay.bhu@gmail.com)

Quanwen WAN
Hubei Provincial Museum, Wuhan 430077, China
(wqwhbsb@yahoo.com.cn)

Changsui WANG
Graduate School of Chinese Academy of Sciences, Beijing 100049, China
(cswang@gucas.ac.cn)

Zhankui WANG
Shaanxi Provincial Institute of Archaeology, Xi'an, Shaanxi 710054, China
(zhankuiwang@yahoo.com.cn).

François WIDEMANN
Centre de Recherche et de Restauration des Musées de France (C2RMF) UMR 171 du CNRS Palais du Louvre, porte des lions 14, quai François Mitterrand, 75001 Paris, France
(francois.widemann@culture.gouv.fr)

Pengfei XIE
College of Urban and Environmental Sciences, Peking University Beijing 100871, China
Institute of Historical Metallurgy and Materials, University of Science and Technology Beijing, Beijing 100083, China
(xiepengfei@yahoo.com.cn)

Tianjin XU
School of Archaeology and Museology, Peking University, Beijing 100871, China
(xtj@pku.edu.cn)

Junchang YANG
Head of Scientific Department of Conservation and Restoration, Shaanxi Archaeological Institute, 31 Leyou Road, Xi'an 710054, China
(junchangy@yahoo.com)

Zhihui YAO
Faculty of History, Zhengzhou University, No.1, Kexue Road, Zhengzhou, Henan Province, 450001 China
(yzhmz@zzu.edu.cn)

Liangren ZHANG
27 Wangfujing Dajie, Institute of Archaeology, Chinese Academy of Social Sciences, Beijing 100710, China
(zhlr31@hotmail.com)

Weirong ZHOU
China Numismatic Museum, 17 Xijiaominxiang St., Beijing 100031, China
(chinumis@yahoo.com.cn)

Wenli ZHOU
School of Archaeology and Museology, Peking University, Beijing 100871, China
UCL Institute of Archaeology, University College London, 31–34 Gordon Square, London WC1H 0PY, UK
(Juliazhouwl@gmail.com)

Introduction

The 6th International Conference on the Beginnings of the Use of Metals and Alloys (BUMA VI), held 15–20 September 2006 at the Institute for Historical Metallurgy and Materials, University of Science and Technology Beijing, received over 70 paper abstracts covering a wide range of topics. About 60 delegates gave oral presentations, while a number of papers were presented as posters. These presentations were organised into eight sessions as follows: *Metallurgy and interaction across Eurasia; Ancient iron and steel technology; Reconstruction of metallurgical processes; Bronze casting; Metallography and metalworking crafts; Ancient gold, silver and lead; Metallurgy and culture; Metallurgy and civilisation.* On 16 September 2006, at a meeting of the BUMA VI International Advisory Committee, a decision was made that the Local Organising Committee should begin planning the publication of the BUMA VI proceedings at the end of the conference. It was also agreed that a coherent theme for the proceedings would be essential and necessary. A call for papers to be published within the BUMA VI proceedings was then announced at the closing session. *Metallurgy and Civilisation: Eurasia and Beyond* was suggested as the theme in order to encourage the submission of papers that would be related to Asia in one way or another.

After a long process of peer review and revision, 23 manuscripts were finally accepted for publication and are presented in this book. These papers are organised into four equally strong sections, namely *Early metallurgy across Eurasia; Bronze casting technologies in ancient China; Ancient iron and steel technologies in Asia;* and *Ancient metallurgical and manufacturing processes.* These sections indicate the wide range of topics covered in this book, as well as its emphasis on Eurasia and Asia.

The first section, *Early metallurgy across Eurasia*, contains six papers focusing on cultural interactions or connections across Eurasia at different moments in time, stretching from the Bronze Age to the present. In his paper on 'Ancient metallurgy in the Eurasian steppes and China: problems of interactions', Evgenij Chernykh offers an insightful overview of early interactions between the Eurasian steppes and China. He reminds us that while the north-west influence of the pastoral nomads on Chinese metal production has been noted widely, it is crucial to also recognise the independent and distinctive characteristics of ancient Chinese metallurgy. His view is echoed by Jianjun Mei's review of recent studies of early metallurgy in China, which indicates that although new research work carried out in north-west China has revealed an increasing body of evidence for cultural connections between this region and the Eurasian steppes, it still remains unclear whether the beginning and early development of bronze metallurgy in the Central Plains of China received any stimulation from north-west China, or even from the Eurasian steppes. Based on morphological and compositional analyses, Liangren Zhang suggests that communities of the Seĭma-Turbino Phenomenon in the Eurasian steppes had intimate trade connections with those of the Abashevo and Sintashta cultures. This hypothesis points to a wider cultural context for the emergence of the Seĭma-Turbino Phenomenon. The common interest of the above three papers in the 'large picture' of early Eurasian metallurgy represents a new research trend that has emerged with the expanding archaeological discoveries in the Eurasian steppes since the 1990s.

The other three papers in the first section focus on more specific cultural and metallurgical connections between Europe and Asia, and within Asia itself. François Widemann's paper on *baitong*, or *paktong*, an ancient copper-nickel alloy, presents an interesting case that seems to confirm the existence of early cultural connections between south-west China and India, though many questions remain to be answered. Mainly based on the examination of ancient texts, Alessandra Giumlia-Mair *et al.* show that significant trade contacts via land or sea between Europe and Asia can be traced back to ancient times, and continued throughout the medieval period. They suggest that systematic examination of the textual evidence for ancient trade in metals could reveal some details that 'are fundamental for the philological reconstruction of the history of ancient metallurgy and could never be obtained from metallurgical analyses only'. 'The black bronzes of Asia' by Paul Craddock *et al.* concentrates on diverse metalworking traditions of black-patinated copper alloys in different countries such as Japan, China, Korea, Thailand and Myanmar, as well as their relationship to the European-inspired and distributed *sawasa* wares. The issue concerning the origins of black bronzes in the Orient is also addressed.

The ongoing progress in the studies of bronze casting technologies in ancient China is well represented by the six papers in the second section, which are mostly based on recent archaeological finds in China and contain considerable primary data and new ideas. Based on her first-hand study of the ritual vessels excavated at Yinxu in Anyang, Henan province, Yu Liu explores the development of bronze casting technology during the Shang dynasty, focusing on such issues as the section-mould division, separate casting, casting-on and spacers. One of the most significant archaeological discoveries for the Western Zhou period in recent years is the unearthing of three foundry sites in Zhouyuan, Shaanxi province, the most important region for the origin and development of Zhou culture. In the paper by Wenli Zhou *et al.*, scientific examinations of metal objects, furnace fragments and slags from these three sites have been reported for the first time in English, revealing important new evidence for reconstructing the bronze casting technology in the Zhouyuan region.

The question of when lost-wax casting technology began to be used in China has become a hotly debated issue among a number of scholars in recent years. Previously, it was widely held that lost-wax casting technology came into use in China no later than the late Spring and Autumn period (*c.* 5th century BC), as evidenced by the bronze *jin* vessel excavated at Xiasi in Xichuan, Henan province, as well as by the bronze *zun* and *pan* vessels unearthed from the tomb of Marquis Yi of State Zeng (曾侯乙) at Leigudun in Suixian, Hubei province. This view now faces a serious challenge from Weirong Zhou *et al.*, who, in their paper, argue that all those bronze vessels previously believed to be cast by using lost-wax techniques were actually made by using the clay piece-mould method. Furthermore, they even maintain that lost-wax casting technology was not in use at all in Bronze Age China. Such a radical claim has already drawn some serious counter-arguments, indicating that the debate will continue for many years to come.

The following two papers concern the development of ancient bronze technologies in south-west China. Tzehuey Chiou-Peng's paper offers an updated discussion on the beginning of metallurgy in Yunnan on the basis of the most recent archaeological discoveries at the Haimenkou site. Having examined the Haimenkou bronzes from both typological and metallurgical perspectives, it concludes that 'perceptible signs of incipient copper-based metallurgical activities at 2nd-millennium sites in western and northwestern Yunnan point to steppe-related cultures as the source for inspiring metallurgical activities at the Haimenkou site, which commenced during the latter part of the 2nd millennium BCE'. The paper by Zhihui Yao and Shuyun Sun focuses on the formation of a particular mottled pattern, also referred to as 'pattern-etched' or *hu-ban-wen* (虎斑纹), on the surface of some bronze weapons excavated in the Ba-Shu region (now Sichuan province) and dated to the 5th–2nd century BC. Through scientific examination of three samples, they found that the main compositional elements of the mottled layer are SnO_2 and $Cu_{41}Sn_{11}$. In their opinion, hot-tinning followed by annealing should have been a suitable method to create the mottled pattern.

In the final paper of the second section, Katheryn M. Linduff highlights a very important archaeological discovery of 25 ceramic models for making bronze-casting moulds in a Qin dynasty tomb found in the northern suburbs of Xi'an in 1991. In her opinion, the fact that these models were clearly used for casting 'steppe-style' bronze plaques not only confirms that the Chinese were producing objects for their pastoral neighbours, but also suggests that the metal industry formed a significant feature of the Qin state-level, as well as the local, economy and of its 'foreign policy'.

The theme of the third section is *Ancient iron and steel technologies in Asia*. Six papers are included in this section, covering a variety of regions and topics. There are two papers on early iron objects found in China. The paper by Rubin Han and Hongmei Duan reports their research results on early iron objects excavated in southern Shanxi, pointing out that the earliest cast iron objects found so far in the world are two iron fragments recovered at the Tianma site of the Qucun area in Shanxi province. It concludes that the ancient Jin cultural region of the Shanxi province played an important role in the development of cast iron- and steel-making in ancient China. Wu Guo's paper focuses on early iron objects found in Xinjiang, as well as their cultural background. After carefully reviewing the archaeological evidence available so far, it comes to the conclusion that iron artefacts appeared in Xinjiang no later than the 9th–8th century BC, and they seemed to be introduced into Xinjiang from north-west Iran. This is consistent with the fact that there are some finds in Xinjiang pointing to the existence of cultural connections with north-west Iran during the early 1st millennium BC.

The following two papers concentrate on Indian materials, one on ancient high carbon iron alloys, and the other on traditional iron technology. Sharada Srinivasan *et al.* report their research results on iron and steel artefacts recently uncovered at the site of Kadebakele, Karnataka in southern India, which has been radiocarbon dated between 800 and 400 BC. They have found evidence for the use of high carbon steels, which leads to comparative studies dealing with surface finds of crucibles related to high carbon steel production at Mel-siruvalur. The paper by Vibha Tripathi and Prabhakar Upadhyay is representative of a different perspective, in which the value of traditional Indian iron technology is emphasised. It systematically introduces the ironworking of the Agarias, an ethnic community known to consist of iron smelters in the Sonbhadra-Sidhi region in the state of Madhya Pradesh. It is suggested that the survival of the traditional iron technology in this community is a heritage of the past and is of significance for understanding the ancient metallurgical processes with greater clarity.

Japanese swords are famous for the high quality of their material. Masahiro Kitada's paper presents a detailed analysis of an old sword excavated in 1885 near Tokyo, from a tomb of the Yayoi period (AD 200–600), which offers important information concerning its microstructures, non-metallic inclusions and mechanical properties. Such information is useful for understanding the original raw materials and fabrication techniques used for making ancient Japanese swords.

The development of iron and steel technology in the 1st millennium AD across the Old World is an issue of long-standing interest for many scholars. In his thorough paper, Brian Gilmour provides a detailed review of the current understanding of the issue, pointing out that the early development of steel-making is still poorly understood both in Europe and China. Against the widely held assumption that steel initially was produced primarily by a carburisation or cementation process (getting bloomery iron to absorb carbon), he argues that there is a small but growing body of evidence – archaeological and textual – indicating that from an early period steel was made as part of the bloomery process. Focusing on written sources from about the 3rd century AD, he also examines the manufacture and use of crucible steel in Central Asia and the Middle East during the 1st millennium AD, and finds that there is hardly any evidence to support the most persistent myth concerning 'Damascus steel', that posits Damascus as a great steel- and sword-making centre in the medieval and pre-medieval periods, stretching back into the 1st millennium AD.

The last section consists of five papers with a focus on *Ancient metallurgical and manufacturing processes*. The first paper by Christopher J. Davey presents an overview of the early history of the lost-wax process in many parts of the Old World, including China. It points out that complexity of shape cannot be taken as conclusive evidence for the use of the lost-wax process. This idea echoes in a way the view offered by Weirong Zhou *et al.* in a previously introduced paper. Having discussed various types of evidence associated with lost-wax casting, such as a mould from the Tell edh-Dhiba'i collection, the Akkadian copper head and the Nahal Mishmar hoard, the paper concludes that lost-wax casting began in the Chalcolithic period, before 4000 BC, and the attestation of the technique at Mehrgarh points to an origin possibly on the Iranian plateau.

Bastian Asmus' paper deals with the design and use of natural draught furnaces for non-ferrous casting purposes. Employing an experimental approach, he explores problems related to crucible handling, fuel charge and gas flow in the suggested furnace design. This research demonstrates the possibility to operate such a furnace without artificial oxygen induction, and thus offers an alternative way for interpreting features related to metal casting traces at archaeological sites.

The last three papers are all related to ancient silver metallurgy in Japan and China. Eiji Izawa discusses the beginning of the liquation process for extracting silver from complex copper-silver ores in Japan. In his opinion, the liquation process was transferred from Europe to Japan by the end of the 16th century AD and was applied successfully for silver extraction from argentiferous copper, originating from copper-bearing silver ores. During the 17th and 18th centuries AD, the liquation process became the core technology not only for silver production, but also for the production of high quality copper.

So far, very little scientific research has been carried out on ancient gold and silver objects in China. It is in this light that the paper by Junchang Yang *et al.* is very welcome. The paper reports the results of the examination of 22 samples taken from ancient silver objects, mostly dated to the Tang dynasty, but with one dating to the Warring States period and seven to the Western Han dynasty. The examination shows that almost half the samples had silver contents of 97% or more. Evidence for the use of mercury gilding has also been found. This research addresses one of the gaps in the study of ancient silver technology in China.

Research on the ancient silver-lead smelting process is also rather rare. In the last paper in this section, Pengfei Xie and Thilo Rehren present their results on two sets of lead-silver slag samples from northern and south-west China. Such new and original data, while barely scratching the surface of this topic, are an important first step and useful in contributing to our understanding of ancient and historic smelting practices for lead and silver.

In summary, the 23 papers published in this book are, to some extent, representative of the wide range of presentations to the BUMA VI conference, with a clear emphasis on cultural and technological connections or interaction across Eurasia, as demonstrated by the papers on the origins of bronze metallurgy in China, the Seĭma-Turbino cultural phenomenon, the early use of iron, steppe connections, lost-wax casting, *paktong*, 'black bronze', crucible steel or 'Damascus steel', liquation for silver extraction, and so on. It is important to note that many papers are based on the most recent archaeological discoveries, thus offering valuable information about some important aspects of Asian archaeometallurgical studies. A number of issues raised in this book, such as the beginnings of bronze metallurgy in China, the early history of lost-wax casting in Asia, the development of early steel-making technology in Europe and Asia, and the role of the steppe influence in metalworking in Eastern Zhou China, will continue to attract substantial research interest. We are delighted to see that among the 23 papers published here, about one-third are contributed by young scholars, evidence that the field of archaeometallurgical studies is growing and will become even stronger in the years to come.

Collectively, the papers presented here provide a good overview of the breadth and depth of current archaeometallurgical research relevant to Asia and beyond. The important role that casting technologies play in wide parts of Asia is immediately evident, as is the importance of studying technical aspects not in isolation, but as parts of complex and multidimensional cultural developments. These papers show how long-distance cultural connections and technological inspirations remained active over many millennia. Importantly, this exchange typically had multiple directions rather than being a one-way system. The movement of actual material – often through many hands and over vast distances – could provide stimuli for developments as much as the movement of actual artisans could help cross-fertilise practices within a widely non-literate sector of society. The richness of past and present interactions, the ever-expanding archaeological knowledge, and an increasing cooperation across disciplinary, geographical and political boundaries visible in this volume are impressive.

Also impressive, however, is the potential for new research still ahead of us. Despite significant effort over the last century or so, we are only just beginning to understand and appreciate the range of smelting technologies employed to produce the key metals that maintained the fundamental economic and cultural developments of the majority of humankind for the last five millennia. What is the range of ores exploited to produce copper east of the Urals? Is there a more-or-less linear sequence from using native copper to smelting rich secondary ores before the large but typically poor primary sulphides were exploited, as we sometimes see it in western Asia and Europe? What role did arsenical copper play in Asia, and complex ores containing metals such as zinc, silver and antimony, in addition to copper? How did furnace design and the overall organisation of mining, beneficiation, fuel provision and distribution of the raw copper respond to changes in the ore base, socio-political developments, and changing volumes of production? Similar questions have to be asked for other key metals such as tin, gold and silver, and of course iron, but the answers are likely to be specific and different for each of these metals, and with possibly very different answers across time and space. The vastly differing cultural and ecological parameters of Siberia, Central Asia, the Great Plains of China, South and South-East Asia, or Japan, to name but a few key areas, will have resulted in quite different solutions to primary metal production and will have led to specific manufacturing traditions. Generations of archaeologists and art historians have accumulated the knowledge that allows us to allocate most artefacts to specific cultural units based on stylistic features; but how about the underlying technological diversity? Much work is still ahead of us.

Interactions between the technologies used for producing different metals need to be explored, most obviously related to the two large-scale cast alloys, bronze and cast iron. Superficially, they both require similar installations to handle the liquid metal. However, the technical requirements for the alloy production just before the casting step are quite different: highly reducing single-ore smelting for cast iron, alloying well-defined amounts of pre-smelted copper and tin (and lead) for bronze. Does this reflect in different furnace designs? Did cast iron smelting furnaces develop from earlier bronze casting furnaces, and how did the blast furnace necessary to smelt cast iron influence later copper smelting practice? Similar relationships and interactions exist among such metallurgical couples as lead-silver, copper-tin, copper-zinc, gold-silver, and will be fruitful to research.

Finally, exploring the interactions between the main high-temperature industries – metallurgy, ceramic production, and glass- and pigment-making – will reap rich new rewards. What is the relationship between the development of highly fired ceramic wares – notably proto-porcelain and porcelain – and the emergence of refractory technical ceramics suitable for blast furnaces? Is there more to the use of lead-barium minerals for Chinese glass-making, and of copper-barium minerals for making Han blue and purple? After all, copper, lead and barium occur frequently together in nature in hydrothermal veins, the most important type of ore deposit exploited in ancient metallurgy. Did the glassy and often colourful slags of early metal furnaces provide an inspiration for this genuinely Chinese glass technology?

This is not the place to set out a detailed research plan for Asian archaeometallurgy; however, the potential is so huge that it is almost overwhelming. How much – or rather how little – do we know about early iron production in Central Asia, in South-East Asia, in Siberia or north-east Europe? We should not be scared by the sheer volume of what we don't know, but build on what we have. After all, a journey of a thousand miles starts under one's feet.

Jianjun Mei and Thilo Rehren
May 2009

Early metallurgy across Eurasia

Ancient metallurgy in the Eurasian steppes and China: problems of interactions

Evgenij Chernykh

ABSTRACT Since the 2nd millennium BCE the extensive Steppe Belt that was occupied by the Eurasian pastoral nomads quite often played either an exclusive or primary role as the sole mediator between the Western and Eastern worlds. Undoubtedly the first of these complex interactions was reflected in the metals used. In many respects it seems impossible to deny the influence of the pastoral nomads on Chinese metal production that led to a whole series of imitations of weapons. Basically, however, ancient Chinese metallurgy appears as an independent, rather distinctive, and particularly dazzling phenomenon.

The worldwide view on ancient Eurasian metallurgy has led to the conclusion that it is possible to distinguish at least two general models of the formation and development of metallurgical production: Western and Eastern. The vast majority of the leading metallurgical and metalworking centres connected with the steppe pastoral nomadic cultures have been included in the broad circle of the Western technologies. Chinese archaeometallurgy, on the other hand, embodied the Eastern type. The beginnings of the Western model date back earlier than the Eastern form, and the dominant vector for the spread of this technology across Eurasia was quite definitely from West to East. For this reason, many archaeologists and historians in their studies often came to the conclusion that early metallurgy in China was secondary, almost completely dependent upon the earlier Western impulse.[1] But the evidence, in the author's opinion, suggests that this situation is not at all as simple as it initially seemed.

In trying to clarify the confusion surrounding the subject of interconnections of ancient metallurgy among the peoples of China and the Eurasian steppes, a series of wide-ranging difficult questions is frequently encountered. From the very beginning, the long-debated problem concerning single or multiple centres of sources and distribution of metallurgy in Eurasia is inevitably recalled. However it is the author's belief that this problem is no longer a subject of debate. In all probability, there existed numerous centres for the beginnings of one of the most important technological inventions in the history of mankind. It seems to be extremely doubtful that this conclusion could be challenged in light of recent hypotheses.

In this respect, one more aspect of this problem is important. Developments of metallurgy can also be characterised by two basic models. The first – more customary and traditional for archaeological practice – is an evolutionary one. In general, it goes back to the theoretical constructs of Oskar Montelius, a classic figure in archaeology. If a researcher could reconstruct a typologically consecutive and morphologically linear progression from the simplest to the most complex, this would reveal their technological development.

The second model could be called an 'explosive' one. It assumes that, in a place without a centuries-old preparatory base, a developed technology of metallurgical production could suddenly erupt. It is often very difficult to understand and to interpret concrete manifestations of this model, but it seems indisputable that this latter model is more efficient and effective in understanding the progressive development of cultures. For this reason, this model promises to be of greater interest to researchers of historical and archaeological problems. A striking example of this 'explosion' of mining and metallurgical production has been reconstructed for the cultures of the northern Balkans and Carpathian basin during the Copper Age of the 5th millennium BCE (Chernykh 1992: 48–53). We are faced with a similar 'explosive' possibility when studying the interaction between Chinese and Eurasian steppe archaeometallurgy.

This paper, unfortunately, cannot be comprehensive. In essence it has a preliminary character, and there is no opportunity to present a full-scale canvas of detailed argumentation. For this reason, the discussion is limited to a few briefly formulated suggestions.

The watershed between the two 'worlds' of Eurasian mining and metallurgical production – Western and Eastern – ran across the Sayan-Altai mountain system (Fig. 1) where an original, and in all probability independent, metallurgi-

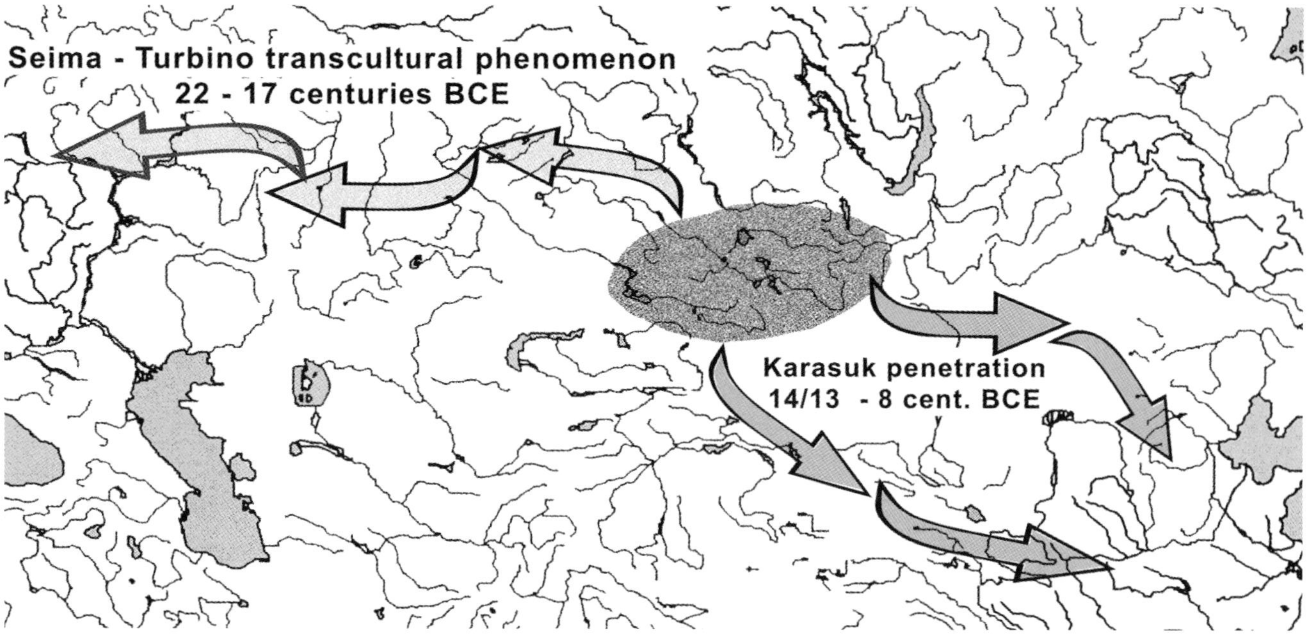

Figure 1 Sketch map of Seĭma-Turbino and Karasuk metallurgical models of expansion spreading.

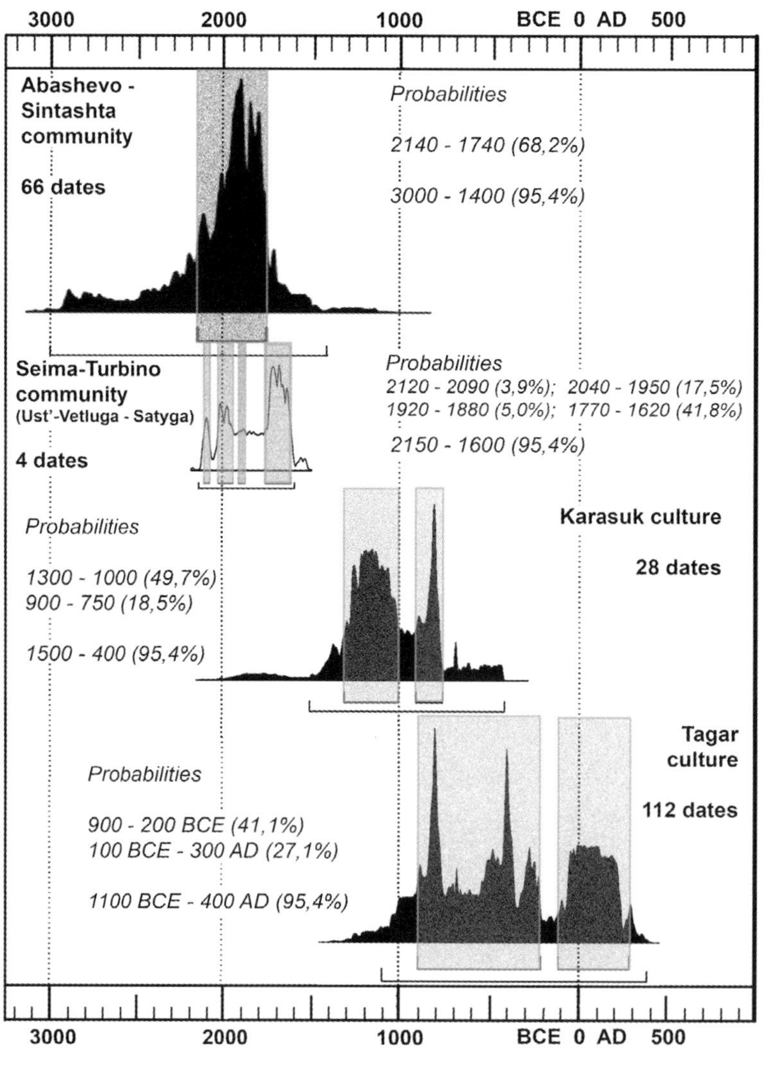

Figure 2 Sum probabilities of Abashevo-Sintashta and Seĭma-Turbino communities, Karasuk and Tagar cultures. Notes: the calibrated dates were obtained using OxCal vers. 3.10. Organic remains in Seĭma-Turbino burials were found to be rare and this is the main reason for the small number of radiocarbon dates. The close interrelations of Seĭma-Turbino groups with the Abashevo-Sintashta community determine a comparison of the chronological position of these two cultural associations.

cal centre appeared. In this region, the Seĭma-Turbino type metallurgy (Chernykh and Kuz'minykh 1989), which was characterised by its surprisingly high technological level, emerged spontaneously. Detachments moved west and swiftly overcame the many thousands of kilometres of the almost impenetrable forested expanses of northern Eurasia. Surprisingly, we find far to the west – as far as Moldova and the Baltic basin – that the metal weapons made in the eastern European workshops were part of this transcultural phenomenon.

Another but later product (according to radiocarbon determinations (Fig. 2)) of the Sayan-Altai centre is the phenomenon of the Karasuk-style metallurgy. Unlike the Seĭma-Turbino occurrence, the general direction of expansion of the producers of these forms and their carriers was the opposite – to the east and south-east. In the following text, the Karasuk (or what can be more precisely termed in certain respects, the Karasuk-Tagar)[2] metallurgical production type is compared with the bronze industry of the Shang and Western Zhou periods, and more particularly with the materials that were concentrated in the Yellow River (Huang Ho) basin,[3] with a focus on the power, specificity and originality of this metallurgical production.

The mighty technological 'explosion' of metallurgy and metalworking of bronzes during the Shang period of the 16th–11th century BCE cannot but surprise and attract special attention. This 'explosion' is especially striking in light of the earlier periods of the development of mining and metallurgical productive technology. The metals of previous periods – especially of the Longshan and of the early dynastic 'Xia' culture (although the latter is not clearly reflected in its archaeological sites)[4] are poorly expressed both qualitatively and quantitatively (An 2000: 64–5; Mei 2004: 173–5; Wagner 2006: 266–75; Chang and Xu 2005: 85–141; Linduff 2005). Numerous ritual vessels of the highest quality casting and with a surprising variety of forms were excavated from the imperial tombs and burial places of the highest nobility at Anyang (Fig. 3) and from other sites of the Shang and Western Zhou periods, many weighing tens and hundreds of kilograms. They were cast in complex multi-piece compound clay moulds, the internal surfaces of which were covered with the most complicated and elegant carved decorations (Fig. 4). Essentially, it is extremely difficult to find such magnificence of metalworking in the previous cultures of the Yellow River and Yangtze basins. Nothing similar can be found in the inventories of the metal-bearing cultures of the Western world.

Morphologically, the Karasuk and later Tagar metallurgy differs markedly from the ancient Chinese metallurgy of the Shang and Western Zhou. Metal weapons are most important for the inhabitants of the Sayan-Altai area, especially the famous Karasuk curved single-bladed knives with figured handles (Fig. 5) and, more rarely, daggers. Compared to these categories of metal products, other forms are less distinctive – such as decorations or mysterious objects supposedly representing 'models of yokes' for two-horse chariots (Fig. 5: 22). The northern steppe[5] forms or, more likely, their imitations are also present in the Shang imperial funeral complexes that are mostly found in the Royal Necropolis of Anyang (Fig. 6).

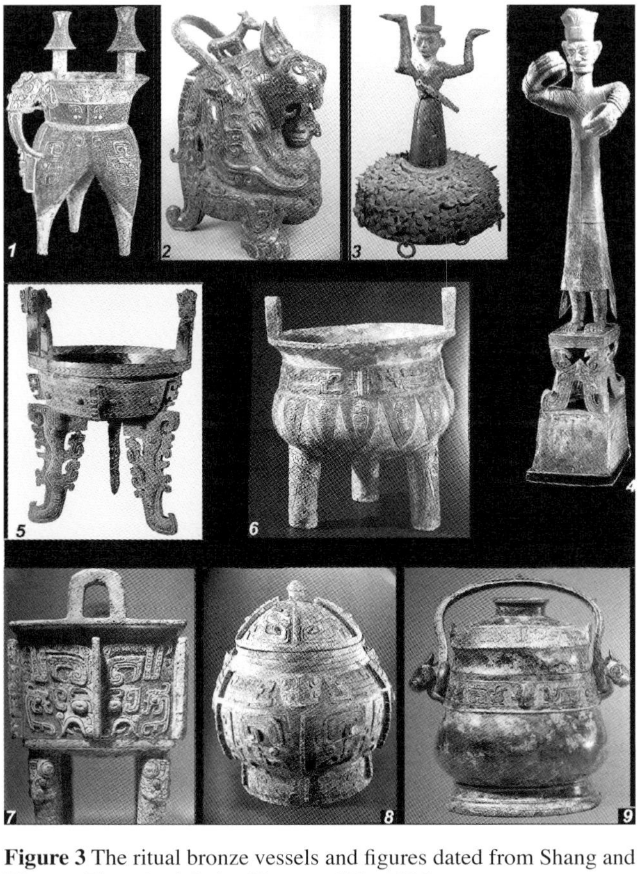

Figure 3 The ritual bronze vessels and figures dated from Shang and Western Zhou (mainly by Chang and Xu 2005).

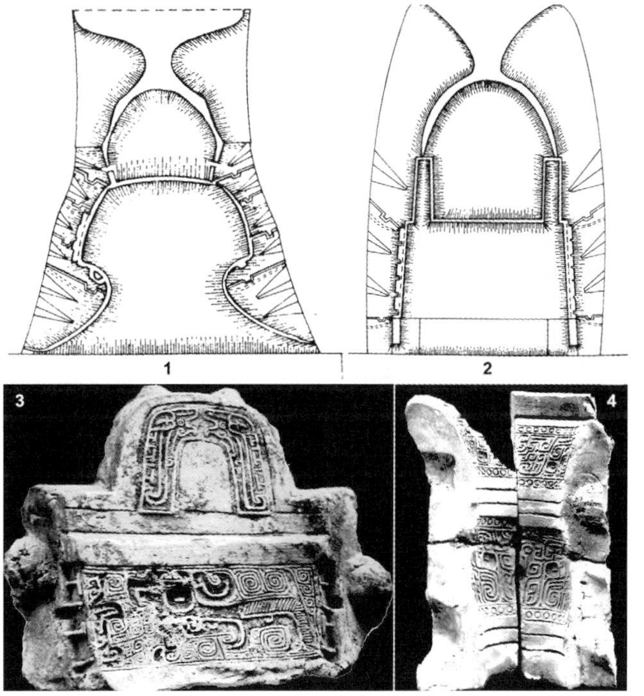

Figure 4 The complete casting moulds and their reconstructions: Shang period (1 and 2 from Chang and Xu 2005: 145 (fig. 6.2); 3 and 4: Anyang Museum).

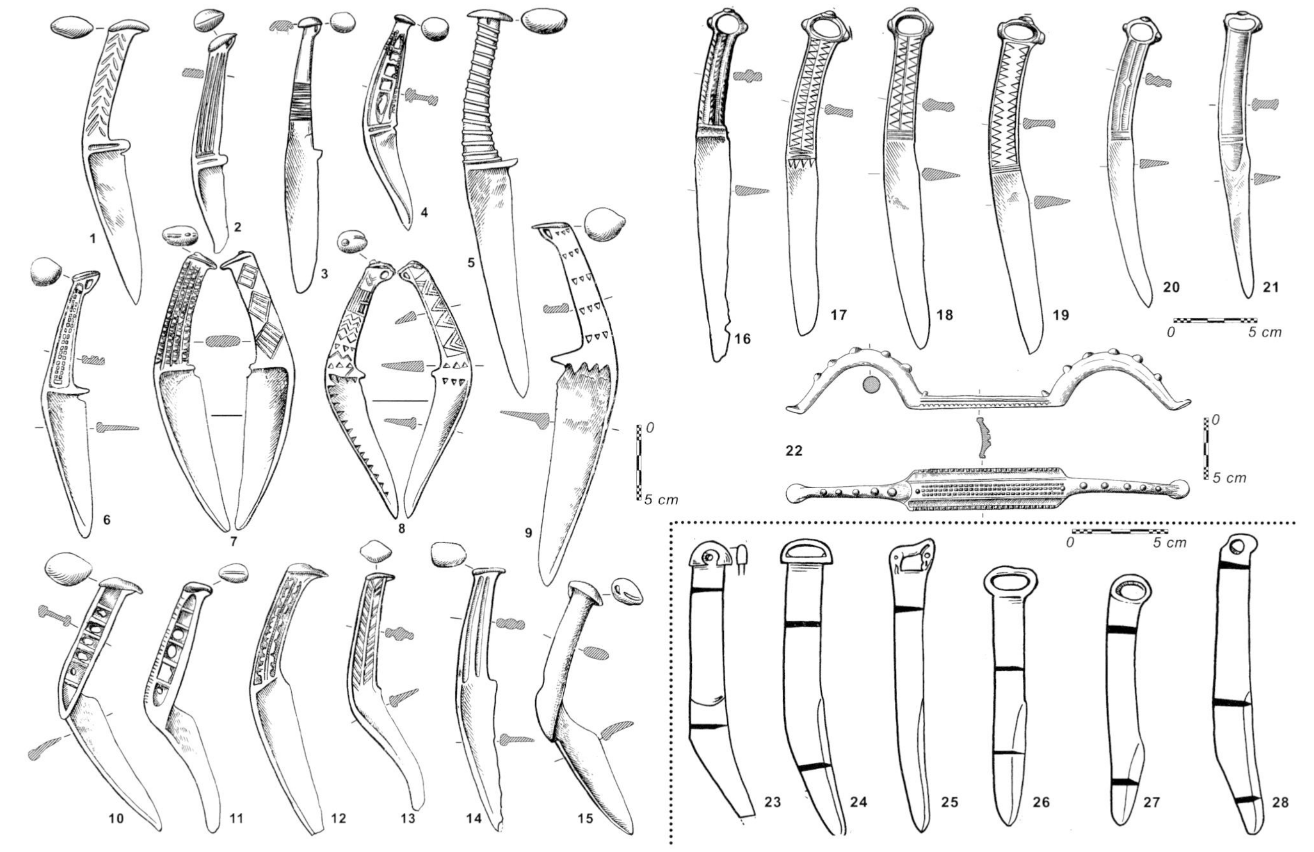

Figure 5 The Karasuk (1–21) and Karasuk-Tagar (23–28) curved single-bladed knives; 'models of yokes' for two-horse chariots (22). All materials dispersed mainly in the Khakassia-Minusink region, Western Sayan mountain system (Chlenova 1972: tables 1–5, 57, 58).

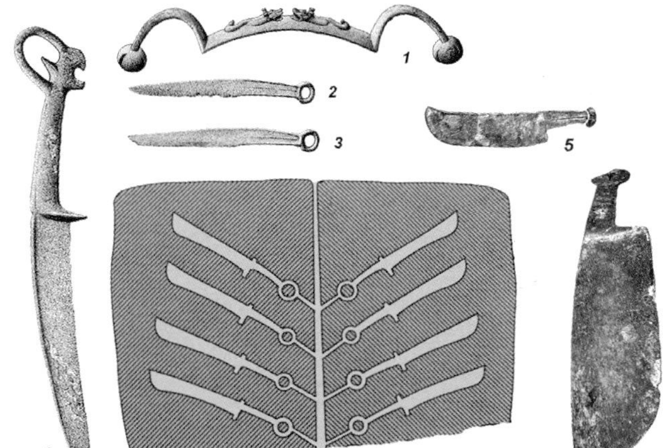

Figure 6 Chinese imitations of Karasuk bronze forms and complete casting mould for the curved single-bladed knives (1–4 from Müller-Karpe 1979: 44–7 (Abb. 56–58, 71); 5 and 6: Anyang Museum; 7 from Chêng Tê-kun 1960: fig. 7).

It is obvious that the broken chain (a break of nearly 3,500 km in a straight line) of the Karasuk finds (or essentially similar Karasuk prototypes) stretched far to the east: from the Sayan-Altai region through Xinjiang (Mei 2004), Mongolia (Erdenebataar 2004), northern China (including Inner Mongolia), right up to the Liao Ho basin and almost up to the Liaodong gulf (Wagner 2006: 101–276). Another line of the distribution of the 'steppe' forms extends to the south and south-east (Fig. 1). Similar products are known to us in the semi-arid and arid foothills of the Altun-Shan and Shanxi-Shaanxi plateau. They appear near the territory that was under the control of the Shang governors on the Yellow River (Chinese Archaeology 2003: 585–90).

It is also necessary to note another rather essential difference of these proposed models. The lion's share of the bronzes of the Shang and Western Zhou metallurgical production was certainly directed to the sacral purposes connected, first of all, with diverse, ritually magical ceremonies. The Karasuk or, more broadly, Karasuk-Tagar metallurgy (Fig. 5) is incomparably simpler and more rational. Its general orientation is completely different. To all appearances, its major efforts were directed towards the creation of metal weapons. The decorations on the curved knives with figures of animals or with certain ornamental patterns did not change the basic character of these northern handicrafts. Plaques, pendants and even mysterious objects supposedly representing 'horse yokes' were certainly secondary or had a less significant effect on the production cycle of the steppe smelters and smiths.

It is generally known that during many millennia, aggressive steppe pastoral nomads were a source of trouble for the cultures and civilisations of ancient China. Distinct and, probably, initial traces of this steppe aggression can be detected in the appearance of northern weapons in the inventories of the Shang period assemblages; however, it is much more difficult to distinguish such manifestations for earlier periods. Concerning the widely held view on the derivative nature of ancient Chinese metallurgy, assertions have been made and hypotheses advanced on powerful and decisive western or northern steppe impulses leading to the appearance of corresponding types of military horse-drawn chariots and metal weapons (Vasiliev 1976: 265–79).

To a significant degree, this hypothesis seems to recall the well-known discussions between Confucians and legal scholars (Perelomov 1981: 221–306). Supporters of Confucius's views rejected any chance of borrowing anything from the steppe 'barbarians', who were deeply despised by the native Chinese 'Xia' people. More practical legal scholars insisted that in spite of their undoubted contempt, it was necessary to think pragmatically about adopting from these barbarians the things in which they were more highly skilled; namely, first and foremost, military matters. While Confucius lived and worked at the time of the Tagar culture, the idea of proud self-isolation had most likely been widespread in China much earlier, including the Shang period. After all, the late phase of this period coincided with the activities of the Karasuk steppe nomads. However, as can be seen, borrowings from the north and north-west and imitations of their forms (Fig. 5) were already reliably under way during the second half of the 2nd millennium BCE. At that time, it is probable that the predecessors of Confucius and his followers had begun to struggle against such actions.

At the same time, analysis of the 'steppe' penetrations and adoptions does not allow us to exaggerate the role of such impulses. The weapons found in the Huang He basin are not represented exclusively by the Karasuk curved knives.[6] Bronze spearheads (Fig. 7: 5–8) have been documented there in far greater numbers than the knives, but this weapon is not so characteristic of Karasuk metallurgical production. Also, heavy flat poleaxes and distinctive 'arm-tooth' or *ge*-halberds (Fig. 7: 1–4), which were found in the royal complexes of Anyang, are not known in the steppe inventory.

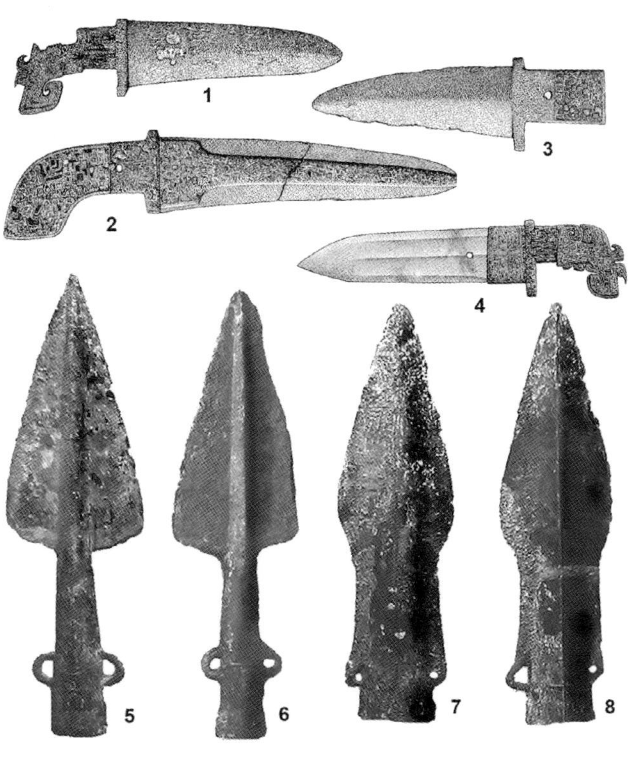

Figure 7 Some forms of Chinese bronze weapons of the Shang period (1–4 from Müller-Karpe 1979: 47 (Abb. 66–68); 5–8: Anyang Museum). Note the nephrite blade (4).

Steppe influences are more obvious at spaces located to the west and to the north of the indigenous areas of the ancient Chinese state – in Inner Mongolia, the Xinjang-Ujgur region, Qinghai, the Ordos, and even in the north-east of present-day China in Liaoning province (Chinese Archaeology 2003: 535–645, figs 8-20, 8-33, 8-41, 8-60).

The considerable abundance of Shang period bronze spearheads is a reminder of the hypothesis, which was first articulated years ago, concerning Seĭma-Turbino influences in the formation of the complex of metal Chinese weapons of the Shang dynasty (Kiselev 1960: 264–5). However, Kiselev did not notice any direct parallels with the Seĭma-Turbino spearheads or with the socketed axe-celts in the Chinese collections of metal of this period. This fact contrasts drastically with those quite obvious Karasuk analogies observed for the metal of Shang times. A unique and very long spearhead with a 'divaricate' rod (not clearly expressed) was found recently in the arid mountainous region of Qinghai province north of Tibet (Molodin and Komissarov 2001; Linduff 2005: 30, fig. 3). This obvious representative of the Seĭma-Turbino bronzes was an isolated find located more than 1000 km distant from the most northern and eastern complexes of this transcultural phenomenon. Consequently, any discussion of a probable chance of the Seĭma-Turbino influence in the formation of the weapons complex of Shang times is scarcely promising.

In such a light, the author identifies both the major problems and the nature of interrelations between two drastically opposed models of metallurgical production. It is important to note that since the 2nd millennium BCE, the extensive Steppe Belt that was occupied by the Eurasian pastoral nomads quite often played either an exclusive or primary role as the sole mediator between the Western and Eastern worlds. Undoubtedly the first of these complex interactions was reflected in the metals used. In many respects it seems impossible to deny the influence of the north-west pastoral nomads on Chinese metal production that led to a whole series of imitations in the form of the weapons. Basically, however, ancient Chinese metallurgy appears as an independent, rather distinctive and particularly dazzling phenomenon. Finally, it is necessary to note that these conclusions should for the most part only be considered as preliminary due to the lack of an exhaustive database on the ancient metals of China of the 2nd millennium BCE and to the fragmentary data concerning their chemical compositions.

Acknowledgements

I am grateful to Dr Liangren Zhang for translation of some Chinese texts and to Professor Philip L. Kohl for correcting my English manuscript of this paper.

Notes

1. See, for example, the review of the discussion on this theme in the work of L.S. Vasiliev (1976: 260–79) and also the author's arguments in favour of the dependence of early Chinese technologies on Western influences (Chernykh 2005: 56–59; 2006). The same questions were discussed in the works of Hua Jueming and An Zhimin (Hua 2000: 51–9; An 2000: 63–81). Cf. Linduff et al. 2000; Linduff 2004; Mei 2004.
2. It is quite probable that it is also reasonable to use the concept of the 'Karasuk-Lugav' metallurgy type. On the other hand, N.L. Chlenova was inclined to recreate, for example, a rather complex picture of the introduction and interaction of multi-ethnic tribes within the area of the Minusinsk depression. On this canvas the carriers of the Lugav culture are represented by successors of early local tribes (Okunevo type), and the Karasuk people were newcomers. According to Chlenova, the Tagar culture owes its origin to a much greater degree to the indigenous Lugav people (Chlenova 1967: 211; 1972: 117–31).
3. Note the intentional reference to the metallurgical production of Shang dynasty times as an industry.
4. For example, it is not quite clear if a rather rich collection of metal products in one of the most well known of the ancient Chinese sites – Erlitou and some others (Chang and Xu 2005: 145–51) – belong to the 'Hsia' or 'Xia' culture. This discussion is still far from settled. Is it possible to relate many metal-bearing complexes to the earlier Xia culture or is it more reasonable to correlate them with the Shang period?
5. Of course, the concept of 'steppe metallurgy' is used here conditionally. The centres of metallurgical production of this type covered not only the steppes, but also mountain taiga, semi-desert, and even deserted regions of Central Asia.
6. The general morphological pictures of the Karasuk curved knives (Fig. 5) and other metals have not been included as they are widely known from the above-mentioned works such as Chlenova (1972). In this paper, particular attention is paid to the ancient Chinese imitations (or Karasuk imports?) in the funeral complexes of Shang times.

References

An Zhimin (2000) 'Some problems concerning early Chinese copper and bronze artifacts', in K.M. Linduff, Han Rubin and Sun Shuyun (eds), *The Beginnings of Metallurgy in China,* 29–46. New York: Edwin Mellen Press Ltd.

Chang Kwang-chih and Xu Pingfang (eds) (2005) *The Formation of Chinese Civilization: An Archaeological Perspective.* New Haven/ London: Yale University Press and New World Press.

Chêng Tê-kun (1960) *Archaeology in China, Vol. 2.* Cambridge: Heffer.

Chernykh, E.N. (1992) *Ancient Metallurgy in the USSR: The Early Metal Age.* Cambridge: Cambridge University Press.

Chernykh E.N. (2005) 'Puti I modeli razvitiya arkheometallurgii v Starom I Novom Svete' ('Ways and models of development of archaeometallurgy in the Old and New Worlds'), *Rossiyskaya arkheologia* 4: 49–60.

Chernykh, E.N. and Kuz'minykh S.V. (1989) *Drevnaya metallurgiya Severnoy Evrasii: seiminsko-turbinskiy transkulturnyi fenomen* (Ancient Metallurgy in the North Eurasia: the Seĭma-Turbino Transcultural Phenomenon). Moscow: Nauka.

Chlenova, N.L. (1967) *Proiskhozhdenie I rannaya istoriya plemen tagarskoy kultury* (The Origin and Early History of the Tagar Culture Tribes). Moscow: Nauka.

Chlenova, N.L. (1972) *Khronologiya pamiatnikov karasukskoy epokhi* (The Chronology of the Karasuk Period Monuments). Moscow: Nauka.

Chinese Archaeology (2003) *Xia and Shang.* Archaeological Monograph Series, Type A, No. 29. Beijing: China Social Sciences Press [in Chinese].

Erdenebataar, D. (2004) 'Burial materials related to the history of the Bronze Age in the territory of Mongolia', in K.M. Linduff (ed.), *Metallurgy in Ancient Eastern Eurasia from the Urals to the Yellow River,* 189–222. New York: Edwin Mellen Press Ltd.

Hua Jueming (2000) 'On the origin of metallurgy in China', in K.M. Linduff, Han Rubin and Sun Shuyun (eds), *The Beginnings of Metallurgy in China,* 51–62. New York: Edwin Mellen Press Ltd.

Kiselev, S.V. (1960) 'Neolit i bronzovyi vek v Kitaie' ('The Neolithic and Bronze Age in China'), *Sovetskaya arkheologia* 4: 244–66.

Linduff, K.M. (ed.) (2004) *Metallurgy in Ancient Eastern Eurasia from the Urals to the Yellow River.* New York: Edwin Mellen Press Ltd.

Linduff, K.M. (2005) 'Kak daleko na vostok rasprostranyalas' evrasiyskaya metallurgicheskaya traditsiya?' ('How far east does the Eurasian metallurgical tradition extend?'), *Rossiyskaya arkheologia* 4: 25–35.

Linduff, K.M., Han Rubin and Sun Shuyun (eds) (2000) *The Beginnings of Metallurgy in China.* New York: Edwin Mellen Press Ltd.

Mei Jianjun (2004) 'Metallurgy in Bronze Age Xinjiang and its cultural context', in K.M. Linduff (ed.), *Metallurgy in Ancient Eastern Eurasia from the Urals to the Yellow River,* 173–88. New York: Edwin Mellen Press Ltd.

Molodin, V.I. and Komissarov, S.A. (2001) 'Seiminskoye kopie iz Tsinkhaya' ('Seĭma-type arrowhead from Qinghai province'), *Problemy arkheologii, etnografii, antropologii Sibiri I sopredel'nykh territoriy. Tom VII. Novosibirsk*: 374–81.

Müller-Karpe, H. (1979) 'Das Grab der Fu Hao von Anyang', in H. Müller-Karpe (ed.), *Beiträge zur allgemeinen und vergleichenden Archäologie, Bd. 1,* 31–68. Munich: Beck'sche Verlagsbuchhandlung.

Perelomov, L.S. (1981) *Konfutsianstvo I legizm v politicheskoy istorii Kitaya* (The Confucius Views and Legalism in the Political History of China). Moscow: Nauka.

Vasiliev, L.S. (1976) *Problemy genezisa kitayskoy zivilizatsii* (The Problems of Genesis of Chinese Civilization). Moscow, Nauka.

Wagner, M. (2006) 'Neolithikum und Frühe Bronzezeit in Nordchina vor 8000 bis 3500 Jahren', *Archäologie in Eurasien, Bd. 21,* 1–355. Mainz: Verlag Philipp von Zabern.

Early metallurgy in China: some challenging issues in current studies

Jianjun Mei

ABSTRACT This paper first reviews major recent progress in the studies of early metallurgy in China, particularly the growing body of evidence pointing to the existence of early cultural interactions between north-west China and the Eurasian steppe. It then highlights four challenging issues in current studies of early metallurgy in China, namely the nature of steppe influence, the early use of arsenical copper, the significance of regional interactions, and the invention of piece-mould casting technology. Finally, it argues that an interdisciplinary approach is essential for an in-depth exploration of the role of bronze metallurgy in the growth of early Chinese civilisation.

Introduction

The beginnings of the use of metals and alloys in China is one of the central issues in the exploration of the rise and evolution of early Chinese civilisation, attracting constant interest among both archaeometallurgists and sinologists over the past decades. Before the 1950s, it was widely assumed, especially among Western scholars, that metallurgy was introduced into China from the outside world (Bishop 1942: 14; Loehr 1949: 129). From the 1960s to the early 1980s, on the basis of new archaeological evidence of early metal in China, especially from the Gansu-Qinghai region, more and more scholars began to argue that metallurgy had indigenous origins in China (Barnard 1961, 1983; Barnard and Sato 1975: 1–16; Cheng 1974; Ho 1975: 177–221; Sun and Han 1981), while some still favoured the diffusion theory and found 'it impossible to believe that the basic ideas of metallurgy were so easy to come by *ad nuovo*' (Smith 1977: 81; also cf. Jettmar 1981; Watson 1985).

In 1986, on the occasion of the BUMA II conference in Zhenzhou, China, James Muhly (1988: 16) made the following observation on the beginnings of metallurgy in China:

> What work during the past decade has accomplished is the solid documentation of the long-missing formative state in the development of Chinese copper and bronze metallurgy. The beginnings of metallurgy in China have now been shown to be not so much different from what can be seen elsewhere throughout the ancient world. This neither proves nor disproves the independent invention of metallurgy in east Asia, but it does move the argument into a more comprehensible frame of reference and should eliminate once and for all the argument that Chinese metallurgical technology developed in ways totally different (and thus apart from) anything known in the West.

Twenty years have passed since this statement was made. I wonder what new observations can be offered now – are there any new developments in our understanding of the beginnings of metallurgy in China?

In my view, at least four major developments can be highlighted here. The first is a fresh understanding of the early use of metals in the Gansu-Qinghai region, north-west China, based on a number of new archaeological finds and metallurgical analyses; the second is the initiation of systematic examination of early copper and bronze metallurgy in Xinjiang, west of Gansu, a region that was virtually ignored in previous studies of early metallurgy in China; the third consists of the studies of newly excavated metallurgical remains at the Erlitou (二里头) site of the first half of the 2nd millennium BC in Henan, the Central Plains of China; the fourth is the archaeometallurgical research on early metals of the Zhukaigou (朱开沟) and Lower Xiajiadian (夏家店下层) cultures found in the northern border areas of China.

In this paper, I would like to offer a brief review of these new developments and to see what contributions they have made to our current understanding of early metallurgy in China. In view of a growing body of evidence pointing to the existence of early cultural interactions between north-west China and the Eurasian steppe, I shall then address several challenging issues concerning the early development of copper and bronze metallurgy in China. Finally, I shall give a brief discussion of some potential directions for future research.

Early metallurgy in China: recent progress

The past 20 years have witnessed three major changes in the study of early metallurgy in China: first, an increasing number of early copper and bronze objects have been excavated in four regions, namely Gansu-Qinghai, Xinjiang, the Central Plains and the Northern Zone; secondly, extensive scientific analyses of these early metals have been carried out and substantial new analytical data have been obtained, which provide us with crucial evidence for characterising early Chinese metallurgy; thirdly, cultural connections or interactions between north-west China and the Eurasian steppe have been increasingly recognised among scholars, leading to a better understanding of steppe influence in the growth of early metallurgy in north-west China.

Early copper and bronze objects recovered in Gansu and Qinghai

The importance of the Gansu-Qinghai region for the study of early metallurgy in China became clear in the early 1980s, when dozens of metal objects associated with the cultures of Majiayao (马家窑) (3300–2650 BC), Machang (马厂) (2300–2000 BC) and Qijia (齐家) (2200–1800 BC) were found to be made of copper, tin bronze or leaded tin bronze (Sun and Han 1981: 296–300). Since then, a number of new discoveries of metal objects have come to light in the region, including copper and bronze objects found at the sites of Zongri, Zongzhai, Shenna, Qijiaping, Xinzhuangping, Xinglin, Donghuishan, Huoshaogou, Yingwoshu, and Ganguya. They are mostly small implements, weapons and ornaments, such as knives, axes, awls, spearheads, bracelets, buttons, plaques, rings and mirrors (Figs 1 and 2), and belong to two cultures, namely Qijia and Siba (四坝) (1950–1550 BC) (Mei 2003a: 3–13; Li 2005: 240–51). Many of these objects show a clear similarity in typology to metals found in the Eurasian steppe, leading scholars to consider the probability of some cultural connections between them (Fitzgerald-Huber 1995: 38–52; Mei 2003a: 5–13). The spearhead excavated at the Shenna site in Qinghai is, perhaps, the most outstanding example for such typological comparison. It is 61 cm in length, with a downward hook and a small loop on either side of the socket. Such a distinctive form closely parallels the spearheads recovered in southern Siberia and associated with the Seĭma-Turbino context (Fig. 3; Chernykh and Kuzminykh 1989: 69–70; Chernykh 1992: 215–33; Takahama 2000: 114; Mei 2003b: 32–5).

Another profound development in the study of early metallurgy in Gansu in the past 20 years is the identification of a significant number of arsenical copper artefacts among the metal objects associated with the Siba culture. According to the examinations carried out by Sun Shuyun and others, among 15 copper and bronze objects from the Donghuishan site, 12 are made of arsenical copper, while among 46 examined pieces from the Ganguya site, 10 of arsenical copper have been identified (Sun 1998: 191–5; Sun and Han 1997: 81). Previous analyses of the Huoshaogou metals found no evidence for the use of arsenical copper. However, a recent analysis of 26 Huoshaogou objects has revealed the existence of six made of arsenical copper and two of copper-tin-arsenic alloys (Sun *et al.* 2003: 86). This new evidence for the use of arsenical copper along the Hexi Corridor (河西走

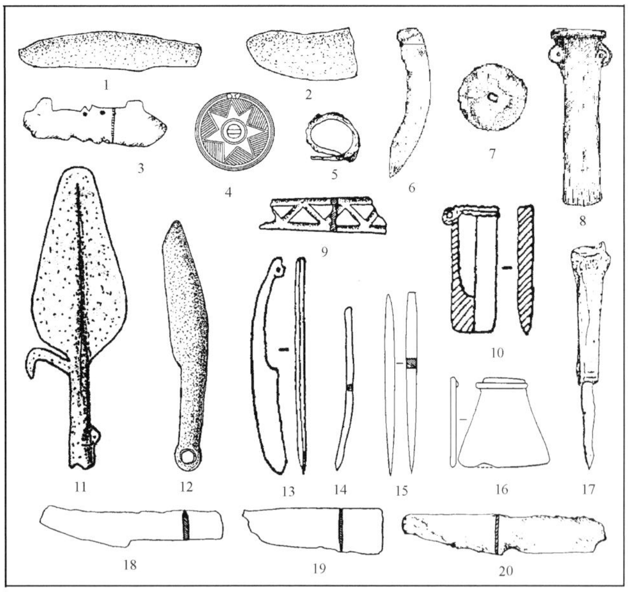

Figure 1 Copper and bronze artefacts recovered at the sites of Majiayao, Machang and Qijia cultures in Gansu and Qinghai: 1: knife from the Linjia site of the Majiayao culture; 2: knife from the Jiangjiaping site of the Machang culture; 3–20: objects from various Qijia culture sites. 1–3, 6, 12, 13, 18–20: knives; 4, 7: mirrors; 5: ring; 8, 10: axes; 9: handle; 11: spearhead; 14, 15: awls; 16: axe-shaped object; 17: knife with a bone handle (1, 2 after Sun and Han 1997, figs 1 and 3; 3–10, 13–20 after Debaine-Francfort 1995, figs 49, 61, 71; 11 and 12 drawings after photos).

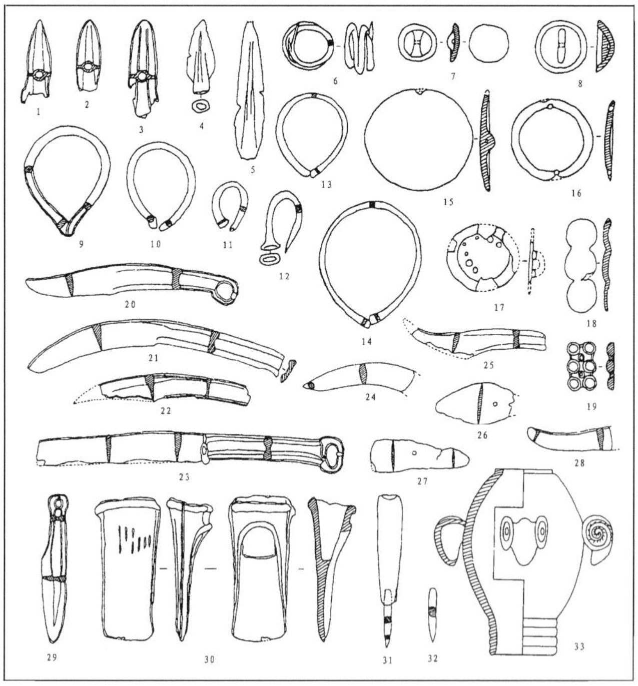

Figure 2 Copper and bronze artefacts of the Siba culture in Gansu: 1–5: arrowheads; 6: ring; 7, 8: buttons; 9–14: earrings and rings; 15–19: ornaments; 20–29: knives; 30: socketed axe; 31, 32: awls; 33: macehead (after Bai 2002, fig. 3).

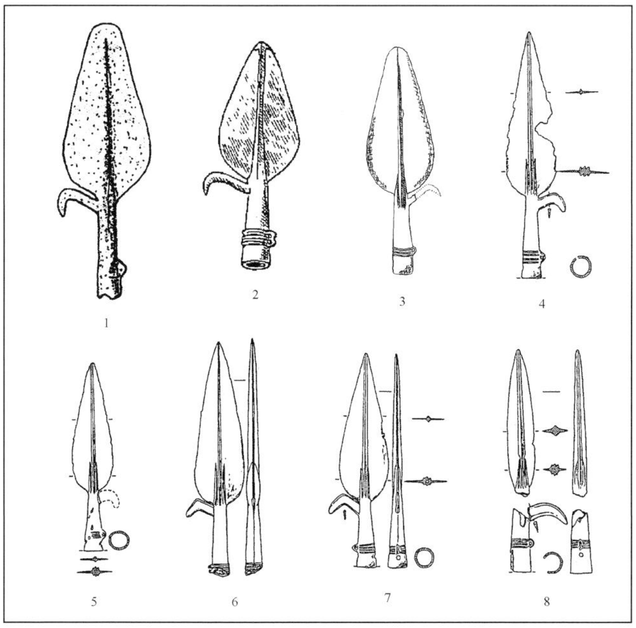

Figure 3 Copper and bronze spearheads recovered in China and Siberia: 1–3: from China; 4–8: from southern Siberia; 1: from the Shenna site in Qinghai; 2: in the collection of the Shaanxi Museum in Xi'an; 3: in the collection of Shanxi Museum, Taiyuan (1, 3. drawings after photos; 2, 4–8. after Takahama 2000, fig. 3).

廊) in Gansu has led to a fundamental change in our understanding of the early metallurgy in the region (Li and Shui 2000). It also offers further evidence to support Muhly's argument: 'The beginnings of metallurgy in China have now been shown to be not so much different from what can be seen elsewhere throughout the ancient world' (Muhly 1988: 16).

Early metallurgy in Xinjiang

Twenty years ago we knew almost nothing about early metallurgy in Xinjiang, but now extensive archaeological and metallurgical data have been accumulated. To date, more than 400 ancient copper and bronze objects recovered in Xinjiang have been subjected to metallurgical examinations. Among them, about 150 can be dated to the first half of the 2nd millennium BC and most of them were found at the Tianshanbeilu (天山北路) cemetery in Hami, eastern Xinjiang. The copper and bronze objects recovered at the Tianshanbeilu cemetery represent a wide range of types, including knives, awls, axes, tubes, earrings, bracelets, mirrors, beads, buttons, plaques, daggers and arrowheads (Figs 4 and 5). Furthermore, a small number of gold and silver ornaments were also found there (Mei 2000: 38–41; Lü *et al.* 2001: 179–84; Qian 2006: 40–41).

On the basis of scientific examinations, a tentative but entirely new understanding of early metallurgy in Xinjiang has been established. First, tin bronze came into use in Xinjiang no later than the early 2nd millennium BC, and it was the predominant metal in Xinjiang throughout the Bronze Age (the 2nd millennium BC). The emergence and early use of tin bronze in Xinjiang may have been subject to cultural influences from bronze cultures in the Eurasian

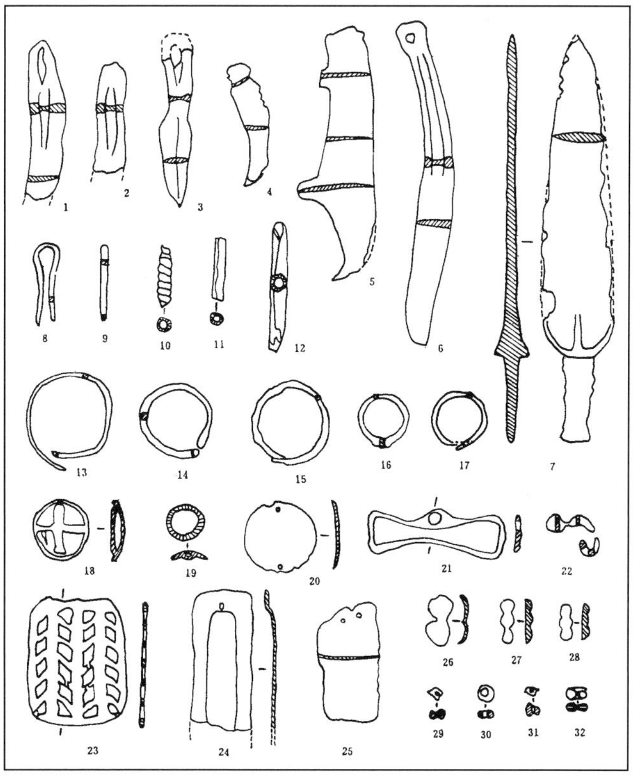

Figure 4 Copper and bronze artefacts found at the Tianshanbeilu cemetery, Hami, Xinjiang: 1, 2, 4–6: knives; 3, 7: daggers; 8–32. rings, plaques, buttons, tubes and other small ornaments and implements (after IHMM 2001, fig. 1).

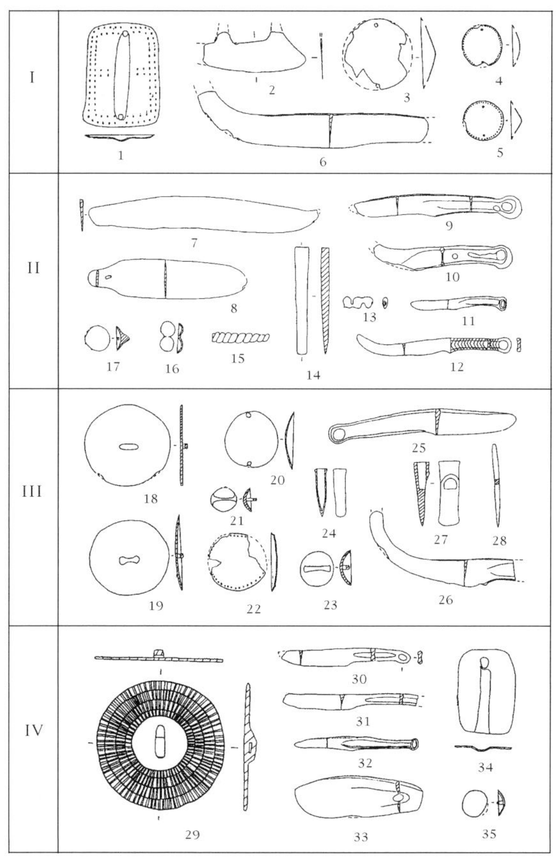

Figure 5 Copper and bronze artefacts found at the Tianshanbeilu cemetery, Hami, Xinjiang (periods I to IV in chronological order): 1, 3–5, 13, 15–23, 34, 35: plaques, buttons and other ornaments; 2, 6–12, 25, 26, 30–33: knives, sickles and knife-dagger; 14, 24, 27, 28: awls and axes; 29: mirror (after Lü *et al.* 2001, 182–3, figs 15–18).

steppe, such as the Andronovo culture (Mei *et al.* 1998; Mei and Shell 1999; Mei 2000: 38–9). Secondly, arsenical copper is another major copper alloy that was used mainly in eastern Xinjiang from the early 2nd millennium BC. The first evidence for the use of arsenical copper in Xinjiang was provided by the examination of two copper objects from Wupu (五堡) cemetery in eastern Xinjiang, which showed that they contain 3–4% arsenic (Mei 2000: 40). Then, analysis of 89 Tianshanbeilu metal items showed that 61 were of copper-tin, 11 copper, 9 copper-arsenic, 4 copper-tin-arsenic and four other alloys, confirming the predominant place of tin bronze as well as the significant presence of arsenical copper (IHMM 2001). It seems most likely that the use of arsenical copper in eastern Xinjiang was somehow related to what happened in the Hexi Corridor in Gansu, although the context for such a connection remains rather obscure, as the excavation reports of two major archaeological sites, Huoshaogou and Tianshanbeilu, have not yet been published.

It has become increasingly clear that Xinjiang played an important part in the beginnings of early metallurgy in north-west China, especially in view of its crucial role in providing connections between the Hexi Corridor and southern Siberia (Mei 2003b). It has been suggested that a two-way traffic of cultural influence could have existed along the prehistoric 'Silk Road' (the Hexi Corridor); while painted pottery spread westward from Gansu into Xinjiang, bronze technology was transmitted in the reverse direction (Debaine-Francfort 2001: 59–60; Mei 2003a: 24–5). Since the similarities and connections in the use of early copper and bronze in both Xinjiang and Gansu-Qinghai are so strong, it has become necessary to consider north-west China as one cultural entity in order to discuss its early metallurgical relationship with the Central Plains of China.

Erlitou: early bronze metallurgy in the Central Plains of China

Although there are a number of archaeological discoveries of metallurgical remains in association with the Yangshao (仰韶) and Longshan (龙山) cultures that can be dated from the fifth to the 3rd millennium BC, the most significant evidence for early bronze metallurgy in the Central Plains of China is still seen in the finds from the Erlitou site in Henan province, which can be roughly dated to the 18th–16th century BC. These finds include bronze objects such as ritual vessels, knives, awls, dagger-axes, arrowheads, plaques and buttons, as well as casting moulds, crucibles and slag (Fig. 6; Bai 2002: 31–2; Li 2005: 256–62). The emergence of bronze vessels and piece-mould casting technology clearly marked a new era in the development of bronze metallurgy in China.

Since the late 1990s, considerable scientific research on Erlitou metallurgical remains has been carried out in order to reveal their technical characteristics. Qu and Zhang (1999: 399) examined 32 metal objects from the Erlitou site and found that they were made of various metals, such as copper-tin-lead (14), copper-lead (9), copper-tin (5), copper (3) and lead (1). Jin (2000: 56–8) reported analytical results of 13 copper objects from the Erlitou site, showing the use of copper-tin-lead (4), copper-tin (4), copper (4) and copper-arsenic (1). It should be noted that among these 13 analysed objects, all vessels are made of copper-tin-lead, suggesting that the copper-tin-lead alloy was most likely chosen deliberately for casting bronze vessels. The identification of a copper-arsenic alloy is also worthy of attention, as its presence in the Central Plains seems to be rare and has some implications for tracing cultural connections with north-west China, where it was widely used during the early half of the 2nd millennium BC (Mei 2001: 61–2).

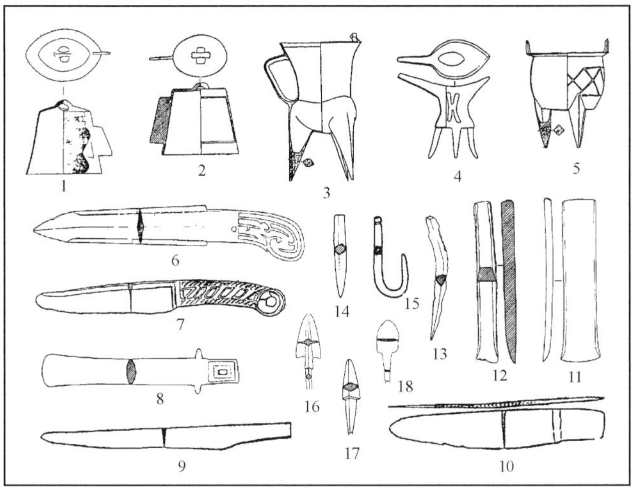

Figure 6 Copper and bronze artefacts of the Erlitou culture found in Yanshi, Henan, central China: 1, 2: bells; 3–5: vessels; 6: dagger-axe; 7, 9, 10: knives; 8: axe; 11–15: small implements; 16–18: arrowheads (after Bai 2002, figs 7–9, 11).

According to Liang and Sun (2004: 35), more than 200 pieces of metal objects have so far been recovered at the Erlitou site. Among them, about 20 are bronze vessels, clearly indicating the rise in importance of ritual vessels in the region. In recent years, Liang Honggang (2004) has carried out a systematic study of various metallurgical remains recently excavated at the Erlitou site. The research results (due to be published soon) are expected to shed new light on how and why bronze metallurgy at Erlitou took a path different from other regions in the development of early metallurgy.

Early metallurgy in the northern border areas of China

In comparison to the situation in the mid-1980s, significant progress has also been seen in the studies of early metal objects recovered in the northern border areas of China, namely the Northern Zone region, over the past 20 years. They include scientific examinations of metal objects from two major sites, Zhukaigou and Dadianzi (大甸子) in Inner Mongolia. Both sites can be dated to the 19th–15th century BC.

Five phases have been identified at the Zhukaigou site, with the first phase being dated to the late Longshan period, and the second, third and fourth phases to the early, middle and late Xia periods. Copper and bronze objects began to occur from the third phase (18th century BC) and in total 25 pieces were recovered from the third and fourth phases. These metal objects are mainly small implements and ornaments, such as pins, chisels, awls, arrowheads, bracelets, finger rings and earrings (Fig. 7; Bai 2002: 29–30). Li and Han (2000: 424–9) examined 13 Zhukaigou metal objects and found that they are made of copper (5) and copper-tin or copper-tin-lead (8), with tin contents being mostly 8–10% and lead contents less than 3%. While the chisel, awl, arrowhead and bracelets were made by casting, the pin, finger ring and earring were all hot-forged into the shape, some with cold-working too. These results provide further analytical evidence for the development of bronze technology in the Northern Zone of China.

Previous research already took note of the discoveries of copper and bronze objects in a number of sites ascribed to the Lower Xiajiadian culture in Inner Mongolia, Liaoning, Hebei, Tianjin and Beijing (Sun and Han 1981: 295; An 1981: 275). However, scientific analysis of these objects was rather limited until recently. So far, more than 100 copper and bronze objects have been recovered from various Lower Xiajiadian sites, including earrings, finger rings, knives, awls and arrowheads (Fig. 8; Bai 2002: 29–30). The Dadianzi site is obviously the most important as it yielded 57 copper and bronze objects, mostly earrings and finger rings, plus one gold and two lead objects. Li *et al.* (2003: 79–82) analysed 41 samples from these objects and found that they are all made of copper-tin or copper-tin-lead alloys, with tin contents being mostly over 8% and lead contents below 6% (generally less than 3%). It is interesting to note that all earrings were made by hot-forging, while all finger rings were made by casting, indicating a deliberate selection of a suitable working process to fashion different types of ornaments.

Some challenging issues in current studies

As can be seen from the evidence presented in the preceding section, our knowledge of early metallurgy in China has advanced immensely over the past 20 years. Three observations may be made: first, archaeological and metallurgical evidence from north-west China (including Gansu, Qinghai and Xinjiang) has confirmed that copper and bronze metallurgy emerged in the region no later than the 3rd millennium BC, and influence or stimulation from the Eurasian steppe may have played a role in this emergence; secondly, the metallurgical remains recovered at the Erlitou site have demonstrated the dominant position of the Central Plains in the development of bronze metallurgy, which is characterised by the invention of piece-mould casting technology; thirdly, the discoveries of early copper and bronze objects in the Northern Zone indicate that the region could have been a major player in regional interactions regarding the transmission of metallurgical knowledge. However, a number of challenging issues still remain. Here, I would like to highlight four of them, namely the nature of steppe influence, the early use of arsenical copper, the significance of regional interactions, and the invention of piece-mould casting technology.

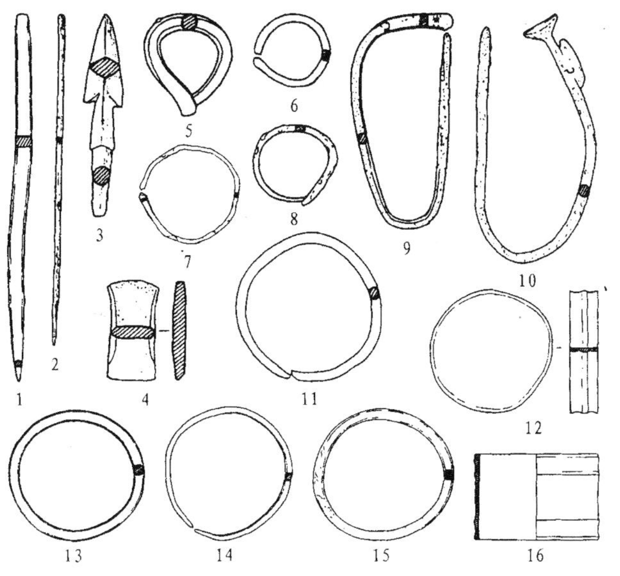

Figure 7 Copper and bronze artefacts of the Zhukaigou culture found in Inner Mongolia, north China: 1: awl; 2: pin; 3: arrowhead; 4: chisel; 5, 6, 8: finger rings; 7, 12–16: bracelets; 9–11: earrings (after Bai 2002, fig. 4).

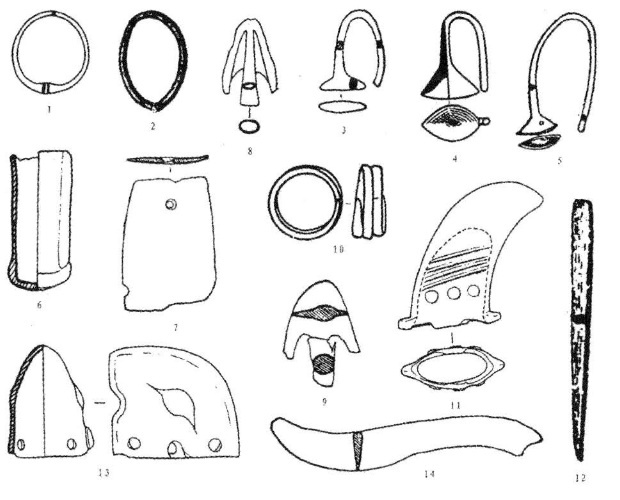

Figure 8 Copper and bronze artefacts of the Lower Xiajiadian culture found along the northern frontier of China: 1, 10: finger rings; 2–5: earrings; 6, 7, 11, 13: ornaments; 8, 9: arrowheads; 12: awl; 14: knife (after Bai 2002, fig. 5).

The nature of steppe influence

While the debate on the independent invention and diffusion of early Chinese metallurgy remains an attractive and important issue, a new trend of research has already emerged putting more emphasis on local or regional developments and interactions, as well as their socio-cultural contexts. One of the major focuses in this new trend addresses the connections between north-west China and the Eurasian steppe. In view of a growing body of evidence pointing to contacts between these two regions, some scholars have already begun to consider the extension of 'steppe metallurgy' into north-west China (Debaine-Francfort 2001) or to see early metallurgy in north-west China as 'part of the

Eurasian tradition' (Linduff 2004). It should be recognised that while steppe influence did exist in the development of bronze metallurgy in north-west China, its nature remains rather obscure. In examining typological characteristics of early metals from north-west China, it has been observed that various steppe elements of different sources co-exist among the early bronze cultures in the region. This preliminary observation seems to suggest that contacts between north-west China and the Eurasian steppe were most likely indirect, occasional and small scale rather than in the form of massive migration (Mei 2003b: 41–2). Further research is clearly needed to clarify the issue.

The early use of arsenical copper

From archaeological discoveries of the past 20 years we know now that arsenical copper was in use in China in the late 3rd and 2nd millennium BC. The most important region for the early use of arsenical copper is north-west China, with a few cases also seen in the Central Plains of China. The emergence of arsenical copper in north-west China can be traced back to the beginning of the 2nd millennium BC or even earlier. It was employed by a number of Bronze Age cultures along the Hexi Corridor in Gansu, as well as in eastern Xinjiang during the first half of the 2nd millennium BC, such as Siba, Qijia and Tianshanbeilu. The number of finds of arsenical copper objects is not in terms of a few, but of dozens (Li and Shui 2000: 40–42; Mei 2001: 59–62). Therefore, the existence of arsenical copper in north-west China is not an accidental phenomenon, but a clear technological development. However, at the current stage of research, we do not know whether these objects were made locally or imported from somewhere else. Although there are some signs suggesting that a local mineral resource is more likely, no hard evidence for this hypothesis has yet been established. On the other hand, typological comparison also shows some cultural connections between north-west China and the Eurasian steppe. Is there any possibility of a stimulating diffusion of arsenical copper technology from the Eurasian steppe into north-west China? Clearly, to answer this question we would need evidence not just from Gansu and Xinjiang, but also from southern Siberia.

The significance of regional interactions

A further challenging issue involves cultural interactions among regional centres, especially between north-west China and the Central Plains (An 1993). Current archaeological evidence indicates that north-west China and the Central Plains are the two major regions where metallurgical remains of the 3rd millennium BC have been found (Linduff 1998). Of great interest is the discovery of a few copper objects at the Taosi (陶寺) site, southern Shanxi, which has been ascribed to the Longshan culture and dated to the late 3rd millennium BC (Liang and Yan 2003). One of the Taosi copper objects has even been identified as arsenical copper, leading some scholars to conjecture that the occurrence of metal objects there was a result of imports from north-west China (Liu and Li 2007: 59). However, such an argument is far from decisive and more solid evidence is clearly needed. It was not until the Erlitou period (18th–16th century BC) that clear signs of the cultural connections between north-west China and the Central Plains appeared. A good example for this is a bronze plaque recovered in Tianshui (天水), eastern Gansu, which is very similar in shape and decorative style to those excavated at the Erlitou site in Henan (Zhang 2002). However, we have to acknowledge that evidence for interactions between the north-west and the Central Plains is not yet plentiful or adequately documented. Therefore, in order to gain a more complete picture of the development of early bronze metallurgy in the Central Plains, regional interaction needs to be further explored and documented.

The invention of piece-mould casting technology

The emergence of piece-mould casting technology and its social background is also a challenging issue in our current studies of early metallurgy in China. We still do not have a clear understanding of the emergence of piece-mould casting technology, which has been widely considered a Chinese-invented technology. If the issue is examined from a technological point of view, it would seem that the technology employed to cast a socketed axe could have led to the birth of piece-mould technology. But it seems to me that the background for the invention of piece-mould casting technology is far more complicated. The evolution of the technology may have been just part of the story. A more important factor could have been a social or institutional one. In my opinion, the ritual practices preceding the technology itself should have been the real driving force behind the full development of piece-mould casting. Before the advent of bronze metallurgy in the Central Plains, institutionalised or regularised ritual practice including both jade and ceramics already existed. When bronze metallurgy was first introduced into the region, it was adapted into the existing socio-cultural customs that confirmed social hierarchies, replacing jade as the primary commodity in elite burials. An increased value placed on bronze vessels and weapons to accommodate ceremonies of ancestral reverence may have resulted in experimentation with piece-mould casting technology (Mei 2009).Therefore, to achieve a better understanding of the role of bronze metallurgy in the growth of early Chinese civilisation, an interdisciplinary approach combining technological and social perspectives is both essential and desirable.

Possible directions for future research

In considering directions for future research, three points may be emphasised: first, technological analysis is fundamental and must be continued, with a greater emphasis on the samples collected from production sites, including slag, ores and furnace materials. Comparative studies based on analytical data should be done in order to characterise regional technological features, which could be used as evidence for understanding patterns of contact or interaction between various regions.

Secondly, regional analysis should include an examination of the socio-cultural context of the metallurgical production, such as ceramic production, ritual preference, burial custom, religious tradition, artistic choice, and so on. More attention should be paid to the exploration of how early metallurgical activities were organised and what they contributed to the increase in social complexity. A combination of technological and archaeological approaches will be necessary.

Thirdly, local innovations and long-distance contacts are two vital issues that are worthy of in-depth exploration by employing an interdisciplinary approach. Several important early metallurgical innovations have been noted, such as piece-mould casting, arsenical copper and leaded bronze, but how they emerged and developed still remains unclear and requires further research. If these issues were examined within a wider Eurasian context, the contributions from both local innovations and outside impetus could be properly highlighted. It is important to understand the scale, patterns and mechanisms of early interaction, including trade and exchange, movement of people, as well as community conflict.

Conclusions

Over the past 20 years considerable progress has been made in our understanding of the development of early metallurgy in China. The new research work carried out in north-west China has revealed an increasing body of evidence for cultural connections between north-west China and the Eurasian steppe. Recent archaeological discoveries have also shown that Erlitou was the most important metallurgical centre in the Central Plains of China, where the earliest bronze industry based on piece-mould casting technology emerged and developed. It is suggested that social requirements of ritual practice may have been the major driving force behind the 'birth' of piece-mould casting technology. Further research is needed to clarify the relationship between the north-west and the Central Plains of China regarding the advent of bronze metallurgy in both regions.

Acknowledgements

I would like to thank Professors Tsun Ko, Han Rubin, Sun Shuyun, Kathy Linduff, Li Shuicheng, Jessica Rawson, Thilo Rehren, Mayke Wagner, Christopher Cullen, Li Xiao, Liu Xuetang and Bryan Hanks for their encouragement, support and help with this research. I am especially grateful to Raul Carstocea and John Moffett for their great help with revising this paper.

This research received financial support from the project entitled 'Exploring the Origins of Chinese Civilization' (Project number: 2006BAK21B03) and from a research grant awarded by the Chiang Ching-kuo Foundation.

References

An, Zhimin (安志敏) (1981) 中国早期铜器的几个问题 ('Some issues concerning early copper and bronze objects in China'), 考古学报 (*Acta Archaeologica Sinica*) 3: 269–84.

An, Zhimin (1993) 试论中国的早期铜器 ('On early copper and bronze objects in China'), 考古 (*Archaeology*) 12: 1110–19.

Bai, Yunxiang (白云翔) (2002) 中国的早期铜器与青铜器的起源 ('The origins of early Chinese copper and bronze objects'), 东南文化 (*Southeast Culture*) 7: 25–37.

Barnard, N. (1961) *Bronze Casting and Bronze Alloys in Ancient China*. Monumenta Serica Monograph XIV. Canberra: Australian National University.

Barnard, N. (1983) 'Further evidence to support the hypothesis of indigenous origins of metallurgy in ancient China', in D.N. Keightley (ed.), *The Origins of Chinese Civilization*, 237–77. Los Angeles: University of California Press.

Barnard, N. and Sato, T. (1975) *Metallurgical Remains of Ancient China*. Tokyo: Nichiosha.

Bishop, C.W. (1942) *Origin of the Far Eastern Civilizations: A Brief Handbook*. Washington, DC: Smithsonian Institute.

Cheng, T.K. (1974) 'Metallurgy in Shang China', *T'oung Pao* 60(4/5): 209–29.

Chernykh, E.N. (1992) *Ancient Metallurgy in the USSR: The Early Metal Age*. Cambridge: Cambridge University Press.

Chernykh, E.N. and Kuzminykh, S.V. (1989) *Drevnyaya Metallurgiya Severnoi Evrazii* (Ancient Metallurgy in Northern Eurasia). Moscow: Nauka.

Debaine-Francfort, C. (1995) *Du Néolithique à l'Âge du Bronze en Chine du Nord-Ouest: La culture de Qijia et ses connexions*. Paris: Éditions Recherche sur les Civilisations.

Debaine-Francfort, C. (2001) 'Xinjiang and northwestern China around 1000 BC: cultural contacts and transmissions', in R. Eichmann and H. Parzinger (eds), *Migration und Kulturtransfer*, 57–70. Bonn: Rudolf Habelt GmbH.

Fitzgerald-Huber, L.G. (1995) 'Qijia and Erlitou: the question of contacts with distant cultures', *Early China* 20: 17–67.

Ho, P.T. (1975) *The Cradle of the East: An Inquiry into the Indigenous Origins of Techniques and Ideas of Neolithic and Early Historic China, 5000–1000 BC*. Hong Kong: Chinese University of Hong Kong.

IHMM (Institute of Historical Metallurgy and Materials, University of Science and Technology Beijing) (2001) 新疆哈密天山北路墓地出土铜器的初步研究 ('A preliminary study of copper and bronze objects excavated at the Tianshanbeilu cemetery in Hami, Xinjiang'), 文物 (*Cultural Relics*) 6: 79–89.

Jettmar, K. (1981) 'Cultural and ethnic groups west of China in the second and first millennia BC', *Asian Perspectives* 24(2): 145–62.

Jin, Zhengyao (金正耀) (2000) 二里头青铜器的自然科学研究与夏文明探索 ('A scientific study of the Erlitou bronzes and an exploration of the Xia civilisation'), 文物 (*Cultural Relics*) 1: 57–64.

Li, Shuicheng (李水城) (2005) 西北及中原早期冶铜业的区域特征及交互作用 ('The regional characteristics and interaction of early metallurgical industries in the northwest and the Central Plains'), 考古学报 (*Acta Archaeologica Sinica*) 3: 239–77.

Li, Shuicheng and Shui, Tao (水涛) (2000) 四坝文化铜器研究 ('A study of the copper and bronze objects of the Siba culture'), 文物 (*Cultural Relics*) 3: 36–44.

Li, Xiuhui (李秀辉) and Han, Rubin (韩如玢) (2000) 朱开沟遗址出土铜器的金相学研究 ('A micrographic study of copper and bronze objects excavated from the Zhukaigou site'), in Inner Mongolian Institute of Archaeology (ed.) 朱开沟—青铜时代早期遗址发掘报告 (*Zhukaigou-Excavation Report of an Early Bronze Age Site*), 422–46. Beijing: Wenwu Chubanshe.

Li, Yanxiang (李延祥), Jia, Haixin (贾海新) and Zhu, Yanping (朱延平) (2003) 大甸子墓地出土铜器初步研究 ('A preliminary study of copper and bronze objects excavated from the Dadianzi cemetery'), 文物 (*Cultural Relics*) 7: 78–84.

Liang, Honggang (梁宏刚) (2004) 二里头遗址出土铜器的制作技术研究 (*A Study of the Manufacturing Techniques of the Copper and Bronze Objects Excavated from the Erlitou Site*). Unpublished PhD dissertation, University of Science and Technology Beijing.

Liang, Honggang and Sun, Shuyun (孙淑云) (2004) 二里头遗址出土铜器研究综述 ('A review of the studies of copper and bronze objects from the Erlitou site'), 中原文物 (*Cultural Relics from the Central Plains*) 1: 29–39.

Liang, Xingpeng (梁星彭) and Yan, Zhibing (严志斌) (2003) 襄汾陶寺新石器时代遗址 ('The Neolithic site at Taosi, Xiangfeng'), 中国考古学年鉴-2002 (*The Yearbook of Chinese Archaeology 2002*), 138. Beijing: Wenwu Chubanshe.

Linduff, K.M. (1998) 'The emergence and demise of bronze-producing cultures outside the Central Plains of China', in V.H. Mair (ed.), *The Bronze Age and Early Iron Age Peoples of Eastern Central Asia*, 619–43. Washington, DC: Institute for the Study of Man.

Linduff, K.M. (2004) 'How far does the Eurasian metallurgical tradition extend?', in K.M. Linduff (ed.), *Metallurgy in Ancient Eastern Eurasia from the Urals to the Yellow River*, 1–14. New York: Edwin Mellen Press Ltd.

Linduff, K.M., Han, Rubin and Sun, Shuyun (eds) (2000) *The Beginnings of Metallurgy in China*. New York: Edwin Mellen Press Ltd.

Liu, Xuetang (刘学堂) and Li, Wenying (李文瑛) (2007) 中国早期青铜文化的起源及其相关问题新探 ('A new exploration of the origins of early Chinese bronze cultures and relevant issues'), 藏学学刊 (*Journal of Tibetology*) 3: 1–63.

Loehr, M. (1949) 'Weapons and tools from Anyang and Siberian analogies', *American Journal of Archaeology* 53: 126–44.

Lü, Enguo (吕恩国), Chang, Xi'en (常喜恩) and Wang, Binghua (王炳华) (2001) 新疆青铜时代考古文化浅论 ('On the Bronze Age cultures in Xinjiang'), in Su, Bai (宿白) (ed.), 苏秉琦与当代中国考古学 (*Su Bingqi and Archaeology in Contemporary China*), 172–93. Beijing: Kexue Chubanshe.

Mei, Jianjun (梅建军) (2000) *Copper and Bronze Metallurgy in Late Prehistoric Xinjiang: Its Cultural Context and Relationship with Neighboring Regions*. BAR International Series 865. Oxford: Archaeopress.

Mei, Jianjun (2001) 关于中国冶金起源及早期铜器研究的几个问题 ('On the origins of metallurgy and the studies of early copper objects in China'), 吐鲁番学研究 (*Turfanological Research*) 2: 57–68.

Mei, Jianjun (2003a) 'Cultural interaction between China and Central Asia during the Bronze Age', *Proceedings of the British Academy* 121: 1–39.

Mei, Jianjun (2003b) 'Qijia and Seĭma-Turbino: the question of early contacts between northwest China and the Eurasian steppe', *Bulletin of the Museum of Far Eastern Antiquities* 75: 31–54.

Mei, Jianjun (2009) 'Early metallurgy and socio-cultural complexity: archaeological discoveries in northwest China', in B. Hanks and K.M. Linduff (eds), *Monuments, Metals and Mobility: Trajectories of Social Complexity in the Late Prehistoric Eurasian Steppe*. Cambridge: Cambridge University Press.

Mei, Jianjun and Shell, C. (1999) 'The existence of Andronovo cultural influence in Xinjiang during the second millennium BC', *Antiquity* 73(281): 570–78.

Mei, Jianjun, Shell, C., Li, Xiao (李肖) and Wang, Bo (王博) (1998) 'A metallurgical study of early copper and bronze artefacts from Xinjiang, China', *Bulletin of the Metals Museum* 30: 1–22.

Muhly, J.D. (1988) 'The beginnings of metallurgy in the Old World', in R. Maddin (ed.), *The Beginning of the Use of Metals and Alloys*, 2–20. Cambridge: MIT Press.

Qian, Wei (潜伟) (2006) 新疆哈密史前时期青铜器及其与周边文化的关系 (*The Prehistoric Copper and Bronze Objects and their Relationships with Neighbouring Cultures*). Beijing: Zhishi Chanquan Chubanshe.

Qu, Changzhi (曲长芝) and Zhang, Riqing (张日清) (1999) 二里头遗址出土铜器X射线荧光分析 ('XRF analysis of copper and bronze objects excavated from the Erlitou site'), in Institute of Archaeology, Chinese Academy of Social Sciences (ed.) 偃师二里头 (*The Erlitou Site in Yanshi*), 399. Beijing: Zhongguo Dabaike Quanshu Chubanshe.

Smith, C.S. (1977) 'Review of Barnard & Sato 1975 and Ho 1975', *Technology and Culture* 18(1): 80–86.

Sun, Shuyun (1998) 东灰山遗址四坝文化铜器的鉴定及研究 ('The examination and study of the copper objects from the Siba culture site at Donghuishan'), in Gansu Provincial Institute of Cultural Relics and Archaeology and Jilin University (eds), 民乐东灰山考古：四坝文化墓地的揭示与研究 (*Archaeology of the Donghuishan Site in Mingle: Excavation and Study on the Cemetery of Siba Culture*), 191–5. Beijing: Kexue Chubanshe.

Sun, Shuyun and Han, Rubin (1981) 中国早期铜器的初步研究 ('A preliminary study of early Chinese copper and bronze artefacts'), 考古学报 (*Acta Archaeologica Sinica*) 3: 287–301. For an English translation, see Linduff *et al.* 2000: 129–51.

Sun, Shuyun and Han, Rubin (1997) 甘肃早期铜器的发现与冶炼制作技术的研究 ('Discovery of early copper and bronze artefacts in Gansu and studies of their smelting and manufacturing techniques'), 文物 (*Cultural Relics*) 7: 75–84. For an English translation, see Linduff *et al.* 2000: 175–93.

Sun, Shuyun, Qian, Wei and Wang, Hui (王辉) (2003) 火烧沟四坝文化铜器成分分析及制作技术的研究 ('A study of the compositions and manufacturing techniques of copper objects of the Siba culture from the Huoshaogou site'), 文物 (*Cultural Relics*) 8: 86–96.

Takahama, Shu. (2000) 'On several types of copper objects of the first half of the second millennium BC from Central Eurasia', in Research Centre for the Silk Road (ed.), *Metals and Civilisation*, 111–23. Nara: Research Centre for the Silk Road [in Japanese].

Watson, W. (1985) 'An interpenetration of opposites? Pre-Han bronze metallurgy in west China', *Proceedings of the British Academy* 70: 327–58.

Zhang, Tian'en (张天恩) (2002) 天水出土的兽面铜牌饰及有关问题 ('The bronze plaque with animal face unearthed in Tianshui and relevant problems'), 中原文物 (*Cultural Relics from the Central Plains*) 1: 43–6.

Metal trade in Bronze Age Central Eurasia

Liangren Zhang

ABSTRACT The Seĭma-Turbino Phenomenon is an important conceptual breakthrough in understanding the 442 metal artefacts discovered in an extraordinarily vast territory in Eurasia. Much of the phenomenon, however, has not been well understood. The locations where the metal artefacts were produced, for instance, remain a puzzle. The commonly held assumption that they were locally produced has become questionable in the face of ever-accumulating morphological and compositional data of metal artefacts of neighbouring cultures. Alternatively, the idea of 'metal trade' is introduced to account for them. This paper suggests that communities of the Seĭma-Turbino Phenomenon had intimate trade connections with those of the Abashevo and Sintashta cultures. Compositional analysis eloquently discloses that Seĭma-Turbino communities constantly imported metal ingots from Abashevo and Sintashta communities. Morphological and compositional analyses of actual artefacts further indicate that they also constantly imported ready-made products from the latter. This paper also suggests that Abashevo communities imported both copper ingots and metal products from their Sintashta partners. The resulting conclusion is that ancient communities in Central Eurasia were actively involved in metal trade, and the Sintashta communities were the major producers and suppliers of metal ingots and artefacts. The fundamental impetus underlying this pattern of metal trade is that the highly demanded arsenic- and silver-containing copper ores are concentrated in the Eastern Urals, the territory of the Sintashta culture.

Introduction

In the study of ancient Eurasian metallurgy, the Seĭma-Turbino Phenomenon is a great conceptual breakthrough.[1] Formulated by Chernykh and Kuz'minykh (1989), this concept unites a total of 442 metal artefacts and 30 casting moulds that have been discovered in a vast territory in Eurasia, stretching from the Altai Mountains in the east to Finland in the west (Chernykh 1992: 216). The lion's share of these objects comes from cemeteries at Seĭma, Turbino, Reshnoe, Rostovka and Satyga. Much of the social and economic lives of the Seĭma-Turbino populations, however, remains enigmatic because the five cemeteries are mostly devoid of either human skeletons or ceramic vessels. No settlement has ever been discovered. Only the morphologically distinctive metal artefacts (socketed axes, forked-shank spearheads and double-edged knives) and the prevalence of tin bronze connect these finds into one cultural entity (Chernykh and Kuz'minykh 1989; Chernykh 1992: 218–19).

Chernykh and Kuz'minykh (1989) have carried out thorough morphological and spectral analyses of the aforementioned 442 metal artefacts. On the one hand, they found certain artefact types such as socketed axes and some metal alloys such as tin bronze, common to the entire territory; on the other hand, they have discerned two separate regions: an eastern or Siberian region and a western or European region, which are divided by the Ural Mountains (Chernykh 1992: 220).[2] The Siberian region is characterised by the predominance of ornate socketed axes and tin bronze, whereas the European region is characterised by a greater variety of artefact types, including shaft-hole axes, spearheads, adzes and daggers, as well as a variety of metal alloys including pure copper, copper-arsenic and silver-copper alloys, and billon.

Chernykh and Kuz'minykh (1989) assume that the metal artefacts were produced locally by their users. The discovery of 30 moulds testifies that some communities of the Seĭma-Turbino Phenomenon were engaged in metal production and were possibly responsible for a part of the 442 items. The ever-accumulating metallurgical data of metal artefacts of neighbouring cultures, however, point to another possibility. These data, in combination with the above-mentioned morphological and compositional analyses, have illuminated the idea that a significant portion of the Seĭma-Turbino metals were produced by external communities. In this paper I will substantiate this idea by delving into the ancient metal trade with a particular focus on the European region and the contemporaneous cultures of Abashevo and Sintashta. The term 'metal trade' used here refers primarily to the cross-regional flow of ingots and actual artefacts; the bulky copper ores were less likely to be involved in long-distance trade. This task requires the use of data on the distribution of copper ores, chemical composition of metal artefacts and metal-working communities; fortunately these data, although far from sufficient, are now available.

Database

Chernykh and Kuz'minykh, once again, have provided most of the data regarding the distribution of copper ores and the chemical composition of metal artefacts. In 1967–68, Chernykh (1970: 35–49) located over 30 Bronze Age mines in the Eastern Urals (Trans-Urals in Russian)

(Fig. 1). He was even able to define the compositional characteristics of the copper ores to be found in a number of them. The Elenovka and Ush-Katta mines yield ores that contain some tin (up to 1%) and nickel (up to 0.1%). This is a possible source for tin bronze. Other possible tin sources are located some considerable distance away in central Kazakhstan and further east in the Altai Mountains (including eastern Kazakhstan). The Tash-Kazgan and Nikol'skoe mines produce high arsenic and high silver ores, respectively. These are the sources for arsenic-copper and silver-copper alloys.[3] Ores that are rich in gold and silver but with less arsenic have been found at Ishkino (Zaykov *et al.* 2002). Overall, the Eastern Urals is a rich source of polymetallic ores.[4] Copper ores are also available in the Western Urals (Pre-Urals in Russian) in the valleys of the Kama River and its tributaries Belaia and Ufa, but they are all sandstones, which have very low contents of minor metals – arsenic, silver, nickel, lead and iron; antimony and gold are completely absent here (Chernykh 1970: 48). This is the source for chemically pure copper. It follows that if a community in the Western Urals wished to produce arsenic and tin bronzes, it was bound to import ores from the Eastern Urals, or central Kazakhstan, or even farther away in the Altai Mountains. In the Donets River valley, particularly the Bakhmut basin, ancient copper ore mines and smelting sites have also been discovered (Tatarinov 1977). The sandstone ores available in this area are good for obtaining pure copper as well, however, they contain some silver and arsenic. As in the Western Urals, ancient communities in this region had to import tin-containing ingots if they wished to produce tin bronze. Certainly, communities in the regions that lack copper ores, such as the Lower Volga (from the Samara Bend to the end of the Volga River) and much of the territory of the Seĭma-Turbino Phenomenon, had to import either ingots or ready-made artefacts.

Information pertaining to Bronze Age metalworking communities comes mostly from settlements, but sometimes from cemeteries as well. Evidence of metal production has been discovered in a large number of settlements and cemeteries not only in the Eastern Urals, but also in the Western Urals, the Middle Volga (from Nizhniĭ Novgorod to the Samara Bend), and the Don and Donets River valleys. It is very likely that Seĭma-Turbino communities had intimate trade connections with contemporaneous metalworking communities of the Abashevo and Sintashta cultures as well as the Alakul' and Srubnaya cultures (Zhang 2007: 223–37). While the Srubnaya culture dates to the Late Bronze Age, roughly to 1690–1390 BC (Chernykh 2002: 125; Görsdorf 2004: 293–4), the Sintashta cultures date to the Middle Bronze Age, approximately 2040–1730 BC (Epimakhov *et al.* 2005). The Seĭma-Turbino Phenomenon thus spans the Middle and Late Bronze Age and dates roughly to 2040–1390 BC. Due to the limitations of space, this paper only deals with the trading activities among Seĭma-Turbino, Abashevo and Sintashta communities.

Metalworking communities have been widely discovered in the territory of the Abashevo culture, evidenced by settlements such as Shilovskiĭ in the Don River valley, Tochka in the Middle Volga valley, and Balanbash in the Western Urals (Priakhin 1976; Sal'nikov 1954: fig. 2). Although they only had geographical access to sandstone ores in the Western Urals, the Middle Volga, and the Donets River valley, they were nevertheless able to acquire the desired metals by importing polymetallic ingots, or actual artefacts, from the Eastern Urals. By comparison, metalworking communities of the Sintashta culture, represented by a number of settlements and cemeteries, were located in the Eastern Urals, and geographically proximate to the polymetallic mines in this region (Fig. 2). It is reasonable to assume that Sintashta communities produced and supplied arsenic- and silver-containing metals

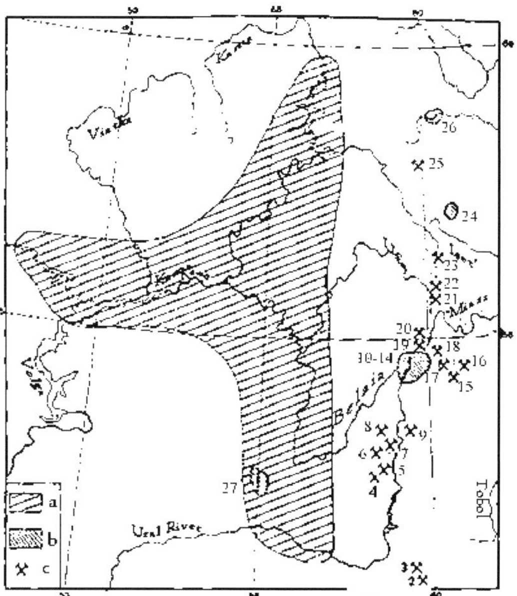

Figure 1 Distribution of copper ores in the Eastern and Western Urals (adapted from Chernykh 1970: 37, fig. 32): (a) area of copper sandstone ores; (b) groups of copper ore mines; (c) copper ore mines: 2. Ush-Katta; 3. Elenovka; 4. Ishkinino; 6. Dergamysh; 8. Bakr-Uziak; 9. Vorovskaia Yama; 17. Tash-Kazgan; 18. Nikol'skoe.

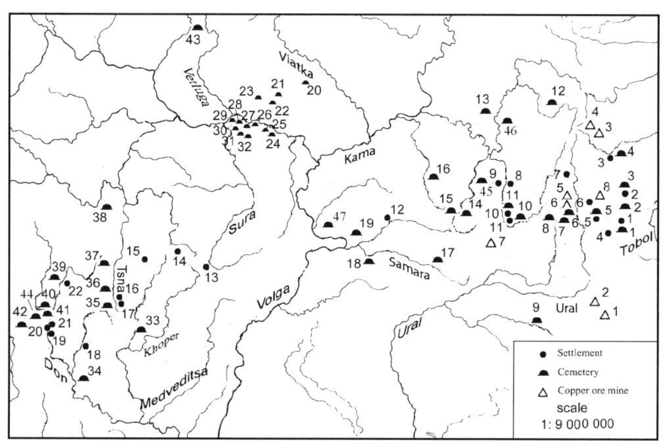

Figure 2 Settlements and cemeteries of the Sintashta and Abashevo cultures and the Seĭma-Turbino Phenomenon (adapted from Priakhin and Khalikov 1987: 126, fig. 23).
Settlements: 1. Sintashta; 5. Arkaim; 9. Balanbash; 10. Beregovskiĭ-I, -II; 12. Tochka; 21. Shilovskiĭ. *Cemeteries*: 1. Sintashta; 2. Kamennyĭ-Ambar-V; 4. Krivoe Ozero; 5. Bol'shekaraganka; 6. Al'mukhametovskiĭ-I; 8. Baishev-IV; 9. Tanabergen-II; 11. Beregovskiĭ-I; 12. Iukalekulevo; 13. Turbino-I, -II; 17. Nikiforovskiĭ; 33. Vlasov-I; 36. Selezni-II; 37. Staroĭur'evo; 38. Reshnoe; 40. Vvedenka; 43. Seĭma; 44. Khokhol'skiĭ; 45. Chukrakly. 46. USt'-Gaiva; 47. Sokolovka; 48. Rostovka. *Copper ore mines*: 1. Ush-Katta; 2. Elenovka; 3. Tash-Kazgan; 4. Nikol'skoe.

Table 1 Chemical composition of metal-related samples from Sintashta and Abashevo communities. Sources of data: Sintashta settlement (SS) (Zaĭkova 1995); Sintashta cemetery (SC) (Chernykh and Kuz'minykh 2005); Arkaim settlement (AS) and cemetery (BK) (Bushmakin 2002); Krivoe Ozero cemetery (KrO) (Degtiareva and Kuz'minykh 2003). * Abashevo (ASH) refers to sites in the Volga and Don River valleys (Chernykh 1971: 207).

Community	Unit	Cu	Cu+As	Cu+Sn	Ag	Cu+Ni	Total
Sintashta	SS	7	60	2	2		71
		9.9%	84.5%	2.8%	2.8%		100%
	SC	27	118	11	11		167
		16.2%	70.6%	6.6%	6.6%		100%
Arkaim	AS	11	12				23
		47.8%	52.2%				100%
	BK		15		2	3	20
			75%		10%	15%	100%
Krivoe Ozero	KrO	3	13	4			20
		15%	65%	20%			100%
Abashevo*	ASH	22	62		3	1	88
		25%	70.5%		3.4%	1.1%	100%

for Abashevo communities. Sintashta communities such as Sintashta (the type site) and Arkaim were in fact specialised in metal production, as is evident from the ubiquitous metalworking remains in their settlements (Gening *et al.* 1992; Zdanovich 1995).

A significant number of settlements and cemeteries of the Abashevo and Sintashta communities have yielded copper ores, slag, moulds and metal artefacts, and thus indicate the presence of metalworking and/or metal-consuming communities. Altogether 88 specimens from Abashevo sites comprising artefacts, ores and slag, have been subjected to spectral analysis, while 301 specimens from Sintashta settlements and cemeteries have been similarly tested (Chernykh 1971; Zaĭkova 1995; Bushmakin 2002; Degtiareva and Kuz'minykh 2003). A total of 167 specimens were taken from the Sintashta cemetery (SC) and analysed by Chernykh and Kuz'minykh.[5]

Overall, the ever-progressing excavation and research endeavours have made it possible to consider the hypothesis of metal trade among Seĭma-Turbino, Abashevo and Sintashta communities. The following text considers in particular two major aspects of this hypothetical trend: (1) the metal trade between Seĭma-Turbino and Abashevo and Sintashta communities and (2) the metal trade between Abashevo and Sintashta communities. The purpose of this focus is to highlight the extensity and intensity of ancient metal trade in Central Eurasia. Compositional and morphological data will be employed to elucidate the two themes.

Metal trade

The hypothesis of metal trade among Abashevo and Sintashta communities is first borne out by the chemical composition of their metal artefacts (Table 1). It must be noted that the factor of recycling has barely been probed in relation to the metal artefacts under discussion, but it cannot be ignored and therefore the points made here are not conclusive. The 301 tested specimens of the Sintashta culture include 167 pieces from the Sintashta cemetery (SC), 71 pieces from the Sintashta settlement (SS), 23 from the Arkaim settlement (AS) and 20 from the Bol'shekaraganka cemetery (BK) associated with Arkaim, and 20 from the Krivoe Ozero cemetery (KrO). It appears that the percentage of objects made of pure copper (derived from sandstone ores), is lower than that of arsenic bronze artefacts, even among Abashevo specimens. This fact denotes a regular trade of arsenic-containing metal ingots and artefacts from Sintashta communities to their Abashevo partners, whether through direct trade of ingots or through indirect trade of recycled metals. Other metals are relatively modest in quantity. It appears that tin bronze and silver were regularly produced and used by Sintashta communities, as they have been found in both their settlements and cemeteries. They were also buried in tombs at Krivoe Ozero and Bol'shekaraganka. Thus, the three silver objects among the Abashevo specimens must have been imported from the Sintashta community. In addition, among the Abashevo specimens is one nickel and copper alloy, which finds three analogies among the specimens from Bol'shekaraganka. The provenance of this alloy might have been Arkaim.

The trade in metal artefacts is a subtle subject. Compositional data are insufficient, because one community may use recycled metals from a third party. It is necessary to introduce the morphological criterion, i.e. because of the complexity of provenancing metal objects, it is assumed that if two objects from two distant sites are both compositionally and morphologically close to each other, one is likely to be an import. One community may acquire metal by recycling metals from source communities, or its products may emulate the forms of metal artefacts of source communities. The likelihood that both situations occur at the same time is considered low. Although many artefacts from the Abashevo and Sintashta communities have been subjected to compositional analysis, many others have not been analysed. The most readily available data are in fact line drawings that show their formal characteristics. Thus, when compositional analysis is done and can be combined with stylistic analysis, it is possible to discern imported artefacts with greater certainty. When compositional data are not available, stylistic analysis is the only means of achiev-

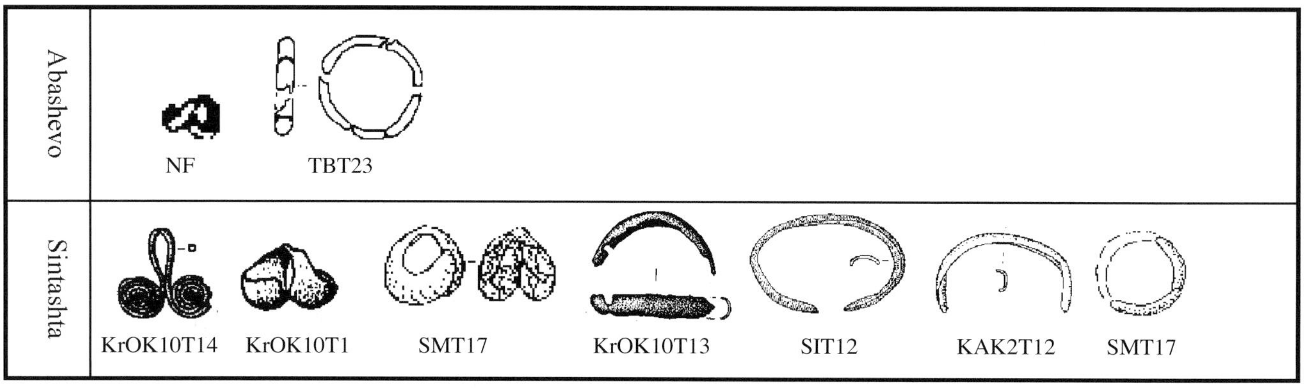

Figure 3 Metal ornaments from Sintashta and Abashevo communities: NF – Nikiforovskiĭ, TB – Tanabergen-II. Sources of data: NF (Vasil'ev and Priakhin 1979: 147, fig. 3: 20); KrOK10T14, KrOK10T1, KrOK10T13 (Vinogradov 2003: 14, fig. 76: 1; 128, fig. 55: 2; 172, fig. 75: 12); SMT17, SMT17, SIT12 (Gening *et al.* 1992: 169, fig. 82: 8, 10; 285, fig. 159: 1); KA-2T12 (Kostiukov *et al.* 1995: 201, fig. 25: 7); TBT23 (Tkachëv 2004: 12, fig. 4: 5).

Table 2 Chemical composition of metal ornaments from Sintashta and Abashevo communities. Sources of data: Sintashta (SM, SI; Chernykh and Kuz'minykh 2005); Arkaim (BK-25; Bushmakhin 2002); Krivoe Ozero (KrO; Degtiareva and Kuz'minykh 2003); Khokhol'skiĭ (KhKh; Chernykh 1971).

	Site	Cu	Sn	Pb	Zn	Bi	Ag	Sb	As	Fe	Ni	Co	Au	Mn
Bracelet	SMT2	core	12	.2		.005	.06	.1	.03	.004	.003			
	SMT2	core	6	.06		.002	.06	.03	.04	.005	.001			
	SMT2	core	6	.1	.02	.004	.2		.03	.07	.001			
	SMT2	core	4	.2	.02	.003	.2	.003	.05	.015	.002			
	SMT13	core	7	.3		.003	.0005	.003	.2	.1	.003		.03	
	KrO	core	6.2	.21	.001	.012	.14	.073	.24	.0037	.0047	.0014	.0001	
	SIT11	core	.006	.0012	.001	.0003	.032	.027	2	.063	.047	.0014	.0015	
	SIT11	core	.0017	.001	.0012	.0003	.027	.027	2.7	.086	.064	.0019	.0019	
	SIT12	core	.0022	.001	.001	.0008	.019	.043	2.7	.13	.047	.0015	.0041	
	SIT12	core	.002	.0022	.001	.001	.089	.029	2.5	.12	.035	.0014	.0037	
	SIT12	core	.0016	.001	.002	.0016	.0048	.0076	1.8	.0037	.052	.0012	.001	
	KhKh	core		.001			.007		4.5	.6	.009	.006		.005
	KhKh	core		.001			.04		3.5	.3	.005		~.001	.001
Earring	SMT2	<40					core						<3	
	SMT2	<3					core						~30	
	SMT13	<1					~50						~50	
	SIT12	>30					core							
	BKK25	11.33			.28	.18	87.13		.82					
	KhKh	10	little	yes	?	little	core	?	yes	little	little			yes

ing this. In this case, it is assumed that local products are stylistically homogeneous and occur repetitively, whereas imported items are incongruent with the styles of local products and are relatively rare. Because moulds were handmade, artefacts from one community should not be expected to be identical. This line of analysis is merely suggestive, and is riddled with uncertainties given the above-mentioned caution that one community may emulate foreign forms.

I will now focus on the types of metal artefacts that are morphologically more sophisticated, and thus more expressive of the aesthetic and technical preferences of the producing communities. The first type to be considered is that of ornaments (Fig. 3). The double-ring pendant from Krivoe Ozero (KrOK10T14) is apparently an import from the Abashevo community because it rarely occurs in tombs of the Sintashta culture but is found frequently in those of the Abashevo culture.[6] On the contrary, the one-and-half-twist earring has been found pervasively at the Sintashta, Arkaim, Krivoe Ozero and Kamennyĭ Ambar-V cemeteries of the Sintashta culture, but it has only been discovered at the Nikiforovskiĭ and Khokhol'skiĭ cemeteries of the Abashevo culture (Vasil'ev and Priakhin 1979; Chernykh 1971: 208–9). The earrings from Sintashta and Abashevo cemeteries are uniform in shape (Fig. 3). The four tested pieces from the Sintashta cemetery (Chernykh and Kuz'minykh 2005) and one from the Arkaim settlement (Bushmakin 2002: 138, fig. 81: 7b) are all made of silver-based alloys (Table 2). This fact and the prevalence of this product at Sintashta cemeteries indicate that the Sintashta community produced and supplied these ornaments as they had better access to the silver-rich mines at Nikol'skoe.

As regards bracelets, most Abashevo specimens were made by bending solid wires (Chernkykh 1964: 107). By contrast, the items from the Sintashta cemetery were predominantly bent from groove-shaped foils (Fig. 3). Two tested specimens from Khokhol'skiĭ were produced locally or by different communities of the same culture, although they were produced of arsenic bronzes, the raw materials of which came probably from the Eastern Urals (Table 2). Five tested samples from Sintashta and Krivoe Ozero contain high percentages of tin (4.2%–8.3%) and a broad range of trace elements (Table 2). They were therefore more likely produced by the Sintashta community. Three analogous bracelets can be found at the Tanabergen-II cemetery of the Abashevo culture (Tkachëv 2004: 12, fig. 4: 5–7). Although the chemical composition of these items is as yet unknown, morphologically they all resemble the Sintashta items.

Daggers have been discovered in much greater quantity in tombs of both the Abashevo and Sintashta cultures. They are morphologically diverse and thus expressive of stylistic preferences of individual communities. Unfortunately, not many specimens have been compositionally tested. They fall into several types (Fig. 4). The distinctive type A specimen, which features a thin and broad leaf-like blade (SMT2), and type B specimen, which is thicker and narrower than type A, occur four times altogether at the Sintashta cemetery

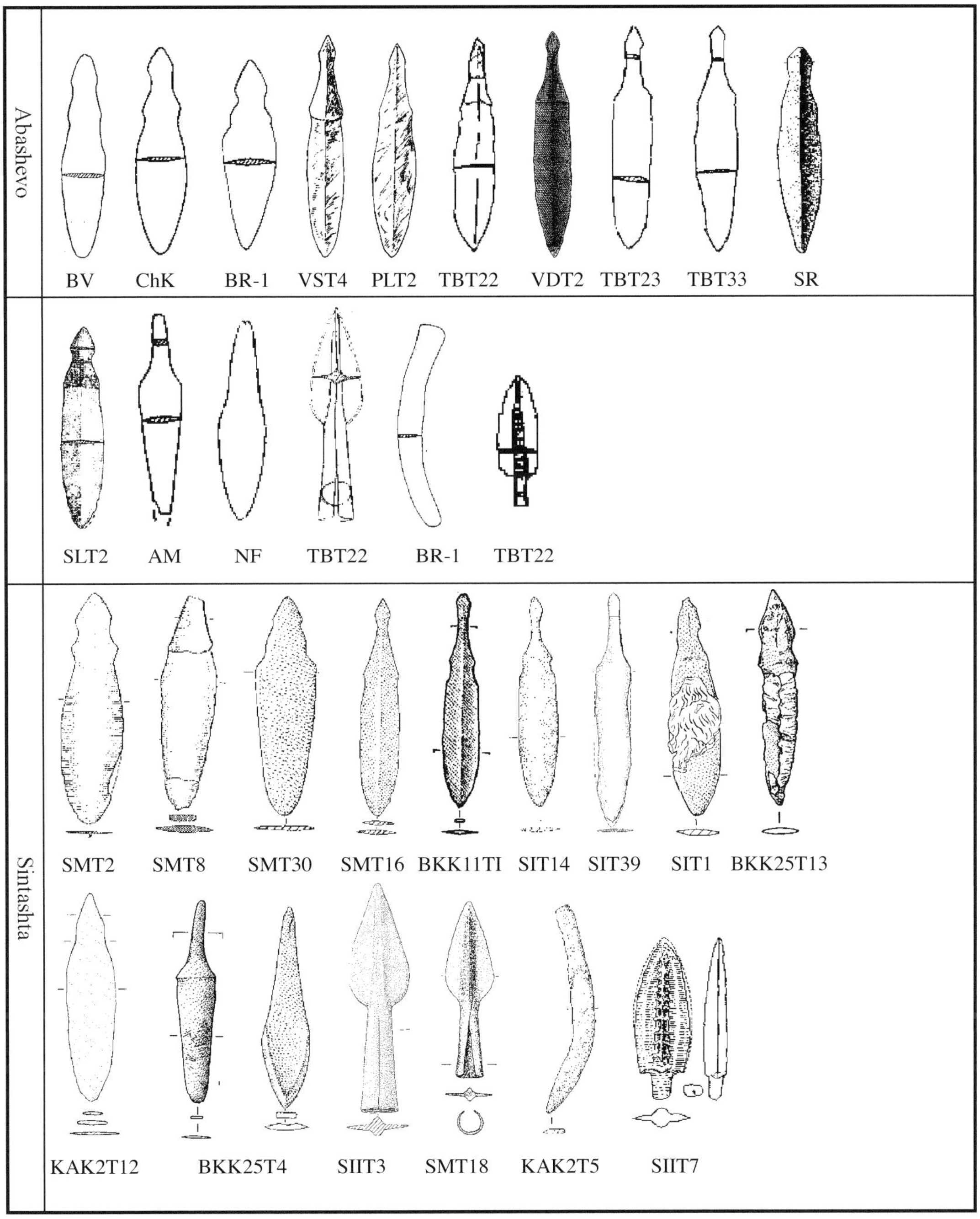

Figure 4 Metal daggers, spearheads, and sickles from Sintashta and Abashevo communities: BV – Baishev-IV; ChK – Chukrakly, BR-I – Beregovskiĭ-I, VS – Vlasov-I, PL – Pavlovsk, TB – Tanabergen-II, VD – Vvedenka, SR – Staroiur'evo, SL – Selezni-2, NF – Nikiforovskiĭ, AM – Al'mukhametovskiĭ, Iuk – Iukalekulevo. Sources of data: BV(Obydënnov and Obydënnova 1996: fig. 34: 1); ChK, BR-I, AM (Gorbunov 1986: 89, fig. 14: 16; 18: 1, 15); VS, PL (Siniuk and Kozmirchuk 1995: 45, fig. 4: 2; 61, fig. 14: 10); VDT2 (Siniuk and Kileĭnikov 1976: 163, fig. 3: 5); SR (Priakhin 1972: 235, fig. 2: 1); SLT2 (Priakhin *et al.* 1998: 13, fig. 6: 1); NF (Vasil'ev and Priakhin 1979: 147, fig. 4: 4); TBT22, T23, T22, T22 (Tkachëv 2004: 9, fig. 2: 7; 12, fig. 4: 11; 9, fig. 2: 1, 8); SMT2, 8, 30, 16, 39, 18, SIT14, 1, 14, SIIT3 (Gening *et al.* 1992: 117, fig. 46: 6; 124, fig. 51: 13; 213, fig. 114: 1; 151, fig. 70: 2; 231, fig. 126: 21; 176, fig. 88: 3; 268, fig. 148: 18; 255, fig. 140: 4; 268, fig. 148: 20; 307, fig. 175: 10); BKK11T1 (Botalov *et al.* 1996: 67, fig. 3: 4); BKK25T13, BKK25T4 (Zdanovich 2002: 60, fig. 35: 2; 34, fig. 16: 1); KAK2T5, KAK2T12 (Kostiukov *et al.* 1995: 197, fig. 21: 1; 201, fig. 25: 1).

(SMT8), and once at the Baishev-IV (BV) and Chukrakly (ChK) cemeteries of the Abashevo culture. The frequency of occurrence attributes the source of these types of daggers to Sintashta communities.

Type C daggers have acute hilt and blade contours and include several variations. The piece from Sintashta (SMT16), the most precisely fashioned variation, finds analogies at the cemeteries of Bol'shekaraganka (BKK11T1) of the Sintashta culture, and Vlasov-I (VST4), Pavlovsk (PLT2) and Tanabergen-II (TBT22) of the Abashevo culture (Fig. 4). They were probably produced and supplied by the Sintashta community as this variation appears at the two neighbouring cemeteries of Sintashta and Bol'shekaraganka, whereas analogous finds are widely dispersed at Abashevo sites. A second variation, which features a slender hilt and long blade, appears three times at Sintashta (only two, SIT14 and SMT39, are shown in Fig. 4) and once at Bol'shekaraganka (BKK11T1, not shown in Fig. 4) of the Sintashta culture, and twice at Tanabergen-II (TBT23, T33) and once at Vvedenka (VDT2) of the Abashevo culture (Fig. 4). This distributional pattern again suggests that they were produced by Sintashta communities.

The type D dagger, featuring a short hilt and a blunt contour, occurs three times at Sintashta (SIT1), once at Bol'shekaraganka (BKK25T13) and Kamennyĭ Ambar-V (KAK2T12) of the Sintashta culture, and once at Staroiur'evo (SR) and Selezni-II (SLT2) of the Abashevo culture (Fig. 4). The item from Krivoe Ozero (Vinogradov 2003: 157, fig. 68: 10) of the Sintashta culture, which has a fragmented body, may also fall within this type. The concentrated finds of this type at Sintashta cemeteries identify the associated communities as the producers/suppliers of this type of dagger. The following two types, however, are not easy to judge in terms of provenance because specimens are isolated. The type E dagger, which has an elongated rectangular hilt, is seen at Bol'shekaraganka (BK-25T4) and Al'mukhametovskiĭ-I (AM) of the Abashevo culture (Fig. 4). The type F dagger, equipped with a long shank and a triangular blade, occurs once both at Sintashta (SIT3), and Nikiforovskiĭ (NF) (Fig. 4).

Other categories of metal artefacts – spearhead, knife and sickle – are limited in quantity. Spearheads, which have robust sockets and wide blades, have been found at Sintashta (SMT18) and Kamennyĭ Ambar-V (KAK2T5) of the Sintashta culture, and Tanabergen-II (TBT22) of the Abashevo culture (Fig. 4); bent sickles have been found at Sintashta (SI14) of the Sintashta culture, and Beregovskiĭ-I (BR-I) of the Abashevo culture (Fig. 4). In addition, the unique metal arrowheads found at Sintashta (SII7) recur at Tanabergen-II (TB22) (Fig. 4). The Tanabergen-II piece imitates the Sintashta items not only in general contour but also in the vertebrae-like stalk. Due to the lack of metallurgical analysis, the provenance of these types of artefacts cannot be determined at present. But their concentration at cemeteries of Sintashta communities suggests that they were produced and supplied by Sintashta communities.

The metal trade between Sintashta and Abashevo communities and Seĭma-Turbino communities becomes apparent when the formal and compositional characteristics of Seĭma-Turbino artefacts are examined. The 442 artefacts comprise 115 socketed axes, 73 spearheads, 126 knives, 21 daggers, 9 shaft-hole axes, 12 adze-chisels, 22 bracelets, 19 rings and 45 other objects (Chernykh and Kuz'minykh 1989: 38). The percentage of the total made from similar alloys was as follows: tin-copper and tin-arsenic-copper alloys (37.2%); arsenic-copper and arsenic-antimony-copper alloys (44%); pure copper (10.8%); copper-silver (3.3%) and silver-copper alloys (4.7%) (Chernykh and Kuz'minykh 1989: 188). The major ore sources were the arsenic- and silver-containing ores from the Eastern Urals and the sandstone ores from the Western Urals and from the Abashevo and Sintashta communities that dominated these regions. Alternatively, they might have been acquired from recycled metal artefacts, which were originally made of metals from these sources. When chemical composition is considered together with formal characteristics, four varieties emerge. First, there are the items that are morphologically absent in cemeteries or settlements of Abashevo and Sintashta cultures, for instance the socketed axes, fork-like spearheads and double-edged knives. There is little likelihood that they were imports from Sintashta and Abashevo communities; rather, they were probably local products or imports from elsewhere. This variety falls further into two different groups. Some are tin bronzes, and the communities might have obtained metals for their production from Elenovka and Ush-Katta, central Kazakhstan and the Altai Mountains; the others are pure copper or non-tin alloys, and they were made of metal ingots imported from Sintashta and Abashevo communities. Secondly, there are those that are morphologically close to specimens from Sintashta and Abashevo communities. This variety likewise falls into two different categories. Some are tin bronzes, and their production locations are ambiguous because Seĭma-Turbino communities could have obtained tin from the Altai Mountains themselves. What deserves our greatest attention are those artefacts that are analogous to finds at Sintashta and Abashevo communities, and are made of pure copper and non-tin alloys. They are the subject of the following discussion.

The spearhead that has rhombic sections from Ust'-Gaiva, and the two from the Sintashta cemetery are similar in shape (Fig. 5), and all were made of copper-arsenic alloys (Chernykh and Kuz'minykh 1989: 285, table 2.1: nos. 4, 6, 7). This is in sharp contrast to the spearheads with fork-like shanks of the Seĭma-Turbino Phenomenon (Chernykh and Kuz'minykh 1989: 285–6, table 2.1: nos. 9–49). While the fork-like spearheads were likely produced by the Seĭma-Turbino community or imported from elsewhere, the spearhead from Ust'-Gaiva was probably manufactured by the Sintashta community. One spearhead from Reshnoe (Chernykh and Kuz'minykh 1989: 285–6, table 2.1: no. 57) has a similar form but with an added side loop made of a copper-arsenic alloy; it might likewise have come from the Sintashta community (Fig. 5). Two daggers from Turbino-I and Seĭma, the hilts of which have triangular tips, resemble the specimen from the Sintashta cemetery (SIT1) (Fig. 5). The one from Turbino-I is a copper-arsenic alloy, and the other from Seĭma contains antimony in addition to this alloy (Chernykh and Kuz'minykh 1989: 293, table 3.1:

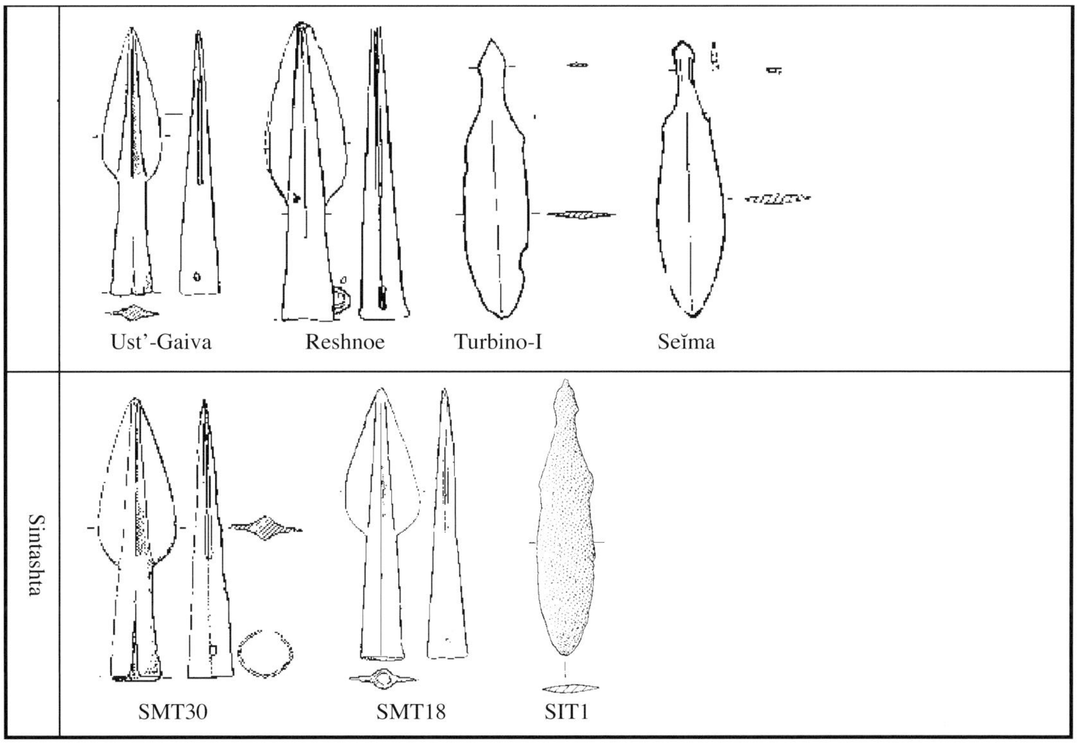

Figure 5 Spearheads and daggers from Sintashta, Potapovka and Seĭma-Turbino communities. Sources of data: Ust'-Gaiva, Reshnoe, SM30, SM18 (Chernykh and Kuz'minykh 1989: 66, fig. 25: 2; 80, fig. 40: 1); Turbino-I, Seĭma (Chernykh and Kuz'minykh 1989: 102, fig. 58: 2, 5); SI1 (Gening *et al.* 1992: 255, fig. 140: 1).

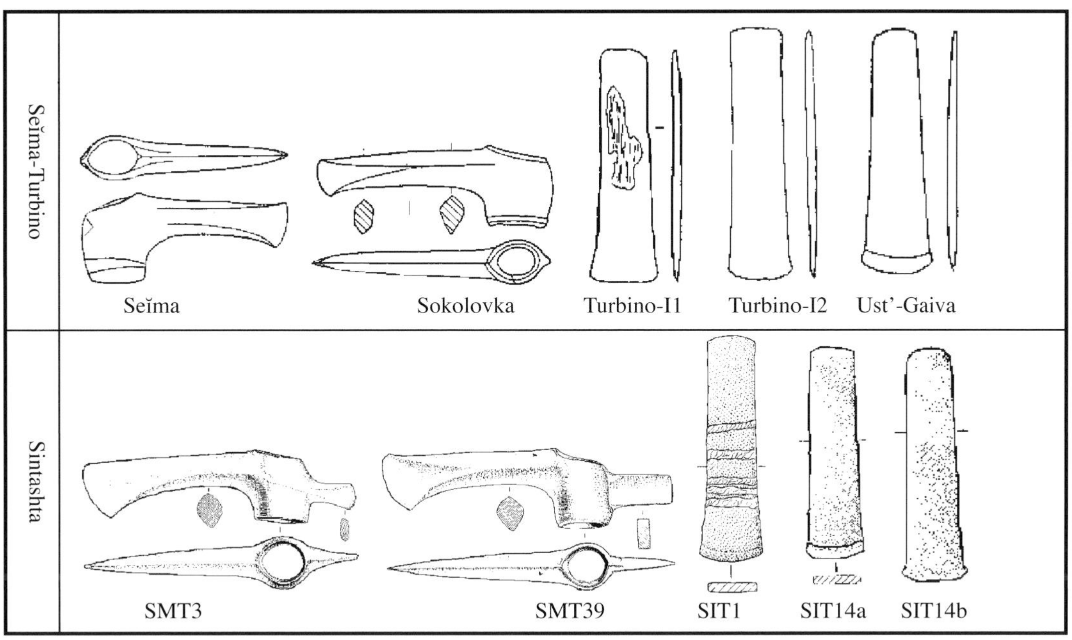

Figure 6 Shaft-hole axes and adzes from cemeteries of the Sintashta culture and the Seĭma-Turbino Phenomenon. Sources of data: Seĭma, Sokolovka (Chernykh and Kuz'minykh 1989: 127, fig. 70: 5, 6); SM3, SM39 (Gening *et al.* 1992: 122, fig. 49: 6; 232, fig. 127: 1); Turbino-I1, Turbino-I2, Ust' Gaiva (Chernykh and Kuz'minykh 1989: 129, fig. 71: 4, 5, 8); SI1, SI14a, SI14b (Gening *et al.* 1992: 255, fig. 140: 7; 268, fig. 148: 15, 16).

nos. 2, 5; Fig. 5). The sources of these daggers must also have been the Sintashta community.

The two shaft-hole axes from Seĭma and Sokolovka morphologically match the two items from the Sintashta cemetery (SMT3, SMT39), except that the latter two have small butts protruding from the shaft barrel (Fig. 6). The former were made of a copper-arsenic-antimony and a copper-arsenic alloy, respectively (Chernykh and Kuz'minykh 1989: 300, table 6.1: nos. 5, 8). The two Sintashta pieces are copper-arsenic-nickel alloys. The chemical compositions of the Seĭma-Turbino pieces are not identical but are close to those of the two Sintashta axes. The three adzes from Turbino-I and Ust'-Gaiva repeat the forms of samples from Sintashta and they chemically belong to the copper-arsenic alloy type (Chernykh and Kuz'minykh 1989: 301, table 7.1: nos. 5, 4, 8; Fig. 6). The formal and compositional

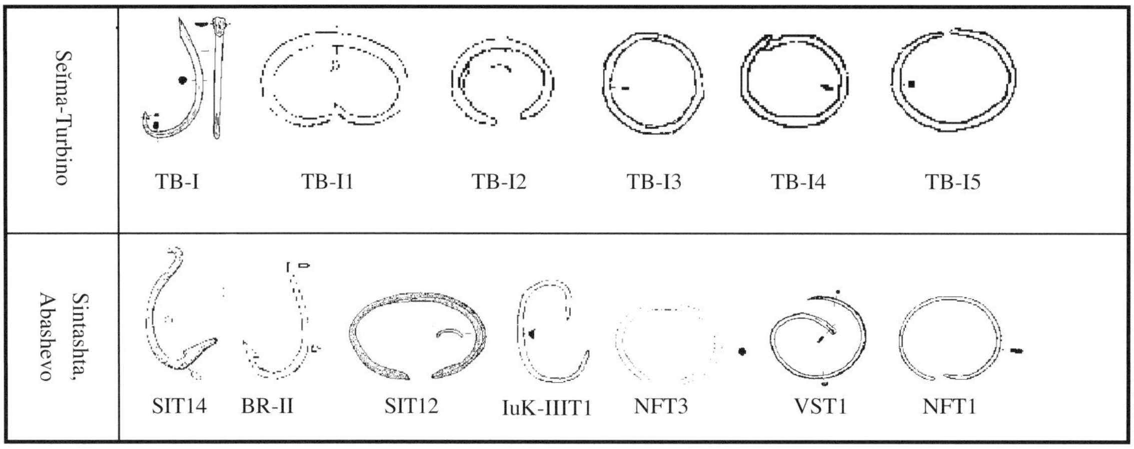

Figure 7 Fishing hooks and bracelets from Sintashta, Abashevo and Seĭma-Turbino communities: TB – Turbino. Sources of data: TB-I, TB-I1, 2, 3, 4, 5 (Chernykh and Kuz'minykh 1989: 131, fig. 72: 24; 133, fig. 73: 17, 21, 15, 10, 8); SIT14, SIT12 (Gening *et al.* 1992: 268, fig. 148: 2; 274, fig. 153: 1); BR-II, IuK-IIIT1 (Gorbunov 1986: 90, fig. 15: 7, 20); NFT3, NFT1 (Vasil'ev and Priakhin 1979: 147, fig. 3: 3, 1); VST1 (Siniuk and Kozmirchuk 1995: 45, fig. 4: 1).

resemblance identifies Sintashta as the producing/supplying community of these shaft-hole axes and adzes.

The fishing hooks from Turbino-I (TB-I) are morphologically common among the finds from Sintashta (e.g. SIT14) and Abashevo (e.g. BR-II) sites; their chemical compositions, unanimously copper-arsenic alloys (Chernykh and Kuz'minykh 1989: 303, table 9.2), indicate that they were imported from Sintashta and Abashevo communities (Fig. 7). In the meantime, among the 29 bracelets (Chernykh and Kuz'minykh 1989: 133, fig. 73: nos. 1–8, 10–28, 30–34) of the Seĭma-Turbino Phenomenon, 25 are made of solid wires that are circular, rectangular and triangular in section. These forms recall specimens of the Abashevo culture. Three pieces from Turbino-I (TB-I3, 4, 5) in particular resemble the four pieces from the cemeteries at Iukalekulevo (IuK-3T1), Nikiforovskiĭ (NFT3, NFT1) and Vlasov-I (VST1) (Fig. 7). One bracelet is made of a copper-silver alloy (Chernykh and Kuz'minykh 1989: 304, table 10.1: no. 15). Two other pieces from Turbino-I (e.g. TB-I1, 2), however, have groove-shaped sections, and match specimens from the Sintashta cemetery (SIT12) (Fig. 7). The two bracelets are made exclusively of copper-arsenic alloys (Chernykh and Kuz'minykh 1989: 304, table 10.1: nos. 19, 22). These bracelets again reflect the vibrant trade from Sintashta and Abashevo communities to their Seĭma-Turbino partners.

Conclusions

Due to the sparseness of compositional data, many of the foregoing hypotheses cannot be confirmed, but the combination of available compositional and morphological data are strongly indicative of dynamic and expansive trading networks that involved Sintashta, Abashevo and Seĭma-Turbino communities. In such networks, not only copper ingots, but also ready-made metal artefacts were circulated. As far as these data go, the major directions of the flow of metal goods were from Sintashta communities to their Abashevo counterparts, and from Sintashta and Abashevo communities to their Seĭma-Turbino counterparts. Sintashta communities were therefore the hubs of these trading networks. The ultimate reason for this pattern of metal trade lies in the concentration of the much-desired polymetallic ores in their territories, the Eastern Urals. The metal types of arsenic-copper and silver-copper alloys from this region were constantly in demand.

This paper therefore not only offers a new understanding of the Seĭma-Turbino metals, but also supplies a new perspective to the issue of social development in eastern Europe. Communities of the Sintashta culture as a whole were economically and, by implication, politically stronger than those of the Abashevo culture. This superiority is manifested in the larger fortified settlements and greater deposits of metal artefacts and other prestige goods in tombs of the Sintashta culture (Gening *et al.* 1992; Botalov *et al.* 1996; Vinogradov 2003; Zdanovich 2002; Epimakhov 2005; Zhang 2007: ch. IV). The probable underlying mechanism is that metals commanded great economic and social value in ancient societies, and the metal producing and supplying communities gained greater economic wealth and political power than the consuming communities. This is not the place to extrapolate this theme at length, but it is worth advancing the point that metal trade may throw new light onto early social evolution.

Notes

1. The term 'Seĭma-Turbino' was first coined by Bader (1964). It was defined as a cultural community that subsumes both the cemeteries at Seĭma and Turbino, which represent two local cultures in the Kama and Oka river valleys.
2. In a recent statistic Chernykh *et al.* (2004: 26) counted 'slightly over five hundred' metal objects belonging to this phenomenon.
3. The other arsenic-rich source is the Caucasian Mountains. Chernykh (1992: 200) argues that active exploitation of this source took place in the time of Pit-Grave, Catacomb and Poltavka cultures, but had ceased in the time of the Seĭma-Turbino Phenomenon. This position is adopted by Kohl (2007).
4. The mine at Dergamysh has been found to contain gold, but the mine at Vorovskaya Yama has not been defined compositionally (Zaykov *et al.* 2002).

5. I would like to express my profound gratitude to E.N Chernykh and S.V. Kuz'minykh for their unmediated generosity in offering me unpublished data, some of which is quoted in this paper.
6. This type of pendant was used widely in Bronze Age eastern Europe, but those from Abashevo communities find the closest match to those of Sintashta communities.

References

Bader, O.N. (1964) *Ancient Metallurgy of the Western Urals*. Moscow: Science Press [in Russian].

Botalov, S.G., Grigor'ev, S.A. and Zdanovich, G.B. (1996) 'Burial complexes of the Bronze Age Bol'shekaraganka cemetery', in G.B. Zdanovich (ed.), *Materials of Archaeology and Ethnography*, 64–88. Cheliabinsk: Kamennyĭ Poias [in Russian].

Bushmakin, A.F. (2002) 'Metal objects from Kurgan 25 of the Bol'shekaraganka cemetery', in D.G. Zdanovich (ed.), *Arkaim: Necropolis*, 132–43. Cheliabinsk: South-Ural Book Publisher [in Russian].

Chernykh, E.N. (1964) 'On metal of the Abashevo culture', in *Sites of the Stone and Bronze Age*, 97–109. Moscow: Nauka [in Russian].

Chernykh, E.N. (1970) *Ancient Metallurgy in the Urals and the Volga River Valley*. Moscow: Nauka [in Russian].

Chernykh, E.N. (1971) 'Metals of the Don variant of the Abashevo culture', in A.D. Priakhin (ed.), *The Abashevo Culture in the Don River Valley*, 205–11. Voronezh: Voronezh University Press [in Russian].

Chernykh, E.N. (1992) *Ancient Metallurgy in the USSR: The Early Metal Age*. Cambridge: Cambridge University Press.

Chernykh, E.N. (2002) 'Absolute chronology of the Late Bronze Age layers of Gornyĭ', in E.N. Chernykh (ed.), *Kargaly, Tome II*, 125–7. Moscow: Languages of Slavonic Culture [in Russian].

Chernykh, E.N. and Kuz'minykh, S.V. (1989) *Ancient Metallurgy of Northern Eurasia*. Moscow: Nauka [in Russian].

Chernykh, E.N. and Kuz'minykh, S.V. (2005) *Chemical Compositions of 167 Specimens from the Sintashta Cemetery* [unpublished data].

Chernykh, E.N., Kuz'minykh, S.V. and Orlovskaia, L.B. (2004) 'Ancient metallurgy of Northeast Asia: from the Urals to the Saiano-Altai', in K.M. Linduff (ed.), *Metallurgy in Ancient Eastern Eurasia from the Urals to the Yellow River*, 15–36. New York: Edwin Mellen Press Ltd.

Degtiareva, A.D. and Kuz'minykh, S.V. (2003) 'Results of analytical investigation of metals from the cemetery Krivoe Ozero', in N.B. Vinogradov (ed.), *The Cemetery of the Bronze Age Krivoe Ozero in the Southern Eastern Urals*, 285–309. Cheliabinsk: South-Ural Book Publisher [in Russian].

Epimakhov, A.V. (2005) *Early Complex Society of North Central Eurasia, Book 1*. Cheliabinsk: Cheliabinsk House of Printing [in Russian].

Epimakhov, A.V., Hanks, B. and Renfrew, S. (2005) 'Radiocarbon chronology of the Bronze Age sites in the Transuralian region', *Russian Archaeology* 4: 92–102 [in Russian].

Gening, V.F., Zdanovich, G.B. and Gening, V.V. (1992) *Sintashta: An Archaeological Site of the Arian Tribes in the Ural-Kazakhstan Steppes*. Cheliabinsk: South-Ural Book Publisher [in Russian].

Gorbunov, V.S. (1986) *The Abashevo Culture of the Southern Western Urals*. Ufa: Bashkir Pedinstitute [in Russian].

Görsdorf, J. (2004) 'C^{14} dates of specimens from Gornyĭ', in E.N. Chernykh (ed.), *Kargaly, Tome III*, 293–4. Moscow: Languages of Slavonic Culture [in Russian].

Kohl, P. (2007) *The Making of Bronze Age Eurasia*. Cambridge/New York: Cambridge University Press.

Kostiukov, V.P., Epimakhov, A.V. and Nelin, D. (1995) 'New sites of the Middle Bronze Age in South Trans-Ural', in I.V. Vasil'ev (ed.), *Ancient Indoiranian Culture in Volgo-Ural*, 156–207. Samara: Samara State Pedagogical University [in Russian].

Obydënnov, M.F. and Obydënnova, G.T. (1996) *Sites of Bronze Age Southern Urals*. Ufa: Press of Eastern University [in Russian].

Priakhin, A.D. (1972) 'A kurgan of the Late Bronze Age at Staroiur'evo village', *Soviet Archaeology* 3: 233–43 [in Russian].

Priakhin, A.D. (1976) *Settlements of the Abashevo Commonwealth*. Voronezh: Voronezh University Press [in Russian].

Priakhin, A. D. and Khalikov, A. Kh. (1987) 'The Abashevo culture', in O.N. Bader, D.A. Krainov and M.F. Kosarev (eds), *The Bronze Age of the Forest Belt of SSSR*, 124–131. Archaeology of SSSR. Moscow: Nauka.

Priakhin, A.D., Moiseev, N.B. and Besedin, V.I. (1998) *Selezni-2: A Kurgan of the Don-Volga Abashevo Culture*. Voronezh: Voronezh University Press [in Russian].

Sal'nikov, K.V. (1954) 'The Abashevo culture in the Southern Urals', *Soviet Archaeology* 21: 52–94 [in Russian].

Siniuk, A.T. and Kozmirchuk, I.A. (1995) 'Some aspects of investigation of the Abashevo culture in the Don valley', in I.B. Vasil'ev (ed.), *Ancient Indo-Iran Cultures in Volga-Ural (Second Millennium BC)*, 72–81. Samara: Samara State Pedagogical University [in Russian].

Siniuk, A.T. and Kileĭnikov, V.V. (1976). 'A kurgan at Vvedenka village', *Soviet Archaeology* 1: 159–69 [in Russian].

Tatarinov, S.I. (1977) 'On the mining-metallurgical center of the Bronze Age in Donbass', *Soviet Archaeology* 4: 192–206 [in Russian].

Tkachëv, V.V. (2004) 'Burial complexes with disc-form cheekpieces in the steppe Western Urals', in A.N. Usacuk (ed.), *Cheekpieces*, 7–30. Donetsk: Donetska Oblasti Folk Culture Museum [in Russian].

Vasil'ev, I.B. and Priakhin, A.D. (1979) 'A non-kurgan Abashevo cemetery at the Nikiforovskii Forestry in Orenburg', *Soviet Archaeology* 2: 145–52 [in Russian].

Vinogradov, N.B. (2003) *The Cemetery of the Bronze Age Krivoe Ozero in the Southern Eastern Urals*. Cheliabinsk: South-Ural Book Publisher [in Russian].

Zaykov, V.V., Yuminov, A.M., Bushmakin, A.P., Zaykova, E.V., Tairov, A.D. and Zdanovich, G.B. (2002) 'Ancient copper mines and products from base and noble metals in the Southern Urals', in K. Jones-Bley and D.G. Zdanovich (eds), *Complex Societies of Central Eurasia from the 3rd to the 1st Millennium BC*, 417–42. Washington, DC: Institute for the Study of Man.

Zaĭkova, E.V. (1995) 'Compositions of metal artifacts from the Sintashta settlement', in G.B. Zdanovich *et al.* (eds), *Russia and the East: Problems of Interaction, Volume V, Book 2*, 152–7. Cheliabsinsk: Cheliabinsk State University [in Russian].

Zdanovich, D.G. (2002) 'Rescue excavations in the zone of the Bol'shekaragan quarry', in D.G. Zdanovich (ed.), *Arkaim: Necropolis*, 110–16. Cheliabinsk: South-Ural Book Publisher [in Russian].

Zdanovich, G.B. (1995) 'Arkaim: Arians on Ural or unstable civilization', in G.B. Zdanovich (ed.), *Arkaim: A Page of the Ancient History of South Ural*, 21–42. Cheliabinsk: Kamennyĭ Poias [in Russian].

Zhang, L. (2007) *Ancient Society and Metallurgy: A Comparative Study of Bronze Age Societies in Central Eurasia and North China*, PhD thesis, University of California, Los Angeles.

Documentary and archaeological evidence for an antique copper-nickel alloy (*baitong*) production in southern China and its exportation to India

François Widemann

ABSTRACT Chemical analyses of Graeco-Bactrian coins and archaeological artefacts from Taxila excavations, made of an alloy of copper and nickel, raised several problems in the history of economy and of metallurgical techniques, not yet understood. Available ancient Chinese, Greek and Roman documents, archaeological data and laboratory studies are presented with a tentative interpretation for the metallurgical technique, its origin and the trade of alloy ingots towards Taxila, in the north of Punjab. Evidence appears in favour of the origin of this technique in Yunnan and Sichuan, where nickel ore was present, and *baitong* (白铜), as mentioned in Chinese documents, was already known in the Spring and Autumn period (722–481 BC). This alloy was not used for coinage in China, with only one, much later, exception. It was considered a precious metal in China, and authors who have come to the conclusion that Bactrian Greek copper-nickel coinage had the same value as copper coinage are probably wrong: the issue of coins made of this alloy by three consecutive Greek kings *c*.195–185 BC can be related to a monetary crisis due to the scarcity of precious metals.

Introduction

At the beginning of the second century BC, the Greeks of Bactria found themselves in a difficult financial and monetary situation described here briefly (for a more developed study of this crisis see Widemann 2000). The secession of Sogdiana, during the siege of Bactra (209–206 BC) by the Seleucid king Antiochus III (Widemann 1989), interrupted the arrival of gold from Altai, and Bactrian gold coinage was no longer issued. The victorious resistance of Bactrian king Euthydemus I led Antiochus to make peace and withdraw through northern India, taking from this very rich country treasures and war elephants. Demetrius, son of Euthydemus, easily conquered, ruined and disarmed kingdoms. He acquired large territories, but the issue by Demetrius as king (*c*.200–195 BC) of silver-plated coinage instead of silver is a clear sign that his glorious conquest did not make him rich (Fig. 1). Demetrius I, who died young, was succeeded by his son Euthydemus II, while still a child. The rule of Indian conquered territories was delegated to Pantaleon. Large and wealthy trading cities in these territories, especially Taxila, had, at least since the Maurya, extensively used a silver coinage of punch-marked ingots (Fig. 2) and coins (Fig. 3). Greek conquerors could not introduce a debased coinage without political risks. A coinage was issued made of an alloy of copper with such a high proportion of nickel that it was white and shone like silver. These monetary issues continued during the reign of three kings – Euthydemus II, Pantaleon and Agathocles – altogether from *c*.195 to

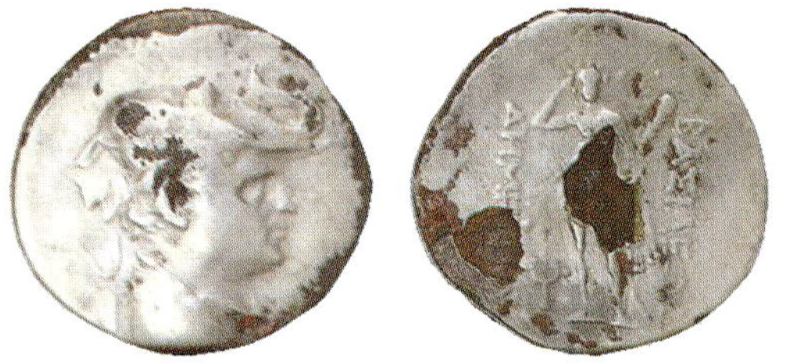

Figure 1 Tetradrachm of Demetrius I, King of Bactria, made of silver-coated bullion. Bust of Demetrios/Heracles. Aman ur Rahman collection no. 124. (Photo: Aman ur Rahman.)

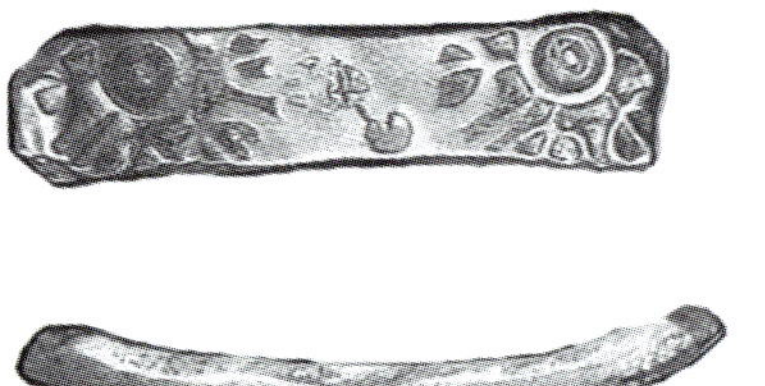

Figure 2 Punch-marked silver ingot 'bent bar'. Northern Indian coinage before the Bactrian Greek invasion, 11.30 g, 36 × 9mm. Aman ur Rahman collection no. 30. (Photo: Aman ur Rahman.)

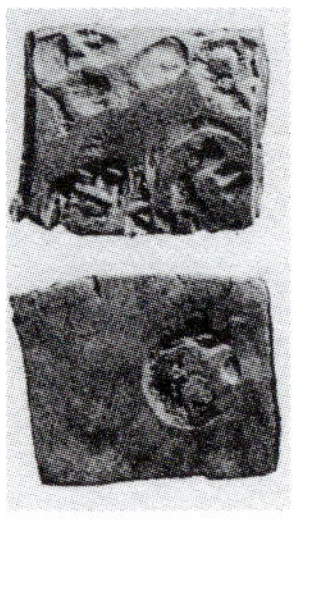

Figure 3 Punch-marked silver coin. Northern Indian coinage before the Bactrian Greek invasion, 3.05 g, 15 × 12mm. First Ai Khanoum coin hoard, no. 477, Kabul Museum.

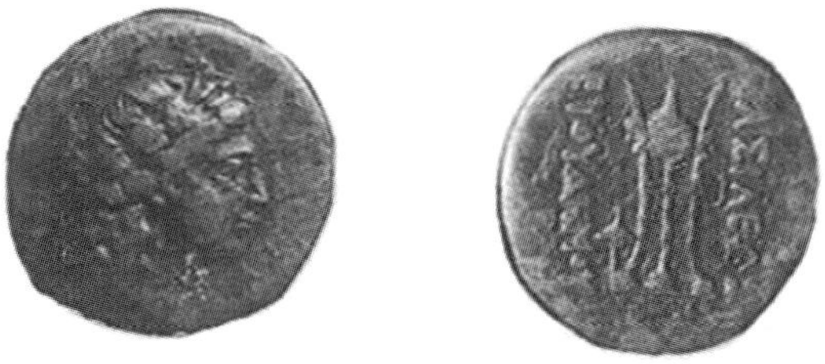

Figure 4 Cupro-nickel coin of Euthydemus II, King of Bactria. Bust of Apollo/Delphic tripod. Cabinet des Médailles, Paris.

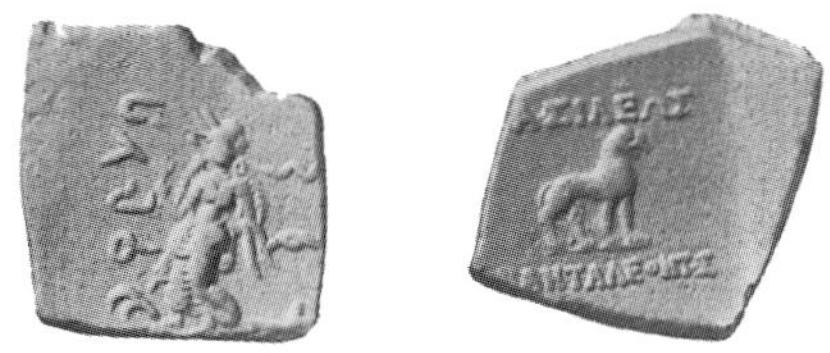

Figure 5 Bilingual bronze coin of Pantaleon, probably issued in Taxila. Indian divinity/lion. Cabinet des Médailles, Paris.

Figure 6 Silver bilingual coin of Agathocles, King of Bactria, with Indian gods on both sides. First Ai Khanoum coin hoard, Kabul Museum.

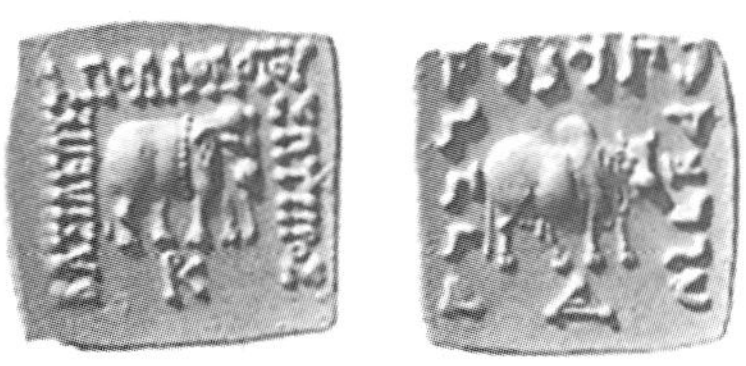

Figure 7 Silver bilingual coin of Apollodotus I in the Indo-Greek drachma module. Elephant/zebu. Cabinet des Médailles, Paris.

186/185 BC. Then production ceased suddenly about 10 years after the first issue.

The so-called nickel coinage introduced under Euthydemus II in Taxila (Fig. 4) can be considered as answering the difficulties met by Greek conquerors when confronted by the wealth and numerical superiority of the Indian population. Their nickel alloy coins bore exclusively Greek religious symbols and legends, suggesting that their main purpose was to pay Greek officers and soldiers in the occupied Indian territories. The same alloy had been used in Taxila for at least three centuries to make precious vases, but never before for coinage – in India or anywhere else. Pantaleon issued the first bilingual bronze coins bearing an Indian divinity on the obverse and a Greek religious symbol on the reverse (Fig. 5). He continued to issue cupro-nickel coinage in his own name after the early death of Euthydemus II. Agathocles, who succeeded Pantaleon, issued similar cupro-nickel and bilingual bronze coins but he also introduced a bilingual silver coinage bearing Indian gods on both sides (Fig. 6). This is interpreted by Paul Bernard as conferring on Hinduism the status of a state religion in the Greek territories of India, however it elicited a strong reaction from the Greek colons. Coin hoards hidden at the end of the reign of Agathocles in northern Bactria show this reign ended in disorder (Audouin and Bernard 1974: 37). Antimachus I and Apollodotus I, his associated king for the Indian territories, ceased the issue of cupro-nickel coins. The local silver city coinage was eventually replaced in the Greek territories of India by a royal bilingual silver coinage, on which any religious figure or symbol other than Greek was prohibited (Fig. 7).

The existence of this very early nickel coinage implies that large quantities of this alloy were available and that the Greek rulers could be certain that a regular supply of alloy would continue in the future. It raises an interesting problem in the field of the history of techniques, mining, metallurgy and trade, summarised by two main questions: (1) where was this metal coming from and (2) how could such an alloy be obtained in antiquity, knowing the rather high fusion temperature of nickel? These technical and possibly distance problems are directly related to the difficulty of obtaining this alloy and then to its price. In conclusion, I shall attempt to answer the question of the value of the 'nickel' currency as compared to currencies made of other metals in the Graeco-Bactrian kingdom.

The use of copper-nickel alloy in antiquity: early studies and hypothesis on its origin

Chemical analysis of a Chinese ingot, said to come from the Huili mines (Sichuan), brought back by a visiting French minister (Levol 1862), and the analysis of a Graeco-Bactrian coin (Flight 1868) brought back from India a few years later by Alexander Cunningham (1884) gave these compositions:

	Ingot	Coin
copper	79.40	77.585
nickel	16.02	20.038
iron	4.58	1.048
		traces of cobalt, tin, sulphur and silver

These surprising (to Europeans) results, obtained using wet chemistry, revealed the ancient use of nickel, an element separated and acknowledged as a new element only in 1751 by the Swedish chemist Cronstedt. The presence of a nickel-rich alloy in ancient India has been inferred by Cunningham from the Roman historian Quintus Curtius (IX, 8) in his mention of the present given by Oxydrachs and Mallians people (south Punjab) to Alexander the Great; among very precious materials were *ferri candidi talenta C*, i.e. one hundred talents (*c*.2.5 tons) of 'white iron'. It is my belief that it may have been copper-nickel alloy, believed to be iron by the Greeks, due to its ferromagnetic property. Other authors[1] interpret this gift as *wootz* steel, an ultra-high carbon steel produced in Tamoul country, in the extreme south of India. A less ambiguous mention of nickel alloy was found by Cunningham in a dedication for a gift of the Greek poet Crinagoras living at the time of the Roman emperor Augustus, that was engraved on a drinking cup: 'an Indian workmanship made in a bronze similar to silver'. Cunningham (1884: 308) gave an erroneous translation from the Greek verse Χαλκεον αργυρεω με πανεικελον, Ινδικον εργον (...) for him 'Indian brass as white as silver'. *Chalkos* means copper, a word that has sometimes been used to refer to bronze but never to brass, which was known in antiquity as *oreichalkos* (ορειχαλκος). Furthermore, it does not state that the *metal* is Indian, but that the *workmanship* of the metal is Indian. This second mistake is misleading about the origin of the alloy.

Later, archaeological excavations of the city of Taxila (in modern Pakistan) led by Marshall between 1913 and 1934, brought to light several artefacts made of copper-nickel alloy dating from 600 BC to 100 BC (Marshall 1951). Chemical analysis, later performed by Prakash and Rawat (1965: tables 5.3 and 5.6) revealed a nickel content (between 10 and 22%) far above the level of the natural nickel impurity in copper ore, which seldom exceeds 1%. No nickel ore is known in the neighbouring regions, and no mention of such an alloy in ancient Indian literature has been found by modern scholars. As Taxila was the only ancient city of northern India in which this alloy was found, there was likely a market for this metal. The Crinagoras mention also emphasises the fact that copper-nickel artefacts were sometimes exported to the Mediterranean basin in Roman times.

A possible Chinese origin for this alloy was first proposed by Cunningham (1884), suggested by the similarity of the ancient copper-nickel alloy with an alloy imported from China since the 17th century AD in rather large quantities, first by England then by various European countries. It is called *paktong* in English and in French, a word derived from the transliteration of the *Guǎngdōng* pronunciation *paăk t'ūng* of the word *baitong* (白铜) according to the accepted pronunciation of this Chinese word in the unified *pīnyīn* transliteration. In Europe, this metal was used mainly to make cheaper silverware (rather than use silver) and small value coins in Belgium in the 19th century. An extensive study of this modern metal was published by Mei Jianjun (1995). However, he only gave a short outline on the question of ancient cupro-nickel alloy, without reference to the archaeological discoveries of artefacts made of this alloy in Taxila, which is the main topic of this paper.

Cheng and Schwitter (1957: table 2) published the first attempt to solve the problem of the origin of the nickel alloy of the Graeco-Bactrian coins. They analysed five copper-nickel coins, archaeological artefacts of various dates and provenances, ores and modern Chinese *paktong*. They also quoted (after Laufer 1919) the *Chi Fu Tsung Shu*, vol. II (in Chinese) by K.L.Y. Ching (ch. 8A, 16), stating that *baitong* had already been mentioned in the dictionary *Kwang ya* (Gwăngyă 广雅), published by Chang I (*Zhāng Yī*, 张揖) in AD 230. Laufer specified that about 1% of arsenic was added to the ancient alloy, giving a more white aspect, a component measured precisely only later (Cowell 1989; Barrandon and Nicolet-Pierre 1989). The authors concluded that the origin of the alloy used by the Greeks was southern China. They found that the modern *paktong* was different from the ancient alloy: about half the copper has been replaced by zinc. The essential characteristics of *paktong* remained the same – high rate of nickel, white silver-looking aspect – but the composition changed slightly. The modern trend was to reduce the level of nickel, sometimes down to 5%, and to add zinc. Therefore, the (now lost) ingot analysed by Levol (1862), which did not contain zinc, may either be ancient or modern but dating before the trend to add zinc. The presence of zinc is the characteristic difference between ancient and modern cupro-nickel, a statement that is also supported by Mei (1995: 47).

Cheng and Schwitter's study has been strongly criticised by the American sinologist Cammann (1958): as no sinologist was associated with the project, several errors have occurred in the transcription of Chinese words and history. In spite of these errors, his contemptuous acrimony was not justified. Cammann's own arguments about the impossibility of production and trade from southern China are not convincing: 'assuming for a moment that there were nickel-copper mines in the still barbarous regions of modern Yunnan, which was not yet a part of the Chinese Empire – although there is no reason to believe that they did have such mines at that time – how could the ore or metal have reached the sea from there?'(Cammann 1958: 410). Without providing firm data on the local situation, Cammann denies the possibility that a 'barbarous' people (which he interprets as being outside the Chinese Empire) could have exploited mines, mastered metallurgical techniques and sold their products. This idea reflects prejudices dating to the 19th century about the level of knowledge and techniques reached by illiterate peoples, which has misled more than one specialist of classical civilisations. Furthermore, Cammann denied the existence of a route from China to India at that time, partly overland and partly on rivers via Burma, a route attested by several ancient references. According to Pirazzoli t'Serstevens (1974: 84): 'The last argument of Schuyler Cammann is that, when Chinese envoys looked for the route from Yunnan to India they did not find it. In fact, this last point does not prove at all the route did not exist: on the contrary, Sima Qian specifies that the access to this route was forbidden by the Kunmings.' Cammann's hypothesis of another origin of the ore, and local fabrication in the Indian subcontinent, was not confirmed.

Ancient Chinese references to nickel alloy[2]

Needham (1974: 232–3) criticised the interpretation of the Guǎngyǎ dictionary based on the identification of a metal called *wù* (鋈) with *paktong* that he called 'a blunt statement'. However, according to the *Hànyū dàcídiǎn* (汉语大词典) (*Great Dictionary of Characters and Expressions of the Han Language*) (vol. 11: 1308 (Shanghai, 1993)), the first meaning of the word *wù* is copper: *tóng* (铜.) But according to the *Shuōwén* 说文 (*Etymological Dictionary of Chinese* (literal translation: *Discourse about the Language*) published *c.* AD 100), '*wù* is white gold (*báijīn*)' (白金) (the character for gold is identical to the character for metal). The Guǎngyǎ is explicit: '*báitóng wèi zhī wù*' (白铜为之鋈) (white copper is called *wù*). In spite of the fact that the word *wù* was used with three different meanings, as Laufer had noted, the existence of *paktong* is really attested in AD 230.

Needham (1974: 229) quoted numerous ancient Chinese references, the first (for its technical precision) being the Thien Kung Khai Wu (*Tiān gōng kāiwù* 天工开物) (*Knowledge of the Things from Heaven* [= Nature] *and Art*), from 1637, by Sòng Yingxīng (宋应星) (1587–?), which describes many alloys, including 'white copper'.

Needham cites several Chinese references for the use of *paktong* in the Middle Ages. This alloy is mentioned in the Chiu Thang Shu (*Jiù Tángshū* 旧唐书) (*Ancient History of the Tang,* AD 945) as used for the decoration of imperial chariots. He also describes several Chinese coinages made of *paktong* (Needham 1974: 231). Each of the series of coins mentioned has been checked by François Thierry (2003), who found that Needham was wrong on this point: all the series of coins were in fact made of other alloys with no nickel content, except for one, quoted in the Hsin Thang Shu (*Xīn Tángshū* 新唐书) (*New History of the Tang*, written in 1060), which mentions a monetary emission of *paktong* coins under the Táng *c.*780. However, to date no single coin of this issue has been found. The text gives the origin of the metal as a mine in the province of Guǎngdōng. A more ancient reference cited by Needham (but not included in his bibliography), is the treatise Hua Yang Kuo Chih (*Huàyáng guózhi* 华阳国志) (*Monograph of the Countries South of Huà* [shān]) written in AD 347 by Chhang Chu (Cháng Qú 常璩). This treatise specifies that *paktong* is produced in the sub-prefecture of Tángláng (螳螂), 150 li (~ 87 km) north of the modern city of Dōngchuān (东川) (Yunnan).This city was in the south close to the river Niúlánjiāng (牛栏江), a tributary of the upper course of the Yáng zǐjiāng (扬子江). Needham (1974: 232, note a) also cites the German traveller Richthofen, who went to China in 1860 and provided a thorough description of the regions he visited.

A still more ancient reference can be found in the *Zhōnghuá dàojiào dàcídiǎn* (中华道教大词典) (*Great Dictionary of Chinese Taoism*): '*wù* is the ancient name of *paktong*. Its secret name *báijīn* 白金 (= white gold or white metal) was already used in the Spring and Autumn period (722–481 BC) and at the time of the Fighting Kingdoms (475–221 BC) in the current chronology. The earliest date of the archaeological cupro-nickel found in Taxila fits the first time interval.

The ores and the technique used to obtain the copper-nickel alloy in antiquity

Most nickel ore deposits consist of pentlandite ($(FeNi)_9S_8$) present with chalcopyrite ($CuFeS_2$). As Mei (1995: 45) stated: 'No ancient reference to the metallurgy of *paktong* has been found so far.' However, it is unlikely that in antiquity, the high temperature needed for the fusion of metallic nickel (1453 °C) could have been obtained. A more reasonable hypothesis assumes that some highly skilled Chinese metallurgists, without recognising nickel as a metal or what could be a eutectic mixing, found an empirical recipe to prepare this alloy, similar to the recipe of the cupro-nickel exported to Europe after the 17th century AD, but without zinc, the latter being known in China only after the 12th century AD or later (for the metallurgical process and modern references see Mei 1995). The secret recipe was only discovered again by R.C. Stanley in 1905 in Canada (Howard-White 1963). Needham (1974) quotes a number of different alloys already known in ancient China; the nickel-copper alloy is only one peculiar case in a country where metallurgical techniques were sufficiently advanced in antiquity to allow for the preparation of many alloys of various metals. Accepting the hypothesis of the production of this alloy in the territory of modern Yunnan, there is the problem of the possibility of its trade with Bactrian Greeks, most probably with Taxila, where the presence of artefacts made of copper-nickel alloy is reported as early as 600 BC, corresponding to the Spring and Autumn period in China, much earlier than the Graeco-Bactrian conquest, without any textual or archaeological alternative evidence for a production in the Indian subcontinent.

Ancient trade routes between southern China and India

The earliest evidence of trade between southern China and the gulf of Bengal was found in the archaeological excavations of tombs of the kingdom of Dian, in modern Yunnan (Pirazzoli-t'Serstevens 1974). Abundant cowrie shells, used as coinage, only found on this seashore, as well as *banliang* coins of the reign of Wendi (179–156 BC), together show the existence before that time of a trade route through the mountains towards Burma and another to northern China, with the early development of trade and a taste for exotic artefacts by the upper classes of this society.

The *Man shu* (*Book of the Southern Barbarians*) written by Fan Ch'o under the Táng, includes itineraries in Yunnan and two itineraries starting from Yung Ch'ang Ch'eng (Yongchang) towards India via Burma (Oey 1961). The first, northern itinerary started from the market city of Yongchang, continued west across Burma, passed the Black Mountains and reached the valley of Brahmaputra (called *Karatoya River* on the map with the book), towards its common delta with the Ganges (*Hêng Ho*). From there it went upstream on the Ganges, then on its tributary the Yamuna River, until it reached the city of Mathura from where an important caravan route continued through Punjab to the major city of Taxila. The second itinerary from Yongchang went south to Longling, then reached the valley of a tributary of the Irrawaddy River to the city of Pyu, capital of an important Burmese kingdom at that time. From there, it continued north-west, across the Black Mountains, and reached the first itinerary in the lower Ganges valley. A third

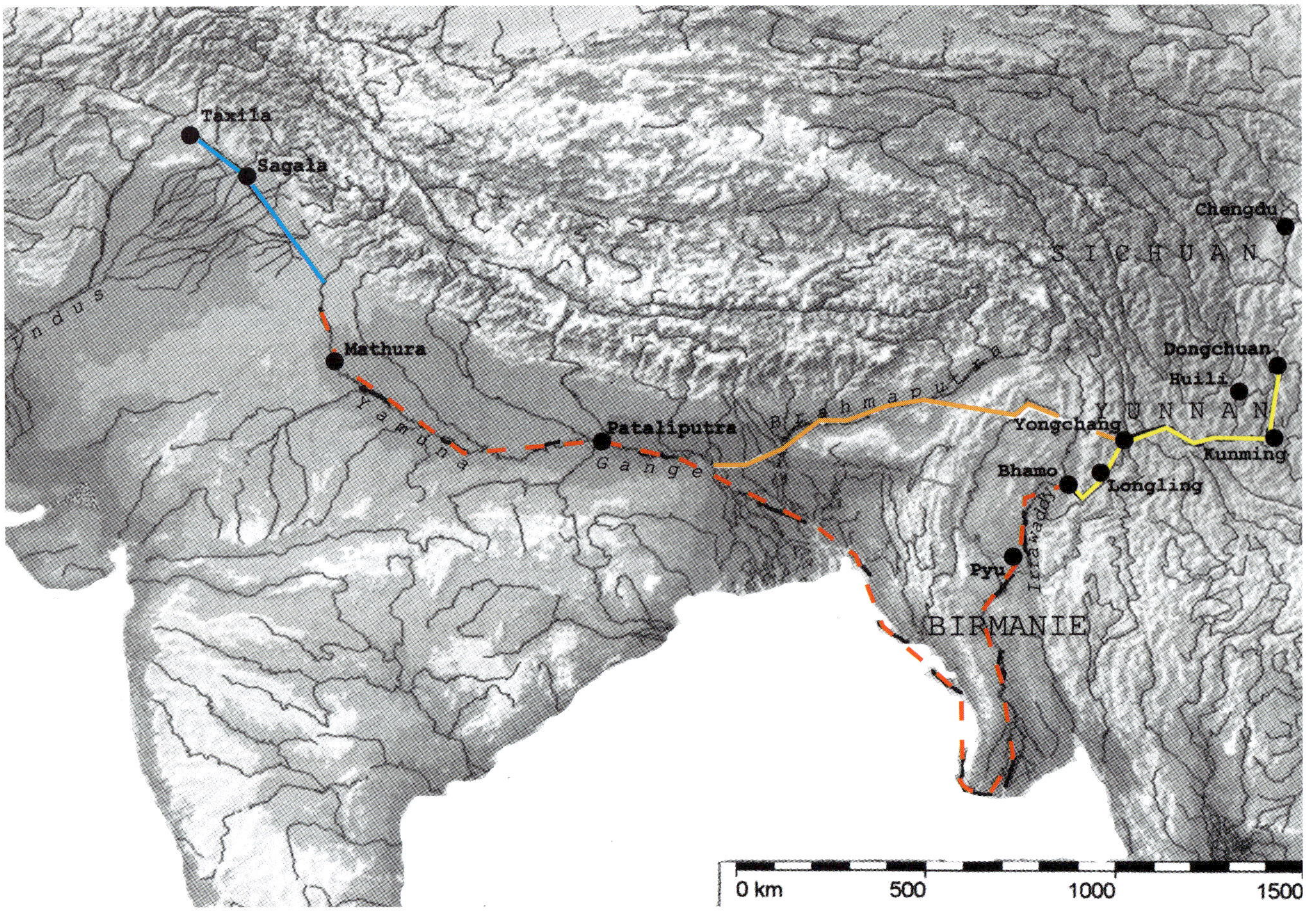

Figure 8 Map of the possible itineraries from the Sichuan and Yunnan nickel mines to Taxila (after the *Man Shu* and other quoted references in the text).

itinerary – longer and perhaps easier and safer – is not mentioned in the *Man shu*, but is indicated by several references: the previously mentioned early cowrie trade; the shipwreck of a Greek vessel off the coast of Bengal (Bernard 1985: 85–95), the shipwreck of a merchant vessel from Taxila (Cunningham 1884: 108); and long-distance ruby trade from Burma (Calligaro *et al.* 1998). Like the second itinerary, it headed south to Longling, but then went on to the city of Bhamo on the upper Irrawaddy. It continued downstream on this river to the gulf of Bengal and then along the coast until reaching the Ganges, following it upstream as the other itineraries until Taxila (Figure 8).

Analytical results since Cheng and Schwitter's study

Howard-White's book (1963), mostly concerned with modern techniques and the uses of nickel and nickel alloys, provided the results of analysis of four Indo-Greek coins, supporting the idea of southern China as the origin of the alloy.

The first analyses of ores from the mines of Limahe, south of Huìlǐ (Sichuān) – close to the Yunnan site of Dōngchuān previously mentioned – published by Wang (1986: 101) are cited in Zhao and Zhou (1998: 201) (Table 1). Nickel is present, mostly between 3 and 4%. Nickel together with copper and sulphur may fit the necessary conditions for the preparation of *paktong*. It would be interesting to know more about the mineralogical nature of the ore and also the analytical techniques that have been used. But even assuming that by chance the two ores were present in the same mine, *paktong* is not a 'natural alloy', as has been suggested (Schwitter 1962).

Table 1 Chemical composition in % of ores from the Limahe mine, south of Huìlǐ (Sichuān) (after Wang 1986).

	Cu	Ni	Fe	S	SiO_2	MgO	CaO
Jin ore n. 1	2.00	3.30	25.40	15.30	24.30	12.70	1.97
Jin ore n. 2	2.05	3.80	22.90	13.78	26.10	13.00	4.80
Fu	1.44	3.68	29.40	18.00	18.30	6.00	6.21
Fu ore n. 1	1.90	3.50	29.04	17.98	16.52	7.11	9.20
U ore n. 2	1.05	2.00	25.74	12.66	19.75	17.11	5.29

Cowell (1989) published a new series of analyses of coins using atomic absorption and gravimetry. He is very negative about the systematic errors and dispersion of Cheng and Schwitter's results, obtained 30 years before using the techniques available at that time. Certainly his measurements are more reliable as he was able to obtain samples from the interior of the coins, without having to take into account the effects of surface corrosion. He rejected the idea of a Chinese origin for the nickel alloy, the only known ancient occurrence of this alloy, without sound argument or suggesting an alternative possibility.

Barrandon and Nicolet-Pierre (1989) analysed 13 Indo-Greek copper-nickel coins of the Cabinet des Médailles of the Bibliothèque Nationale de France in Paris using fast neutron activation and proton activation. These measurements are in general agreement with Cowell's for copper and nickel. However, in one coin of Agathocles (R.3681.58), the nickel (7.5%) and arsenic (0.50%) contents are about half of that measured in other coins. These anomalous values are found together with very high values for lead (5%) and tin (0.46%). In the other coins, lead was undetectable and tin about 100 times lower. Another coin of Agathocles (n. 61), bearing the same control mark ф shows intermediate values between the anomalous and average composition. The authors did not draw any conclusion from this, but it is a significant result. The addition of lead and tin helped to retain the white colour that was the important aspect for this coinage, most likely used for a different value than bronze coinage, as will be seen later. Apart from this interesting result, their measurements include several errors and lacunae. Table 2 (Barrandon and Nicolet-Pierre 1989: 61) gives a silver content of 1.07% for the coin of Euthydemus II (R.3681.51), a very high value and surprising for an element undetected in the other coins, with a detection threshold of 0.001 to 0.01%. Moreover, the level of silver usually strongly correlates with that of gold, which was undetected in this case. For the coin of Agathocles R.3681.60, the authors give a value for lead of 0.005%, lower than their own detection limit. As other authors have measured iron levels to be between 0.4% and 2%, much higher than the detection limit with fast neutrons (40 parts per million in a copper matrix; Beauchesne and Barrandon 1986), it is difficult to understand why iron was not measured, except for four coins in which the iron values were in the usual range. The lack of these measurements is regrettable, because Cowell (1989) has shown that the iron content allows one to distinguish between the copper-nickel coins bearing the two most frequent monograms, ф and Ʀ. In the case of one of the coins whose monogram is illegible, it may have allowed it to be attributed to one of the two main workshops.

All the recent analytical results show the use of 1–2% arsenic, probably as a whitening agent in the cupro-nickel used for Graeco-Bactrian coinage; this contradicts Mei's assumption (1995: 46) that a copper-nickel-arsenic alloy has never been found.

Table 2 gives the available analytical results for seven chemical elements in Graeco-Bactrian cupro-nickel coins.[3]

Table 2 Analytical results for seven chemical elements in Graeco-Bactrian cupro-nickel coins.

King	Monogram	Collection/inv. no.	Analytical technique	Cu%	Ni%	Fe%	Co%	Sn%	Pb	As%	Reference
Euthydemus II	?	BM (destroyed)	WC	77.59	20.04	1.05	0.54	0.038	#	#	Flight 1868
Euthydemus II	?	ANS 2A	XRF	72.9	8.6	1.1	0.4	#	(5.4)	0.2–0.5	Cheng and Schwitter 1957
Euthydemus II	ф	International Nickel Company	XRF	79.3	9.5	1.6	0.6	#	(6.3)	0.2–0.5	Cheng and Schwitter 1957
Pantaleon	?	ANS 4A	XRF	72.0	10.5	1.3	0.6	#	(5.4)	0.2–0.5	Cheng and Schwitter 1957
Agathocles	?	ANS 5A	XRF	66.2	13.8	1.1	0.7	#	(7.8)	0.2–0.5	Cheng and Schwitter 1957
Agathocles	?	ANS 6A	XRF	62.5	13.5	1.8	0.8	#	(8.1)	0.2–0.5	Cheng and Schwitter 1957
Euthydemus II	ф	BM 22 995**	WC	76.7	18.5	1.6	#	#	#	#	Howard-White 1963
Pantaleon	ф	BM 23 002**	WC	77.9	18.6	1.7	#	#	#	#	Howard-White 1963
Agathocles	ф	BM 23 010**	WC	75.5	20.9	1.7	#	#	#	#	Howard-White 1963
Euthydemus II	?	H. de S. Shortt	XRF	O: 83–87 R: 78–82	O: 12–14 R: 16–18	O: 0.5 R: 1.0	O: 0.5 R: 1.0	#	#	O: 0.5 R: 0.5	Howard-White 1963
Pantaleon	?	H. de S. Shortt	XRF	O & R: 75–80	O & R: 19–21	O & R: 1.5	O & R: 1.0	#	#	O & R: 0.5	Howard-White 1963
Agathocles	?	H. de S. Shortt	XRF	O & R: 75–80	O & R: 19–21	O & R: 1.5	O & R: 1.0	#	#	O & R: 0.2	Howard-White 1963
Euthydemus II	ф	BM 22 995**	AA	75.6	18.5	1.1	0.58	< 0.2	0.01	1.80	Cowell 1989
Euthydemus II	Ʀ	BM 22 996	AA	77.4	16.9	0.29	0.34	< 0.2	0.03	0.95	Cowell 1989
Euthydemus II	ф	BM 22 997	AA	71.9	20.3	1.3	0.66	< 0.2	0.01	2.20	Cowell 1989
Euthydemus II	ф	BM 22 998	AA	78.1	17.9	0.78	0.45	< 0.2	0.04	0.70	Cowell 1989

cont'd

Table 2 cont'd

King	Monogram	Collection/inv. no.	Analytical technique	Cu%	Ni%	Fe%	Co%	Sn%	Pb	As%	Reference
Pantaleon	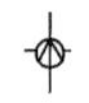	BM 23 001	AA	79.9	17.4	1.2	0.54	< 0.2	0.01	0.80	Cowell 1989
Pantaleon	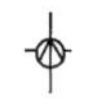	BM 23 002**	AA	76.7	17.8	1.1	0.48	< 0.2	0.01	1.05	Cowell 1989
Agathocles	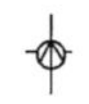	BM 23 005	AA	83.0	12.7	1.6	0.54	< 0.2	0.01	1.80	Cowell 1989
Agathocles		BM 23 007	AA	76.7	19.2	0.75	0.6	< 0.2	< 0.1	1.1	Cowell 1989
Agathocles		BM 23 008	AA	75.6	20.4	0.48	0.57	< 0.2	0.01	0.95	Cowell 1989
Agathocles	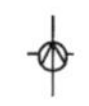	BM 23 009	AA	77.9	18.4	0.94	0.53	< 0.2	0.01	1.00	Cowell 1989
Agathocles	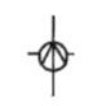	BM 23 010**	AA	75.9	20.3	0.95	0.65	< 0.2	< 0.1	1.85	Cowell 1989
Agathocles	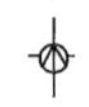	BM 23 011	AA	78.2	19.6	1.4	0.64	< 0.2	0.01	2.00	Cowell 1989
Agathocles		BM 23 012	AA	76.7	19.3	0.39	0.39	< 0.2	< 0.1	1.20	Cowell 1989
Agathocles		BM 23 013	AA	76.8	18.0	0.59	0.45	< 0.2	0.01	0.85	Cowell 1989
Euthydemus II	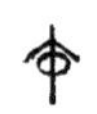	CMP R3681.50	RNA	78.5	19.9	2.36	#	0.006	< 0.04	1.53	Barrandon and Nicolet-Pierre 1989
Euthydemus II	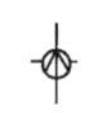	CMP R3681.49	RNA	81.9	16.4	1.24	#	0.043	< 0.03	1.66	Barrandon and Nicolet-Pierre 1989
Euthydemus II	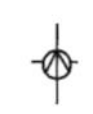	CMP Y.6371	RNA PA	83.8 #	14.5 #	1.35 #	# #	0.0048 #	< 0.02 #	1.25 #	Barrandon and Nicolet-Pierre 1989
Euthydemus II	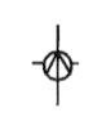	CMP R3681.51	RNA PA	84.5 #	13.0 #	# #	# #	0.013 #	< 0.04 #	1.25 #	Barrandon and Nicolet-Pierre 1989
Pantaleon	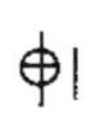	CMP 1974.409	RNA PA	80.8 #	17.4 16.5	# #	# #	0.004 #	< 0.04 #	1.70 #	Barrandon and Nicolet-Pierre 1989
Agathocles		CMP R3681.61	RNA PA	78.9 #	20.0 19.0	# #	# #	0.03 #	< 0.04 #	1.05 #	Barrandon and Nicolet-Pierre 1989
Agathocles	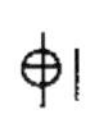	CMP R3681.59	RNA PA	79.3 #	19.0 19.2	# #	# #	0.004 #	< 0.04 #	1.63 #	Barrandon and Nicolet-Pierre 1989
Agathocles	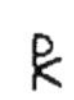	CMP 1974.608	RNA PA	81.4 #	17.6 17.4	0.81 #	# #	0.07 #	< 0.05 #	0.96 #	Barrandon and Nicolet-Pierre 1989
Agathocles	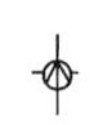	CMP (os) 1974.609	RNA	81.8	16.8	#	#	0.003	< 0.03	1.36	Barrandon and Nicolet-Pierre 1989
Agathocles	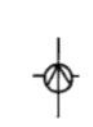	CMP R3681.62	RNA	84.9	14.5	#	#	0.004	<0.015	0.815	Barrandon and Nicolet-Pierre 1989
Agathocles	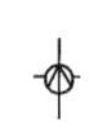	CMP R3681.60	RNA PA	84.8 #	14.0 13.7	# #	# #	# 0.007	0.005 #	1.15 #	Barrandon and Nicolet-Pierre 1989
Agathocles	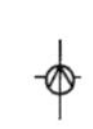	CMP Rollin 61	RNA	86.1	13.7	#	#	0.10	0.06	0.75	Barrandon and Nicolet-Pierre 1989
Agathocles	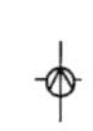	CMP R3681.58	RNA PA	86.5 #	7.5 7.6	# #	# #	0.46 #	5.0 #	0.50 #	Barrandon and Nicolet-Pierre 1989

Key: ? = monogram not indicated in the publication; # = element not detected; BM = British Museum, London; ANS = American Numismatic Society, New York; CMP = Cabinet des Médailles, Paris; ** = coins analysed by two different authors; (os) = overstrike of Agathocles on a coin of Euthydemus II; O = obverse; R = reverse.

Analytical techniques: WC = wet chemistry; XRF = X-ray fluorescence; AA = atomic absorption; RNA = rapid neutron activation; PA = proton activation.

In the Cabinet des Médailles is a copper-nickel coin bearing the monogram ⏀ (Fig. 9), that is an unpublished overstrike of Agathocles struck over a coin of Euthydemus II (Fig. 10). The same control mark appears on the two coins of Agathocles with a low percentage of nickel; it has been replaced by lead and tin at much higher percentages than is found in the other coins of the previously mentioned three kings. A coin from the British Museum in London (Inv. no. 23 005) analysed by Cowell also contains a low percentage of nickel and bears the same monogram. However, lead and arsenic contents are normal, but copper is unusually high. These four independent indications cannot be accidental. Given that the issue of cupro-nickel coins was abolished by Antimachus I, successor to Agathocles, in a turbulent time according to Bernard (see above), it seems that, at least in this workshop, a certain scarcity of nickel alloy before the end of the reign of Agathocles obliged this mint to use either new recipes for alloys or overstrike old coins as an emergency measure. The low content of nickel and the high percentage of lead previously published by Cheng and Schwitter (1957) do not feature in this interpretation. The low nickel contents for all the coins in their analyses were shown by Cowell to result from a calibration lacking precision. Their high lead content was later recognised by Schwitter (1962) to be an experimental artefact.

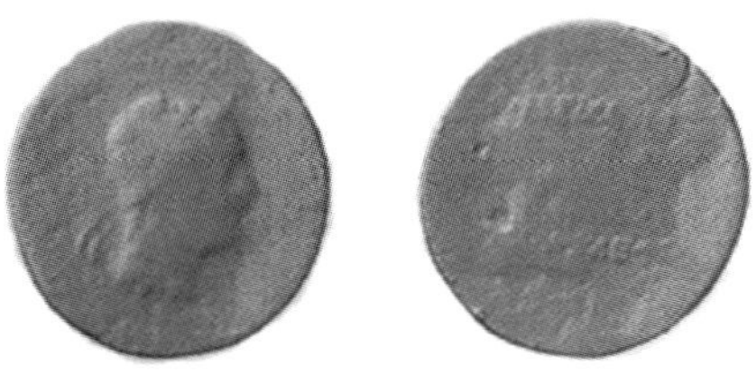

Figure 9 Cupro-nickel coin of Agathocles overstruck on a coin of Euthydemus II. Bust of Dionysos/panther eating grapes. Unpublished overstrike. Cabinet des Médailles, Paris.

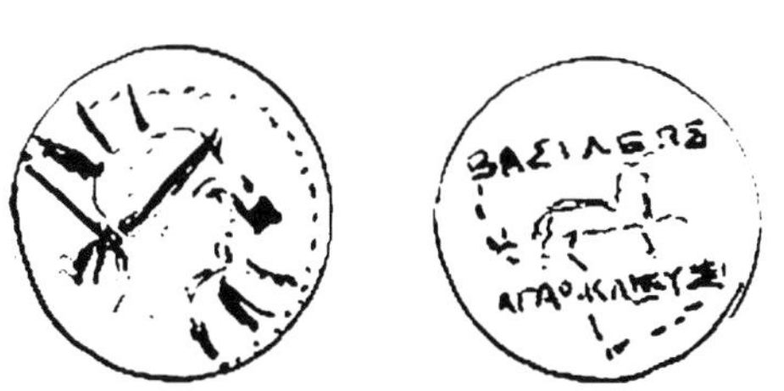

Figure 10 Cupro-nickel coin of Agathocles overstruck on a coin of Euthydemus II. Cabinet des Médailles, Paris. Drawing showing the traces of the first strike.

Another peculiarity of the coinage of Agathocles seems to be the complete lack of issue of silver obolae. Before his reign, obolae were issued by the Graeco-Bactrian kings (Euthydemus I, Demetrius I, Euthydemus II) and an obola is even known of Pantaleon, whose coinage is very rare due to his short reign. After Agathocles, Antimachus I and his successors issued numerous obolae. This complete lack of obolae in the coinage of Agathocles (generally the most common among the three 'nickel kings' probably due to a longer reign), is significant. Contrary to the two other kings, he suppressed the small denomination of silver, giving his subjects no choice other than to use bronze and nickel coins for small denominations. However, if in Taxila copper-nickel was used, it was unknown in other regions of the Graeco-Bactrian Empire and useless for foreign trade. This monetary policy probably contributed to the dissatisfaction that created difficulties for Agathocles at the end of his reign.

Value of copper-nickel coins compared to coins made of other metals

There is no answer, even approximate, to the question of a relative value of metals that would be valid for all countries and all times. For instance, in pharaonic Egypt, silver was twice as precious as gold (Forbes 1964: 209) when usually, in the ancient world, the value of gold was 10–13 times higher than the value of silver. Even within the Greek world the ratio between silver and copper, both used for coinage, varies considerably. An inscription in Athens from 421–416 BC gives the price of 35 silver drachmae (1 Attic drachma = 4.20 g) for one talent of 6000 drachmae of bronze, a ratio of 1:171 (Rébuffat 1998: 139). There is another precise datum for the reign of the Seleucid king Antiochus IV Epiphanes (178–163 BC): a chalcos, a bronze coin weighing 8.78 g, was equivalent to a silver obola of 0.72 g, thus giving a ratio of 1:12.2. In the Parthian Empire, using slightly modified Greek standards, 16 chalcoi of 1.82 g were worth one obola of 0.65 g, hence a ratio of 1:45 (Sellwood 1971: 40). In India the ratio was 1:50 according to Cunningham (1884: 334).

On the subject of the value of Graeco-Bactrian coins made of copper-nickel alloy, relative to coins of other metals, no comparison is possible, as they represent the first coinage made of this material. However, a range of values can at least be estimated for this coinage. The lower limit is the value of bronze coinage. In a time of monetary crisis, it is unlikely that the kings of Bactria would have gone to the expense of making small coins look similar to silver purely for aesthetic reasons. The best way to compensate for silver scarcity would have been to allocate to these coins the same value as silver coins of the same weight, or more. But in Taxila there was a market for this alloy, and so the nickel alloy had a market price in local silver money. Any value in excess of this price would certainly have induced a change in traffic or a massive hoarding of the undervalued silver, disastrous for the economy. The general rule for any coinage made of precious metal is to achieve an official value close to the market price. Usually, a coin made of a precious metal is given an official value slightly greater than the value of the weight of the metal. This difference is called the 'premium'. The premium (usually 5%) minus the manufacturing cost is therefore a profit for the state issuing the coins in exchange for the guarantee of the weight and proportion of precious metal in the alloy (Rébuffat 1998).

If the silver:copper ratio was about 50:50, as attested in India and Iran, the most common value used for small silver coins can be questioned. Double obolae, obolae and half obolae did exist, but the most common by far was the obola. The obola, no longer issued during the reign of Agathocles, was supposed to be replaced by nickel coinage. Nickel coins also existed with three different average unit weights: single 4.20 g; double, 8.40 g; triple, 12.60 g. The most frequently found by far is the double unit. It can be supposed that it was this coin that was used as a substitute for the most frequent silver coin: the obola of 0.72 g, the average value in Graeco-Bactrian coinage. It would give a silver/cupro-nickel ratio of 0.72:8.40 = 1:11.67. In this hypothesis, the ratio between cupro-nickel and bronze, using Cunningham's value, would be 11.67:50 = 1:4.28. This value may seem rather low, considering the cost of long-distance transportation from southern China. If the metal used to make the cupro-nickel coins originated from an important gift given to Alexander, or possibly later to the Graeco-Bactrian kings having conquered Taxila, it was gratis and the available quantity would be sufficient to last a long time. The two and a half tons of metal in the Alexander hypothesis would have allowed the issue of *c*.300,000 coins. Even a much smaller quantity would have allowed the issue of enough coins to make it worthwhile. The corpus of Graeco-Bactrian and Indo-Greek coins has been in progress for about one and a half centuries, and the discovery of numerous coin hoards in the turmoil of the last decades, in particular the huge second hoard of Mir Zakah (Afghanistan), has not provided any evidence to contradict the above scheme of issues. It can be noted that the silver:copper-nickel ratio obtained is close to the silver:bronze ratio under Antiochos Epiphanes. But the large differences between the Seleucid silver:bronze ratio and Parthian or Indian ratios leads one to suppose that the copper coinage was overvalued by the Seleucids. The Seleucid Empire was slowly being conquered by the Parthians and this anomaly may be related to the crisis of this empire under pressure from the Parthians in the east, from Lagid Egypt and the Romans in the west.

Acknowledgements

I would like to express my sincere gratitude to Dr Helen Vetch, sinologist of the CNRS and Dr François Thierry, in charge of the oriental coins in the Cabinet des Médailles of the Bibliothèque Nationale de France for their generous help with the Chinese references and coinage. I would also like to thank Dr Lee Yi-Chuang and Dr Cheng Tung-Ho for providing references on the Chinese analyses of ores from the Limahe mine.

Notes

1. Sharada Srinavasan, pers. comm., 2007.
2. All the transcriptions and translations of ancient Chinese words and titles are by Dr Helen Vetch.
3. Non-quantitative analyses (Flight 1868) are not taken into account. Omitted from the table are: elements searched for in quantitative analyses but not detected (zinc, bismuth); elements that a single author detected (sulphur, silver, chromium, gold) or were detected by several authors at a very low level without any correlation with another element (antimony), because I consider them to be irrelevant results.

References:

Audouin, R. and Bernard, P. (1974) 'Trésor d'Aï Khanoum', *Revue Numismatique* 16: 7–41.

Barrandon, J.-N. and Nicolet-Pierre, H. (1989) 'Analyses de monnaies royales gréco-bactriennes et indo-grecques des IIe – Ier siècles avant J.-C.', *Gazette suisse de numismatique* 155: 57–66.

Beauchesne, F. and Barrandon, J.-N. (1986) 'Analyse globale et non-destructive des objets archéologiques cuivreux par activation avec des neutrons rapides de cyclotron', *Revue d'archéométrie* 10: 75–85.

Bernard, P. (1985) *Fouilles d'Aï Khanoum IV. Les monnaies hors trésors. Questions d'histoire gréco-bactrienne.* Mémoires de la Délégation Archéologique Française en Afghanistan, tome XVIII. Paris: De Boccard.

Calligaro, T., Mossmann, A., Poirot, J.-P. and Querré, G. (1998) 'Provenance study of rubies from a Parthian statuette by PIXE analysis', *Nuclear Instruments and Methods in Physics Research* B 136–138: 846–50.

Cammann, S.C. (1958) 'The "Bactrian nickel theory"', *American Journal of Archaeology* 62: 409–14.

Cháng Qú (常璩) (1965–69) *Huàyáng guózhi* 华阳国志 (*Monograph of the Countries South of Huà* [shān]). Taipei.

Cheng, C.F. and Schwitter, C.M. (1957) 'Nickel in ancient bronzes', *American Journal of Archaeology* 61: 351–65.

Cowell, M.C. (1989) 'Analysis of the cupro-nickel alloy used for Greek Bactrian coins', in Y. Maniatis (ed.), *Proceedings of the 25th International Symposium of Archaeometry*, 335–45. Amsterdam/Oxford: Elsevier.

Cunningham, A. (1884) *Coins of Alexander's Successors in the East.* London: Royal Numismatic Society.

Flight, W. (1868) 'On the chemical composition of a Bactrian coin', *Numismatic Chronicle* 8: 305–8.

Forbes, B.V. (1964) *Studies in Ancient Technologies VIII.* Leiden: J. Brill.

Howard-White, F.B. (1963) *Nickel: An Historical Review.* London: Methuen.

Laufer, B. (1919) *Sino-Iranica: Chinese Contributions to the History of Civilization of Ancient Iran.* Publication 201, Anthropological series 15. Chicago: Field Museum of Natural History.

Levol, M. (1862) 'Analyse d'un cuivre blanc de Chine', *Répertoire de chimie appliquée* 4: 24.

Marshall, J. (1951) *Taxila.* Cambridge: Cambridge University Press.

Mei, Jianjun (1995) 'The history, metallurgy and spread of *paktong*', *Bulletin of the Metals Museum* 24: 43–55.

Needham, J. (1974) *Science and Civilisation in China.* Volume 5, Part II: *Spagyrical Discovery and Invention: Magisteries of Gold and Immortality.* London: Cambridge University Press.

Oey, G.P. (ed.) (1961) *The Man Shu*, G.H. Luce (tr.). Ithaca, NY: Department of Far Eastern Studies, Cornell University.

Pirazzoli-t'Serstevens, M. (1974) *La civilisation du royaume de Dian à l'époque Han, d'après le matériel exhumé à Shizhai shan (Yunnan).* Paris: Ecole Française d'Extrême-Orient.

Prakash, S. and Rawat, N.S. (1965) *Chemical Study of Indian Archaeological Antiquities.* Bombay/New York: Asia Publishing House.

Rébuffat, F. (1998) *La monnaie dans l'Antiquité.* Paris: Picard.

Sellwood, D.G. (1971) *An Introduction to the Coinage of Parthia.* London: Spink.

Schwitter, C.M. (1962) 'Bactrian nickel and Chinese bamboo', *American Journal of Archaeology* 66: 87–92.

Thierry, F. (2003) *Monnaies chinoises. Catalogue des Monnaies du Cabinet des Médailles de la Bibliothèque Nationale de France, Volume II.* Paris: Bibliotheque Nationale De France.

Wang, Deyuan (ed.) (1986) *Huìlǐ nièkuàng zhì* (Annals of the Nickel Mines of Huili). Huili: Huili Nickel Mine Company.

Widemann, F. (1989) 'Un monnayage inconnu de type gréco-bactrien à légende araméenne', *Studia Iranica* 18: 193–7.

Widemann, F. (2000) 'Scarcity of precious metals and relative chronology of Indo-Greek and related coinages (1st century BC–1st century AD)', *East and West* 50(1–4): 227–58.

Zhao, Kuanghua and Zhou, Jiahua (1998) *Zhōngguó kēxué jìshù shǐ, huàxuéjuàn* (History of Chinese Sciences and Techniques: Chemistry). Beijing: Science Press.

Metal trade between Europe and Asia in classical antiquity

Alessandra Giumlia-Mair, Michel Jeandin and Ken'ichi Ota

ABSTRACT This paper presents data which testify to the existence of trade and technology transfer between various European and Asian countries in different periods. The connections between East and West were not always direct, but between the Hellenistic period and the Middle Ages they seem to have been at times quite regular and continuous as ancient texts and circumstantial evidence appear to indicate. The ancient texts describing the trade and exchanges in different periods, the various materials and the trade connections are translated and interpreted.

Introduction

Many exhaustive studies have been carried out on ancient trade and other types of relationships between East and West on themes ranging from parallels between the different philosophical traditions (McEvilley 2002) to the trade of spices in ancient times (Innes Miller 1974), to problems related to a particular region (Vogelsang 1985), and many more. The best studied long-distance trade by far is that of silk (Whitfield and Sims-Williams 2004), a fabric that for centuries remained the secret monopoly of ancient China. However, until now, attention has not focused on the ancient trade in metals between East and West and the ancient Greek and Latin texts which give specific information on details of exchanges of merchandise and know-how in classical times in ancient towns, and in remote places located between Europe and Asia (Fig. 1).

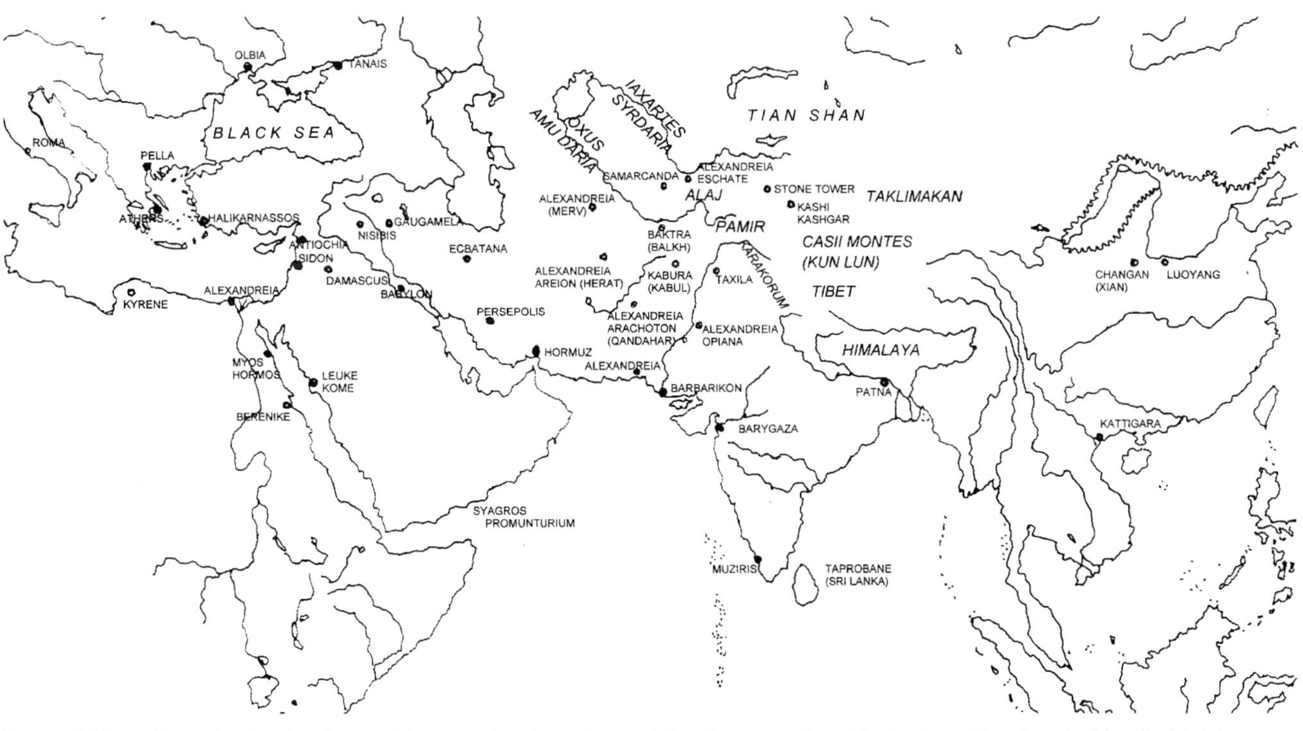

Figure 1 Map of Eurasia showing the most important trade centres and the places mentioned in the text. (Drawing: A. Giumlia-Mair.)

This paper developed from work on ancient metallurgy and on ancient texts on metalworking and metal trade in antiquity. The general picture is of a fascinating history of continuous contacts and relationships between the most important countries and civilisations of the ancient world. They were, in ancient times, not as distant and 'foreign' to each other as they have been at times in some later periods of the history of our common world: the continent Eurasia.

Early metal trade in the Near East and ancient texts

Many exotic names of faraway countries are mentioned in texts dated as early as the 3rd millennium BC as the places of origin of different metals; however, in most cases they are rather difficult to identify (Pettinato 1972; Thapar 1975). The name Meluhha, the country from which (or through which) gold, carnelian, lapis lazuli and tin were imported to Mesopotamia, seems to indicate the southern coast of Pakistan or southern Iran (Reiter 1997: 4; Prange 2001: 11). Dilmun might have been Bahrein, but this interpretation is not accepted by all scholars (Reiter 1997: 8, 24). Zaršu, Hašbar, Yanna and Kusu, mentioned in late Babylonian texts as the countries from which silver came, cannot be easily identified. Even the mines in the silver country Elam are not precisely located, but they might have been in Kerman (Allen 1979: 14, 119; Reiter 1997: 79). The only identification that seems to be generally accepted is that of the legendary copper country Magan with modern Oman (Heimpel 1987; Prange *et al.* 1999; Prange 2001: 29). The origins of early Near Eastern tin have been discussed by many scholars and, until now, the most acceptable hypothesis seems to be that of a tin trade from Afghanistan (Muhly 1985; Stech and Pigott 1986: 47, 56; Giumlia-Mair 2003: 93). In spite of the vague geography there are nevertheless solid proofs of very early intensive trade going to and from areas such as Pakistan and Badakshan to the eastern Mediterranean. Objects such as the vessels made of carved chlorite and decorated with turquoise (Fig. 2) were already being traded from Badakshan through Iran and Syria to the West and to the Indus valley to the East in the second half of the 3rd millennium BC (Carter 1997), while the export of lapis lazuli from Badakshan and Beluchistan (Tosi 1974; Casanova 1993, 1995; Bouquillon and Poirot 1995: 33–4) is documented by its use in the making of jewellery and luxury objects all over the Mediterranean from the 5th millennium BC (Casanova 1995: 17).

Due to space restrictions, this brief paper can only deal with the most important Greek and Latin texts on mines and the metal trade between East and West in the classical, Hellenistic and Roman periods. Many of these texts mention fabulous riches and treasures, often protected by fantastic animals or other rather imaginary details. They were probably inspired by local legends as well as by stories told by merchants who regularly visited remote places to find rare merchandise in an attempt to stave off rival traders. Nevertheless, the metals mentioned in many of these stories can be interpreted as a reflection of actually existing local goods, which took on exaggerated sizes and characteristics through second- and third-hand narrations over long periods of time.

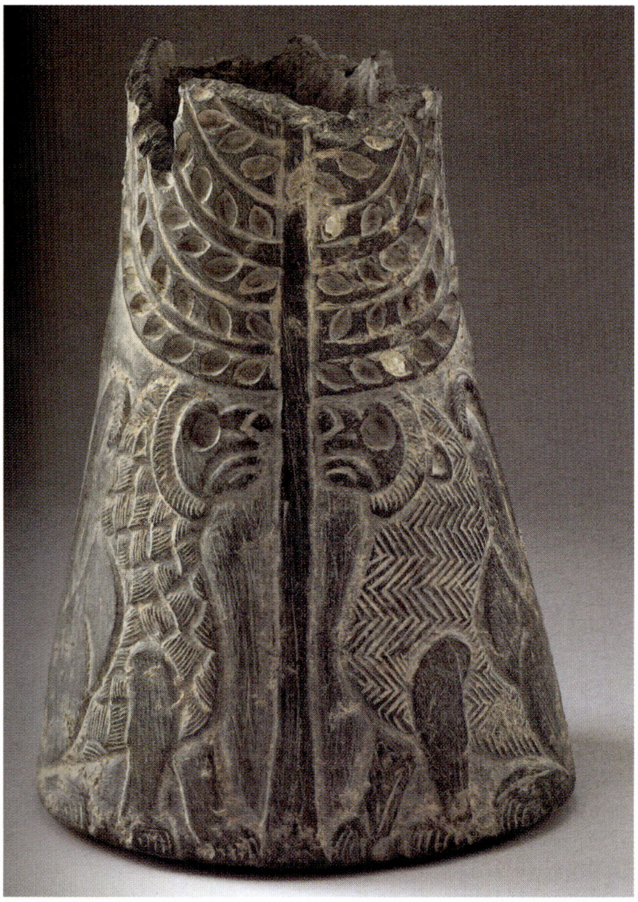

Figure 2 Vessels of carved chlorite decorated with turquoise were already being traded from Badakshan to the East and the West in the 3rd millennium BC (Carter 1997).

We cannot expect to find in the ancient literature detailed descriptions of mining sites, smelting processes and metallurgical production related to the Asian trade, except in very few special cases. However, the information that can be gathered from ancient sources is often valuable because it provides hints on specifics related to myths and beliefs about metal production and metals, and to distinctive features that depended on the idiosyncratic perception by ancient man of the working or production processes. These details are fundamental for the philological reconstruction of the history of ancient metallurgy and could never be obtained solely from metallurgical analyses.

Greek texts

Herodotus

The earliest classical text which gives extensive information on the world behind the Caucasus and the Urals is that of the historian Herodotus, born at Halicarnassus, Asia Minor,

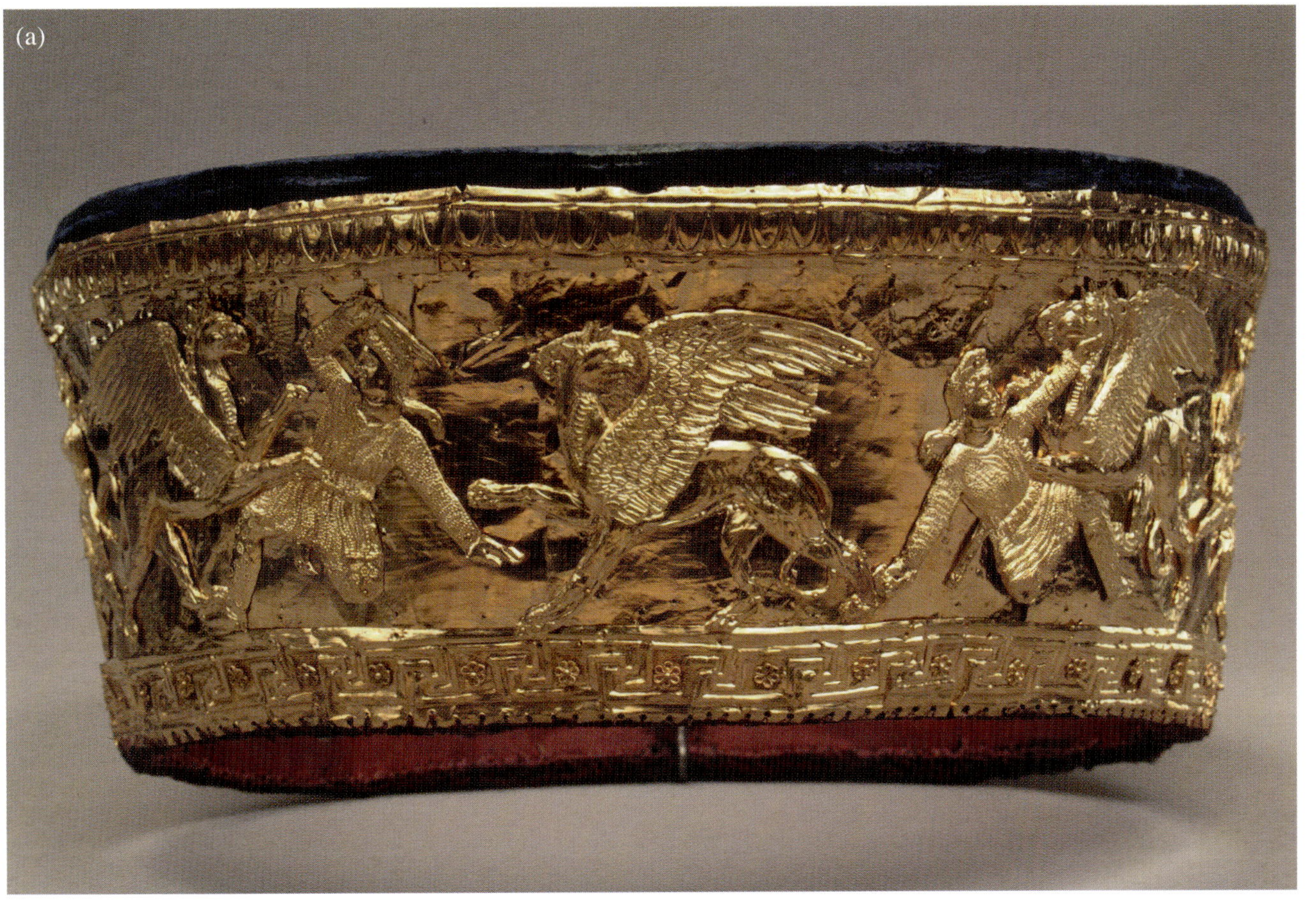

in the 5th century BC (Powell 1939). In the fourth book of his work, *Historiai*, he describes his first-hand impressions of the land and populations of faraway Scythia, where he travelled to the Borysthenes (Dnieper), the Tanai (Don), the palus Maeotis (Maeotian Swamp, now Sea of Azov) and across to the Aral Sea. When describing the populations living on the territories towards the north and east of Scythia he cites the names of tribes taken from the work of Aristeas, a poet from the island of Proconnesus in the Sea of Marmora at the mouth of the Black Sea who lived in the 7th century BC. Aristeas travelled across the Caspian area to the Issedons in Central Asia and wrote a poem, the *Arimaspea,* about the countries he visited: modern-day Kazakhstan, Turkmenistan, Uzbekistan, Tajikistan, Kyrgyzstan and possibly Mongolia.

Herodotus also mentions several of the legends of these countries which reflect the abundance of gold. In several passages he underlines the large amount of gold (Fig. 3a, b) found in the northern regions of Central Asia, among the tribes of the Arimaspoi or Arimaspes (III, 116). From the finds in the kurgans, there is no doubt that this metal was used profusely in this very large area.

In the third book of the *Historiai*, Herodotus mentions the regions of the Achaemenian Empire, up to India and Xinjiang, and enumerates the tributes the Persian king received from the 20 satrapies into which he had divided his empire. It is quite evident that Herodotus can draw on reliable official sources. Most of the tributes were in silver, the quantity depending on the size and the economy of the province. The 9th satrapy for example, which included Phrygia, Paphlagonia and Kylikia, paid 360 silver talents, while Babylonia and Assyria gave 1000 talents. The 20th satrapy along the River Indus paid as much as all the other provinces combined: 360 talents of gold dust. From Herodotus (IV, 44) we also gather that, before the conquest, Darius sent to India a sailor from the West called Scylax of Caryanda, because he wanted to know where the

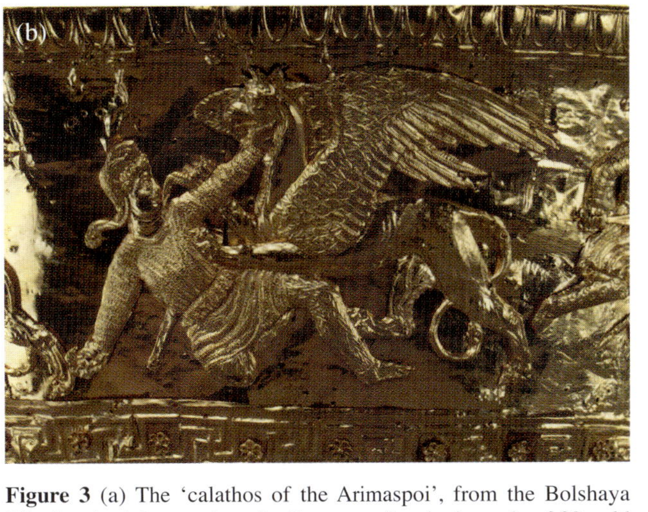

Figure 3 (a) The ‘calathos of the Arimaspoi’, from the Bolshaya Bliznitsa burial mound on the Taman peninsula, is made of 30 gold sheets and decorated with blue enamel. H: 10 cm. (Courtesy of the State Hermitage Museum, St Petersburg.) (b) Detail of the ‘calathos of the Arimaspoi’. The reliefs represent barbarians (Arimaspoi) fighting griffins. The myth of the griffins guarding gold is frequently illustrated in the art of the Bosporan kingdom. (Courtesy of the State Hermitage Museum, St Petersburg.)

Indus flowed into the sea. Scylax left from Caspatyrus in Paktyike (i.e. somewhere in Gandara on the River Kabul). He and his companions 'sailed down the river to the East, toward the sun and to the sea. Once on the sea, they sailed to the West and arrived in the 30th month to the place where the king of Egypt had sent the Phoenicians to circumnavigate Lybia (i.e. Africa)'. The adventure of Scylax, as that of the 'man who first explored Asia', was quite well known in antiquity. In the 5th century BC he was only one of many Western sailors and traders who crossed the Persian Empire and navigated along the coasts of the Red Sea and the Persian gulf to India.

Ctesias

The Greek Ctesias from Cnidus was physician at the court of King Artaxerxes II Mnemon from 405 to 398/97 BC. He took part in the battle of Cunaxa (Xenophon, Anabasis, 1, 8, 26, Photius, 3, 64; Ctesias 2005) as personal physician of the king and after 17 years he returned to Cnidus where he wrote several books about Persia and India. They are now lost, but were quoted by ancient authors such as Diodorus Siculus, Photius, Isocrates, Plato, Aristotle, Theopompos, Ephoros etc. In his books he described silver mines, precious stones and gold mining; however, most of his descriptions tend to be hyperbolised and extravagant, like that of a fountain for example, which 'filled every year with liquid gold', and from which a hundred pitchers of gold (weighing one talent each) were drawn. According to the story, at the bottom of the fountain there was also iron from which Ctesias claimed that he had two swords made – one given to him by the king, the other by his mother Parysatis. This iron could keep off clouds, hail and hurricanes when fixed in the earth, as the king showed him in two different sets of circumstances. Such stories could be termed thaumatology ('the account of miracles').

Ctesias's description of the Persian silver and gold mines is nevertheless rather interesting; he states that:

> the country produces much silver and there are numerous silver mines, not very deep, while those of Bactria are said to be deeper. There was also gold, not found in rivers and washed, as in the river Pactolus, but in many large mountains which are inhabited by griffins. These are four-footed birds as large as a wolf, their legs and claws resembling those of a lion; their breast feathers are red, those of the rest of the body black. Although there is abundance of gold in the mountains, it is difficult to get it because of these birds (Ctesias 2005: 6).

His description might refer to the deep gold mines in Georgia, exploited in the 3rd millennium BC (Hauptmann *et al.* 2006) and belonging, at the time, to the Persian Empire. The griffins were almost certainly taken from local legends and represent a good example of the embroideries which apparently Ctesias could not resist.

Alexander and the Graeco-Bactrian kings

In the 4th century BC, Alexander the Great, king of Macedonia, conquered a wide area from Greece and Egypt to present-day Kazakhstan and India, incorporating all the regions that had previously belonged to the Persian Empire. His armies arrived in Fergana where he founded Alexandria Eschate (now Khojand in Tajikistan) and continued on to Drangiana, Arachosia and Gandara in Pakistan, where he founded more towns: Alexandria Areion (Hera), Alexandria ad Caucasum (Charikar), Alexandria Oxiana (al-Chanum), Alexandria Arachotón (Kandahar), Alexandria Profthasia (Farah), Alexandria-Ghazni, Alexandria-Chenab, Alexandria Margiana (Merv), with a few more in India, for example near Patala, and two on the River Indus. Alexander's conquests opened up new pathways for culture and trade. Suddenly the size of the known world had shrunk (for further details on the influence of Greek culture on the Indian civilisation see, for example, Boardman 1994).

After Alexander's death, the Graeco-Bactrian kingdom was born. Hellenistic Bactria comprised Turkmenistan, Tadjikistan, Uzbekistan, Afghanistan and some parts of Pakistan. On Bactrian history we possess the texts of Strabo (11: 10–12), Polybius (10: 48–9; 11: 34) and fragments of Apollodorus in Strabo (15: 1–3). The main sources of information on the Graeco-Bactrian and Graeco-Indian kings come from coins (Bopearachchi 1991), as no detailed historical account or archive exists. The Graeco-Bactrian kings were the first to issue cupro-nickel coins and much discussion has taken place as to the origins of this alloy. One hypothesis is that of a technology transfer from China, where *baitong* (white copper, i.e. cupronickel) was allegedly in use since the Spring and Autumn period (8th–5th century BC, Widemann, in this volume, pp. 26–34).

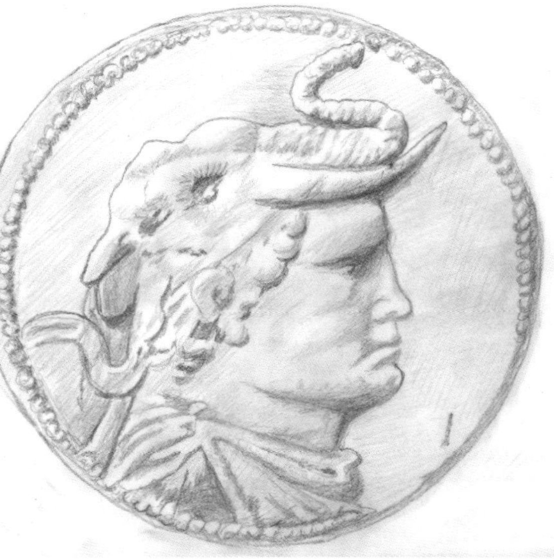

Figure 4 Coin of Demetrius I of Bactria (205–171 BC), the 'Invincible', son of Euthydemus I. He invaded India and arrived at the Ganges and Patna. On the coin he is wearing an elephant headgear, symbol of his conquests in India. (Drawing: A. Giumlia-Mair.)

The foundations of the Graeco-Indian kingdom were laid when in 304 BC Seleucus Nicator invaded Punjab. In 180 BC King Demetrius I of Bactria arrived at the Ganges and Patna (Fig. 4). Menander (155–130 BC) conquered northern India and, as *Milinda*, he became a respected Buddhist sage.

Megasthenes

At the time of the first Bactrian invasion of India, the geographer Megasthenes (350–290 BC) was sent as ambassador by King Seleucus of Syria to King Chandragupta Maurya. He wrote a book (*Indica*) on his travels, but of this only fragments reported by other writers survive.

Megasthenes discusses the riches of the country in several passages: 'while the soil bears ... all kinds of fruits ... it has also underground numerous veins of all sorts of metals, for it contains much gold and silver, and copper and iron in no small quantity, and even tin and other metals' (Oldfather 1935, in Diod, II, 36). His story about ants which dig for gold is rather famous and a parallel, the Pipīlaka gold (i.e. ant's gold) is also found in the Sanskrit epic Mahābhārata (Lal 2005: 52.2–4) (600–500 BC). Megasthenes, reported by Strabo (XV, 1. 44), claims that the 'ants' were ferocious and the size of foxes. They attacked the gold seekers and often killed them and their cattle. Before Megasthenes, Herodotus (III, 102) had also written about gold-digging ants and even the famous Roman scholar Pliny (XI: 111) refers to a similar story. Among other less credible hypotheses, the gold-digging ants have been interpreted as marmots because in the land of the Drok-pa of Ladakh, identified with Megasthenes' *Derdai*, auriferous sand is collected from the burrows of these animals. Their ferocity would again be a story invented to discourage potential foreign gold seekers (Dube 1996: 10). Megasthenes (quoted by Pliny, IV: 81) also knew about Taprobane (Sri Lanka): 'Taprobane is divided by a river, the inhabitants are called Paleogonoi and they are richer in gold and large pearls than the Indians.'

Greek travellers to India (2nd–1st century BC)

At the time of the Graeco-Indian kingdom, the trade contacts with the East were easier and became more intensive, particularly on the sea route. *Portolani* (i.e. pilots' books containing description of the ports) and travel reports multiply. Some fragments of the books *On the Erythraean Sea* with the description of the countries and ports of this area and *Ta katà ten Asìan* (On Asian matters) of Agatharchides of Cnidus (2nd century BC) are preserved in the work of Strabo (14, 2.15) and his description of the gold mines in Nubia is quite famous (Diodorus 3: 12–18) (Burstein 1989; Booth 1814).

The story of the journey of a Greek merchant called Iambulus from Ethiopia to Taprobane ('the fortunate island where people are kind and gentle') by Diodorus (2, 4) is less well known. He remained in Sri Lanka for seven years and returned to Greece via Persia, with the help of the king of the Indian town Polybothia (perhaps a Bactrian king) who 'had a great love for the Greeks and was very studious in the liberal sciences'.

The complex story of Eudoxus of Cyzicus is told by Strabo (2, 3, 4–5) who reports from Poseidonius (135–51 BC). Eudoxus was sent to India by Ptolemy Euergetes of Egypt. He returned with a cargo of perfumes, gold and precious stones, but Euergetes took everything from him. After the death of the king, his wife Cleopatra succeeded him and Eudoxus was sent to India again. On his return he was driven off-course to the south of Ethiopia, where he found a fragment from a wrecked ship with the carving of a horse; he was told that the ship came from the West. When he arrived back in Egypt, the son of Cleopatra, another Ptolemy, was now the pharaoh and he again confiscated Eudoxus' cargo. However, he discovered that the wrecked ship he had found in the south of Ethiopia was from Gades and from this he conjectured that the circumnavigation of Libya (i.e. Africa) was possible. On his return home, he took all his worldly belongings and put to sea bound for India circumnavigating Africa. However, to cut a long story short, apparently after the third attempt and the fifth wrecked ship, he was never seen again.

One of the last Graeco-Indian kings was Hermaeus. From historical Chinese records (48–33 BC) it appears that Hermaeus was their ally in the war against the reign of Jibin. According to the Hou Hanshu (96A) (Hill 2003), he was installed as king of Jibin and vassal of the Chinese emperor. The Graeco-Indian kingdom came to an end when the last territory in Punjab was invaded by Indo-Scythians at the end of the 1st century BC (Tarn 1938).

Roman times

The sea routes and the Periplus of the Erythraean Sea

This famous and very detailed pilot's book describes the navigation and trade from Roman Egyptian ports along the coast of the Red Sea, the Persian gulf and the Indian Ocean. Brass, sheets of copper, weapons, tools and vessels were sold to the locals. Gold and silver plates were reserved for the king (5; 8; 10; 24; 28; 39) (Burstein 1989; Belfiore 2004). Silver coins were also sold as bullion and this apparently created problems for the Roman state. Ancient *drachmae* with Greek letters, possibly from Bactria, were still in circulation at Barygaza (47). At Ozene and Nelcynda copper, tin and lead, and silver and gold coins were much appreciated goods; glass, realgar and powdered antimony were also sold. On the local market, pearls, ivory, silk cloth, transparent stones of all kinds, diamonds and sapphires, and tortoiseshell could be bought (49). In the 1st century AD, Indian iron and steel could be found even on the market on the Dahlak islands (*Alalaios*) in the Red Sea, among all sorts of other exotic goods (5).

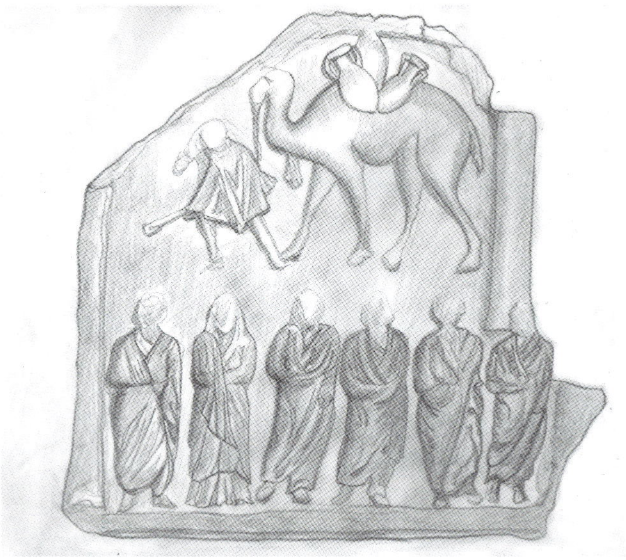

Figure 5 Funerary relief of the Peticii family, Roman merchants, depicting a dromedary as a symbol of trade with the East. Museum of l'Aquila, 1st century BC–1st century AD. (Drawing: A. Giumlia-Mair.)

The land route to Asia

There is considerably less information on the land routes leading to Asia – the traders seem to have kept these secret (Fig. 5). The only extant written document is the *Stathmoi Parthikoi* (Parthic Stations), a geography of the eastern countries by the scholar Isidorus from Spasinus (or Charax Spasinus), at the mouth of the Euphrates, ordered by Augustus when he was planning to go to war against the Parthians (Belfiore 2004: app. B, 205–59). The partially preserved text gives the itinerary of the caravans crossing Parthia in 19 paragraphs. The route and toll stations from Antiochia in Syria to Zeugma on the River Euphrates, to Seleucia, the Caspian Sea and through Turkmenistan, Afghanistan and Pakistan to Alexandria Arachotòn are described in detail. The text gives the distances in *schoinoi*, the linear measure used in Egypt, between 30 and 48 *stadii* (5.2 and 5.6 km). Isidorus' work is dated between the end of the 1st century BC and the beginning of the 1st century AD.

Marinus of Tyros and Claudius Ptolemy

In the 1st century AD, Marinus of Tyros wrote his 'Geography', for which he developed a system of coordinates, with 15 meridians and eight climates. His world map was completed by detailed comments and topographical indications provided by travellers to Asia. Using their first-hand information he was able to correct many of the mistakes of previous geographical works. A sailor called Diodoros of Samos provided details on the sea route to India, and another sailor, Alexandros, described how he came to Kattigara (near Hanoi) in the gulf of Tonkin. Three men – Diogenes, Theophilus and Dioskoros – navigated a route to Zanzibar. Diogenes in particular visited Lake Nyanza (Victoria), and described the Ruwenzori Mountains, the Uganda Nile and a mountain with three peaks (perhaps Kilimanjaro).

However, the most fascinating description is that given to Marinus by Maes Titianos, a merchant who between AD 100 and 120 visited the marketplace at the Λίθινος Πύργος (Stone Tower) (Ptolemäus 1990, I: 17.5), in the country of the Sacha people in Central Asia. Here he and his men exchanged their goods with merchants from all over the known world and in particular with the Seri, the silk people (i.e. Chinese traders) (Alemany i Vilamajó 2002).

The group of traders left Hierapolis on the Euphrates in the direction of Bactria through Edessa, Ecbatana, Rhagai, past the pylae Caspiae, Hecatompylos in Hircania and Antiochia in Margiana. At Bactria they stopped to meet the caravans from India and then continued to the Stone Tower for a longer bivouac. The place was identified (Innes Miller 1974: 242) with Tashkurgan in Xinjiang, because of the name of the town which means 'stone tumulus' or 'stone fortification'. However, the fortification on the site is dated to AD 1277–1367 and the location in Xinjiang does not correspond to the ancient descriptions of a relatively narrow valley between high mountains. There are at least four places called Tashkurgan around the Pamir, but none seems to be acceptable as the location of the ancient Stone Tower. A further hypothesis for the location of the Stone Tower is the area of Qarategin, in Tadjikistan (Belfiore 2004: 251), but again the geography of the place does not correspond to what we know from Maes Titianos.

A location that seems to be much closer to the ancient description is near the town Osh, on the road through the Terek Pass and to the important crossing of Sary Tash (Fig. 6), where it meets the road to the Irkeshtam Pass on the modern border with China. The road from Irkeshtam leads down to Kashi in Xinjiang. Going to the west of Sary Tash through the Kyzyl Suu valley it was possible to reach Sogdiana; to the south a road leads through the Kyzyl Art Pass into the Gorno Badakshan and to the Indus and Gandara. One important reason for choosing this route was that the Terek Pass was open all year and fodder was

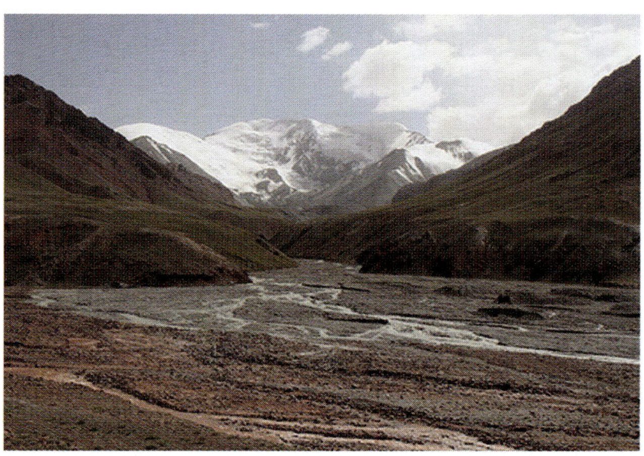

Figure 6 The Gulcho River not far from Sary Tash and the important crossing leading to the east (Xinjiang), to the south (Murghab), to the west (Sogdiana) and to the north (Osh in Fergana). (Photo: J. Barnes.)

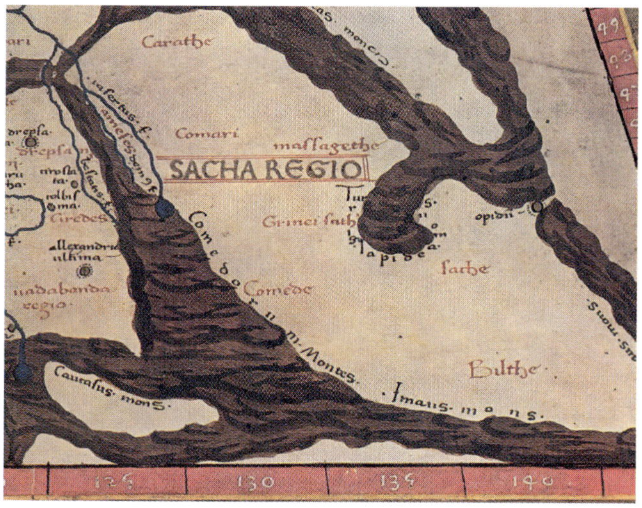

Figure 7 Detail of Ptolemy's *Cosmographia* (1st–2nd century AD). The page representing Central Asia shows the place called *Turris Lapidea Mons* (Mountain Stone Tower in Latin). The writing is placed around a mountain, not a building or a tower, in a valley to the east of Fergana (Ptolemaeus, *Cosmographia, XXII, Asiae Tabula Septima*).

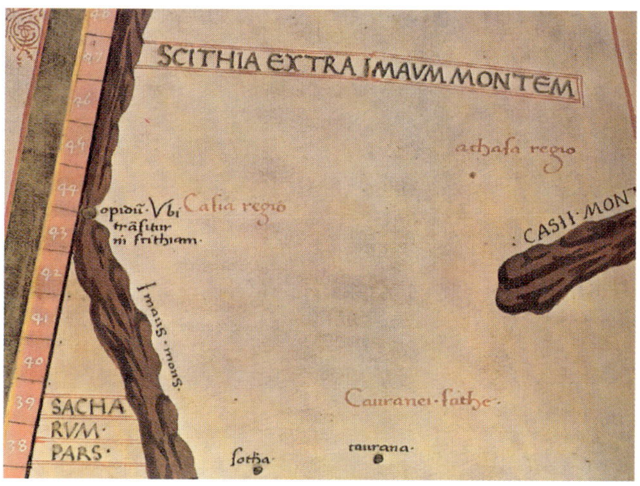

Figure 8 Detail of Ptolemy's *Cosmographia*. The page representing China shows the Casia regio, i.e. the region of Kashi in Xinjiang with the Tian Shan and the Kun Lun Mountains. The note in Latin near the town at the pass into Xingjiang states: 'the town where to pass into Scythia' (Ptolemaeus, *Cosmographia, XXIII, Asiae Tabula Octava*).

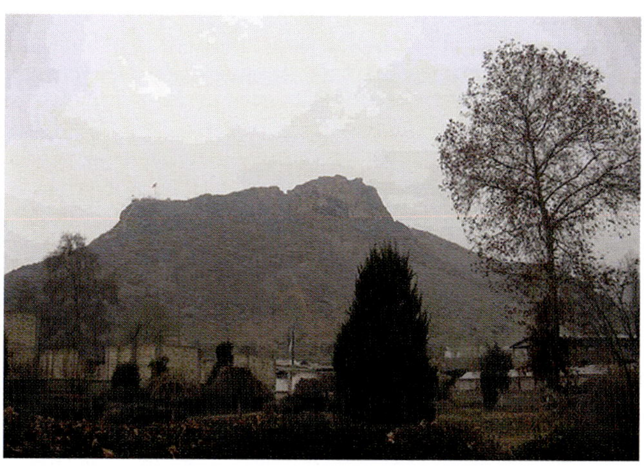

Figure 9 The distinctive Suleyman Mountain at Osh (Kyrgyzstan) might be Marinus' Stone Tower, an important landmark in antiquity. (Photo: W. Liphold.)

available for the animals – no grass grew along the other routes (Stein 1933).

The geographer Claudius Ptolemy relied heavily on Marinus' work and reproduced his writings in his well-known *Cosmographia*. The copy in the National Library of Naples is dated to the 15th century, but it is copied from older versions (apparently from that of Agathodaimon of Alexandria, dated to the 4th century AD) and therefore very close to the original. The precision of the maps representing the Mediterranean and the Arabian coasts is quite amazing, while other parts are not as precise. The Caspian Sea, for example, is horizontal, the triangle of India is too flat and Sri Lanka too large; however, the outlines and the proportions are generally quite recognisable.

The page of Ptolemy's *Cosmographia* representing Central Asia illustrates the place called *Turris Lapidea Mons* (i.e. Latin for Mountain Stone Tower) (Fig. 7). Identification on the map is not easy and is somewhat of a surprise – the letters are placed awkwardly around what is clearly a mountain, not a building or a tower, in a valley to the east of Fergana. Alexandria Ultima (Latin for Alexandria Eschate, Khojand) at the entrance of Fergana is marked behind a mountain range. The site is also without doubt before Xinjiang, because the next page of the *Cosmographia* shows the Casia Regio, i.e. the region of Kashi-Kashgar. Xinjiang is represented as an ample valley with the Tian Shan and Kun Lun Mountains (Fig. 8). The note in Latin jotted next to the town at the mountain pass leading into Xinjiang (Irkeshtam) states: 'the town where to pass into Scythia'. This was the equivalent of Sacha, meaning the regions of the nomadic tribes.

Indeed, the area around Osh is separated from the west of Fergana by a chain of mountains and the shape of the valley is similar to that of Ptolemy's map. In Osh there is a high isolated hill, the venerated Suleyman Mountain (Fig. 9), an important landmark in antiquity as it is now. This mountain might be Ptolemy's *Turris Lapidea Mons*. The Chinese merchants who traded at the market at the Stone Tower were able to give Maes Titianos the exact distance from there to the capital of the silk country, Ch'hangan (present-day Xi'an). From archaeological finds and written sources we know that European countries exported to China worked metal, glass, amber, coral, fruits and dyes (Needham 1984: 89). One of the most appreciated metallurgical products from China was *ferrum Sericum* (Chinese iron), uniformly carburised, obtained from cast iron and decarburised to become steel (Sun 2002: 66–7; Giumlia-Mair and Maddin 2004: 129–30; Wayman and Michaelson 2006). This was a far better metal than forge-welded European iron.

Iron from China and iron from India

In the 1st century AD, Quintus Curzius Rufus wrote a biography of Alexander, the *Historiae Alexandri Magni* in which he states that Alexander received a tribute of 'white iron' from the Oxydrachs and Mallians in Punjab (Curzius Rufus 1977). Some scholars have tried to interpret this passage as an indication of the use of *baitong*, but the text clearly mentions iron (8, 1 *ferri candidi talenta centum*: 100 talents of white iron). There is no reason to interpret it as cupronickel either, as in the 4th century BC most people could distinguish a copper-based alloy from iron. It is not obvious whether cast iron from China or *wootz* from India was meant, but it is clear that it was iron.

According to Pliny (34: 145): '... of all varieties of iron the palm goes to the Seric (Chinese), the second to the Parthian iron'. The reason why Parthian iron was considered the second best is probably too complex and lengthy to be discussed in this paper, but it may also have been cast iron that came to the West from China through Parthia. However, another kind of exotic iron was also in circulation in the Roman Empire: the esteemed *ferrum Indicum* or Indian iron.

Justinian's *Digestum* (AD 527–565) lists the goods imported from the East subject to custom duties. Among spices, precious stones, pearls and ivory, Indian iron is also mentioned. This material, listed also in the *Periplus of the Erithraean Sea* (see above), seems to be *wootz* and is discussed in detail in the work of one of the most famous and important alchemists of Alexandria in Egypt: Zosimos of Panopolis (2nd century AD). He reports a recipe giving the weights and lists ingredients including date peel (as carbon supplier) and 'female magnesia' which in the past, following one of the different translations given for the word 'magnesia' by the great scholar Berthelot (1967: 1, 221, 255), was interpreted as manganese dioxide (Allan and Gilmour 2000: 63–4; Craddock and Lang 2004: 41). However, in this text Zosimos specifies that this material is not the common magnesia, but 'female magnesia of the glass maker'. This material might perhaps be the bone ashes employed by glaziers as stabiliser (see Giumlia-Mair and Maddin 2004: 128–33 for translation and details). If this interpretation of 'female magnesia' as bone ashes is correct, the text might also explain the reason for the presence of phosphorus in the structure of some Damascus blades, which until now has remained a mystery (Gilmour 1996: 120).

It is quite important to note that this recipe belongs to a batch collected by Zosimos from some texts from Persia, but, as expressly stated, 'translated under Philip's reign, King of Macedonia', i.e. under the reign of one of the several successors of Alexander called Philip, between Alexander's death in 323 and the battle of Pydna in 168 BC. This important text confirms the transmission of this recipe to the Mediterranean through Persia (at least in this case), and establishes a date between the 4th and the 2nd century BC for the introduction of crucible steel in the West.

Around the 2nd century BC, pattern-welded blades also begin to circulate in Europe. If, as it seems, crucible steel was already known at this time, perhaps pattern-welded blades might be an imitation of blades made of crucible steel. However, at this time there is no evidence of this, but relatively few blades dated to this time have been studied. Crucible steel blades might have come from the East through different paths – perhaps from India on the sea route, from Persia over the Near East or from Central Asia through the steppes of northern Europe.

The above represents only a concise report on a choice of ancient passages. Apart from the ancient Greek and Latin texts there are more (not less interesting but little known) medieval texts, mainly written by monks and traders, such as Giovanni di Pian del Carpine (1180–1252) and Giovanni di Marignolli (1339–1353), addressing similar issues. There are also a few translated Arabic texts, which conjure up images of a long-lost world and give fascinating insights into the thriving medieval trade between the Mediterranean and Asia. More research and papers on this topic are required in the future.

References

Alemany i Vilamajó, A. (2002) 'Maes Titianos i la Torre de Pedra (I): una font grega sobre els orígens de la ruta de la seda', *Faventia* 24(2): 105–20.

Allan, J. and Gilmour, B. (2000) *Persian Steel, The Tanavoli Collection*, Oxford Studies in Islamic Art XV. Oxford: Oxford University Press.

Allen, J.W. (1979) *Persian Metal Technology 700–1300 AD*. London: Ithaca Press.

Belfiore, S. (2004) *Il periplo del Mare Eritreo di anonimo del I sec. d.C. e altri testi sul commercio fra Roma e l'Oriente attraverso l'Oceano Indiano e la Via della Seta, Memorie della Società Geografica Italiana*, vol. LXXIII. Rome: Società Geografica Italiana.

Berthelot, M. (1967) *Collection des anciens alchimistes grecs*, 1–3. Osnabrück: Otto Zeller [réimpression de l'édition 1888].

Boardman, J. (1994) *The Diffusion of Classical Art in Antiquity*. Princeton, NJ: Princeton University Press.

Booth, G. (1814) *The Historical Library of Diodorus the Sicilian: In Fifteen Books*. London: J. Davis, Military Chronicle Office.

Bopearachchi, O. (1991) *Monnaies Gréco-Bactriennes et Indo-Grecques*. Paris: Bibliothèque Nationale de France (catalogue raisonné).

Bouquillon, A. and Poirot, J-P. (1995) 'Les minéraux et leur origine', in F. Tallon (ed.), *Les pierres précieuses de l'Orient ancient des Sumériens aux Sassanides*, 33–8. Paris: Musée du Louvre, Exposition-dossier des Antiquités orientales 49.

Burstein, S.M. (1989) *Agatharchides of Cnidus, On the Erythraean Sea*. Second Series no. 172. London: Hakluyt Society.

Carter, M.L. (1997) 'The ancient Near East', in J.N. Newland (ed.), *Miho Museum*, 325–34. Nissha: Miho Museum.

Casanova, M. (1993) 'Lapis lazuli beads in Susa and Central Asia: a preliminary study', *South Asian Archaeology* 93: 137–45.

Casanova, M. (1995) 'Le lapislazuli dans l'Orient antique', in F. Tallon (ed.), *Les pierres précieuses de l'Orient ancient des Sumériens aux Sassanides*, 15–20. Paris: Musée du Louvre, Exposition-dossier des Antiquités orientales 49.

Craddock, P. and Lang, J. (2004) 'Crucible steel – bright steel', *Journal of Historical Metallurgy* 38: 35–47.

Ctésias (2005) *La Perse, L'Inde, Autres fragmentes*, D. Lenfant (trans.). Paris: Les Belles Lettres.

Curzius Rufus, Q. (1977) *Storie di Alessandro Magno*, A. Giacone (ed.). Milan: UTET.

Dube, R.K. (1996) 'An analysis of ant's gold in ancient India', *Bulletin of the Metals Museum* 25: 1–13.

Gilmour, B. (1996) 'The patterned sword: its technology in medieval Europe and southern Asia', in *Proceedings of the Forum for the Fourth International Conference on the Beginning of the Use of*

Metals and Alloys (BUMA IV), 113–31. Sendai: Japan Institute of Metals.

Giumlia-Mair, A. (2003) 'Iron Age tin in the oriental Alps', in A. Giumlia-Mair and F. Lo Schiavo (eds), *The Problem of Early Tin*, 93–108. BAR International Series 1199. Oxford: BAR.

Giumlia-Mair, A. and Maddin, R. (2004) 'Iron and steel in the Roman period and in late antiquity', in W. Nicodemi (ed.), *The Civilisation of Iron from Prehistory to the Third Millennium*, 113–43. Milan: Edizioni Olivares.

Hauptmann, A., Bendall, C., Brey, G., Dschaparidze, I., Gambaschidze, I., Prange, M., Steffens, G. and Stöllner, Th. (2006) 'Gold in Georgia: analytical and mining archaeological investigations into the production and distribution of the noble metal in Bronze Age Transcaucasus', in *Metallurgy and Civilisation: 6th International Conference on the Beginnings of the Use of Metals and Alloys, University of Science and Technology, Beijing, BUMA VI, Abstracts*: 57.

Heimpel, W. (1987) 'Das untere Meer', *Zeitschrift für Assyriologie und Vorderasiatische Archäologie* 77: 21–91.

Hill, J.E. (2003) *Annotated Translation of the Chapter on the Western Regions according to the Hou Hanshu*, 2nd draft edition, http://depts.washington.edu/silkroad/texts/hhshu/hou_han_shu.html.

Innes Miller, J. (1974) *Roma e la via delle spezie, dal 29 a.C. al 641 d.C.* Turin: G. Einaudi Editore.

Lal, P. (2005) *Mahabharata*. Volume 2: *Sabha Parva*. Calcutta: Writers Workshop.

McEvilley, Th. (2002) *The Shape of Ancient Thought: Comparative Studies in Greek and Indian Philosophies*. New York: Allworth Press.

Muhly, J.D. (1985) 'Sources of tin and the beginnings of bronze metallurgy', *American Journal of Archaeology* 89: 275–91.

Needham, J. (1984) *Wissenschaft und Zivilisation in China, I, Wissenschaftliche Sonderausgabe*. Frankfurt: Suhrkamp Verlag.

Oldfather, C.H. (trans.) (1935) *Diodorus Siculus*, Loeb Classical Library. Cambridge, MA: Harvard University Press.

Pettinato, G. (1972) 'Il commercio con l'estero della Mesopotamia meridionale nel III millennio a.C. alla luce delle fonti letterarie e lessicali numeriche', *Mesopotamia* 7: 43–166.

Pliny (*Plinius*) (1938–1971) *Natural History* (*Naturalis Historia*) Books I–XXXVII. Loeb Classical Library. Cambridge, MA: Harvard University Press.

Polybius (1922–1927) *The Histories*, vols I–VI, W.R. Paton (trans.). Loeb Classical Library. Cambridge, MA: Harvard University Press.

Powell, J.E. (1939) *The History of Herodotus*. Cambridge: Cambridge Classical Studies 4.

Prange, M. (2001) '5000 Jahre Kupfer in Oman, 2, Vergleichende Untersuchungen zur Charakterisierung des omanischen Kupfers mittels chemischer und isotopischer Analysenmethoden', *Metalla* 8: 1–126.

Prange, M., Götze, H-J., Hauptmann, A. and Weisgerber, G. (1999) 'Is Oman the ancient Magan? Analytical studies of copper from Oman', in S. Young and P. Budd (eds), *Metals in Antiquity*, 187–92. Oxford: Archaeopress.

Ptolemäus (1990) *Cosmographia, Das Weltbild der Antike*. Stuttgart: Parkland Verlag.

Reiter, K. (1997) *Die Metalle im Alten Orient unter besonderer Berücksichtigung altbabylonischer Quellen*. Münster: Ugarit-Verlag.

Stech, T. and Pigott, V.C. (1986) 'The metals trade in Southwest Asia in the third millennium BC', *Iraq* 48: 39–64.

Stein, A. (1933) *On Ancient Central-Asian Tracks*. London: Dover.

Strabo (1917–1932) *Geographia*, 8 vols, H.L. Jones (trans.). Loeb Classical Library. Cambridge, MA: Harvard University Press.

Sun, S. (2002) 'Recent achievements made by the University of Science and Technology Beijing (USTB), China, in the field of Chinese archaeometallurgy', in K. Gyu-Ho, Y. Kyung-Woo and K. Hyung-Tai (eds), *Messages from the History of Metals to the Future Metal Age: Proceedings of the V International Conference on the Beginnings of the Use of Metals and Alloys (BUMA V), 21–24 April, The Korean Institute of Metals and Materials, Gyeongju, Korea*, 61–70.

Tarn, W.W. (1938) *The Greeks in Bactria and India*. Cambridge: Cambridge University Press.

Thapar, R. (1975) 'A possible identification of Meluhha, Dilmun and Makan', *Journal of the Economic and Social History of the Orient* 18: 1–42.

Tosi, M. (1974) 'The lapis lazuli trade across the Iranian Plateau in the 3rd millennium BC, in onore di G. Tucci', *Gururajamanjarika* 1: 3–22.

Vogelsang, W. (1985) 'Early historical Arachosia in south east Afghanistan: meeting-place between East and West', *Iranica Antiqua* 20: 55–99.

Wayman, M.L. and Michaelson, C. (2006) 'The metallurgy of early Chinese wrought iron and steel objects from the British Museum', *Journal of Historical Metallurgy* 40: 95–104.

Whitfield, S. and Sims-Williams, U. (eds) (2004) *The Silk Route: Trade, Travel, War and Faith*. London: British Library.

Xenophon (1989) *Anabasis*, C.L. Brownson (trans.). Loeb Classical Library. Cambridge, MA: Harvard University Press.

The black bronzes of Asia

Paul Craddock, Maickel van Bellegem, Philip Fletcher, Richard Blurton and Susan La Niece

ABSTRACT The many and diverse metalworking traditions of East and South-East Asia of the recent past have been relatively little studied. They include varieties of black-patinated copper alloys containing small amounts of precious metal and inlaid with precious metal, or used themselves as inlays. This paper concentrates on the *sawasa* wares, which were probably made at a number of centres for the European market and the complex inlaid *mylar* lime boxes from Myanmar. Other traditions were – and in some cases still are – practised in China, Korea, Thailand and Japan. Their interrelationships and possible common occidental ancestors have yet to be fully investigated.

Introduction

The first European travellers and traders to the Far East encountered many luxury and exotic products, viewing them both as interesting curiosities and trade goods. These included items of copper alloyed and inlaid with small amounts of precious metals with an attractive black patina. The German doctor Engelbert Kaempfer writing of Japan in the late 17th century noted that 'No Eastern nation is that skilled and able in working, sculpting, cutting and gilding of Sowaas, which is a special kind of precious blackish metal, artificially made of copper with some gold.' It is implicit in this statement that the black-patinated bronzes were also made elsewhere in the East, but unfortunately although there are numerous references to them, there is no single comprehensive description, much less any agreement on a name. Thus the 17th- and 18th-century Dutch and other traders based in the Far East or back home in Europe variously used the terms *savas*, *souassa*, *sowaas*, *suassa* and *swassa* for what might or might not have been the same thing.

From a purely European perspective these black-patinated alloys can be broken down into two groups: items made for Europeans, probably mainly for those domiciled in the East and their contacts, as well as for export to Europe itself, and those made for the indigenous market. Even though the first group include many prestige items – swords decorated with black-patinated hilts presented by Peter the Great to the Elector of Saxony in 1704 for example – they were little known as a group until the 1998 *Sawasa* exhibition at the Rijksmuseum in Amsterdam (de Bruijn and Kist 1998). The second group, with the exception of the Japanese *shakudo* metalwork, remains little known with new discov-

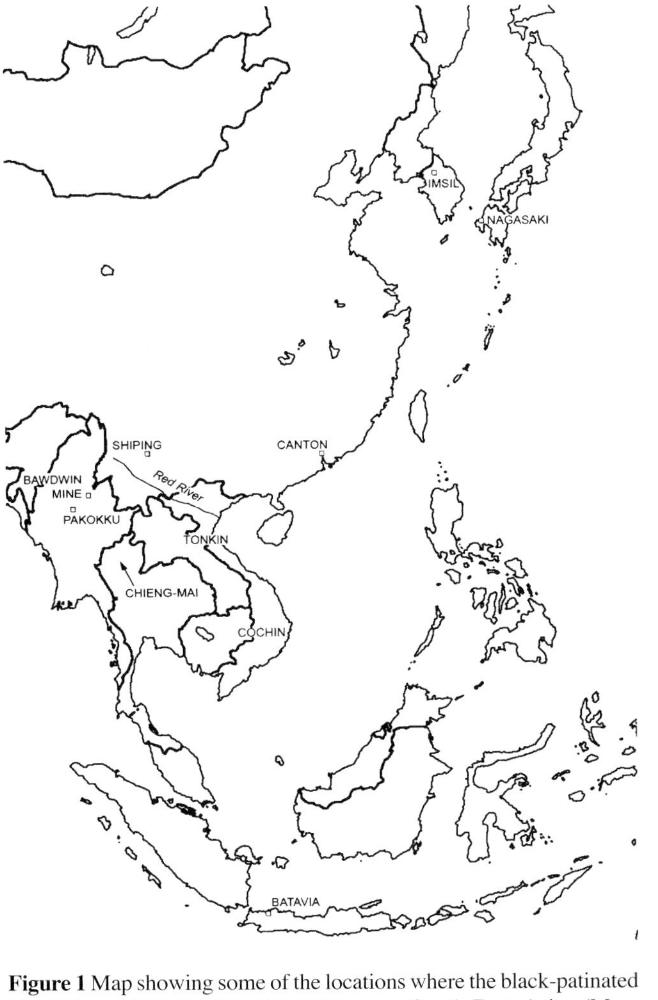

Figure 1 Map showing some of the locations where the black-patinated alloys have been produced in East and South-East Asia. (Map: A. Simpson.)

eries being made and almost certainly more to be made. Items of the black-patinated copper alloys are also known to have been produced for the local market variously in Korea, Yunnan, Thailand, Myanmar, and probably also in Tibet and Indonesia (Fig. 1). They all seem to have flourished through the last three or four centuries and continue in production in Japan and Korea, but their origins and early history remain obscure.

Japanese *shakudo*

This group will be described first as it is the one group of Far Eastern black-patinated metalwork that is reasonably well studied and published in European languages, particularly for the *tsuba* and other sword fittings of the Samurai of the Edo period. Good technical descriptions of the processes by which the whole family of *irogane* alloys, including *shakudo* were produced and patinated were made by Westerners resident in Japan in the 19th century, such as Gowland (1910), and scientific study commenced both within Japan (e.g. Uno 1929; Murakami 1993) and abroad (e.g. Roberts-Austen 1892; Savage and Smith 1979; Notis 1988).

To produce *shakudo*, copper is alloyed with a small amount of gold and sometimes silver, typically between about 0.2 and 3% of each. Although not specified in the recipes, most alloys also have small quantities of arsenic (Roberts-Austen 1892; La Niece 1990). In addition, there is another black-patinated alloy of copper containing just a few percent of arsenic, known in Japan as *kuromido*, literally 'black taste copper'. After the artefact has been made and the surface inlaid with other metals, the surface is carefully polished and cleaned. The type and quality of the final patina is dependent on the nature of the surface and for the Japanese and other patinas a fine mirror-smooth surface is required (the *wu tong* are apparently an exception; all their surfaces would appear to have been finely sanded). The final cleaning and degreasing is done with a variety of vegetable and fruit juices that are mildly acidic and probably commence the oxidisation of the surface. After this procedure the item is placed in a hot aqueous solution containing small amounts of inorganic chemicals such as copper acetate, copper sulphate and alum for 30–60 minutes during which time the patina develops. The dilute solutions are mildly oxidising and acidic. Despite the detailed scientific studies, the exact cause of the distinctive colour, variously described as black, purple black or blue black, is uncertain. The only mineral to be identified by X-ray diffraction (XRD) is the copper oxide cuprite (Cu_2O), which is usually red. It is thought that the gold is precipitated from the solid solution of the body metal as particulate metal finely and evenly distributed within the cuprite layers and these act to absorb much of the red region of the incident light. There is also the possibility of the gold particles in the layers of cuprite promoting interference patterns seen as a purple or blue iridescence.

The origins of *shakudo* are uncertain. Although there are references to items of *shakudo* in inventories from the 8th century AD, this could be taken as the literal translation, just 'red copper', and the earliest surviving pieces are probably sword fittings dating to the late Kamakura period (AD 1185–1333). Harris (1993) suggested that fittings of black lacquer were already widely used by the *shogun* military class, steeped in Zen Buddhism at this time, and the new material, *shakudo*, being both stronger and more attractive, rapidly became popular.

Sawasa: black-patinated wares for the European market

Following the establishment of a trading post for foreigners, including the Dutch East India Company, the Verenigde Oostindische Compagnie (VOC) at Dejima Island, near Nagasaki (Fig. 1), the Japanese began producing items decorated in styles which while still oriental, were more pleasing to European tastes, typically being more ornate. Among these were the distinctive black and gilded copper alloys, recently collectively christened as *sawasa* wares at the 1998 Amsterdam exhibition. Their popularity and distribution seem to have been firmly linked to the fortunes of the VOC, production commencing in the late 17th century, flourishing through the 18th century before ceasing completely around 1800 following the collapse of the VOC. *Sawasa* wares were certainly made in Japan, but may well have been made at other centres where there were European trading settlements, Tonkin being mentioned frequently in the 18th century. This is now the northern part of Vietnam and there were certainly Dutch trading settlements on the Red River that rises in Yunnan, passing Shiping (Fig.1), which was already the main centre of the *wu tong* production (see below).

Fifteen *sawasa* pieces were analysed by X-ray fluorescence (XRF) (Hallebeek 1998: app. 2). The black alloy was of copper with about 0.3–3.5% silver (average 1.5%); 0.3–4.5% gold (average 1.5%); 0.5–1% zinc; 0.1–1% arsenic; not detected–0.4% nickel (average 0.2%); and traces of tin, lead and, rather unusually, chromium. The single example of *sawasa* in the British Museum (OA Reg. 1983 6-16 1) (Fig. 2) has recently been analysed by XRF and found to be of copper with small amounts of gold, silver and arsenic; it also contains traces of chromium similar to the items already analysed (see Table 1). At about this time Japanese copper typically contained around 1% lead (residual from the de-silvering processes, see Craddock 1995: 232–3; Izawa, in this volume, pp. 163–9), but only traces of arsenic and nickel of the order of 0.02–0.05% as exemplified by the Japanese copper ingots from the wreck of the VOC *Waddingsveen* which sank in 1697 (Craddock and Hook 1997). The composition of the *sawasa* is similar to the copper from the mines in Yunnan with the exception of the chromium, suggesting that at least some of the *sawasa* wares were produced in Tonkin or possibly Canton. The gold, silver and arsenic contents of the *sawasa* wares are similar to those found in the *shakudo* and *wu tong* alloys, but the zinc contents are different.

Table 1 XRF results for the black metal compositions (– indicates below the detection limit).

	Cu	Ag	Au	Pb	Zn	Sb	Fe	Cr	Co	Ni	As	Sn
Wu Tong									–	–	–	–
1996,1-15.5b pipe handle	98.4	0.53	0.99	–	0.01	0.03	0.06	0.02	–	–	–	–
1996,1-15.5a pipe handle	97.8	0.84	1.16	0.04	0.03	0.05	0.06	0.01	–	–	–	–
Sawasa												
1983,6-16.1 base	97.0	0.72	1.06	0.19	0.12	0.02	0.09	0.02	–	0.01	0.78	–
1983,6-16.1 lid 1	96.5	0.85	1.21	0.30	–	0.07	0.05	0.01	–	–	0.95	0.04
1983,6-16.1 lid 2	96.5	0.85	1.23	0.23	–	0.05	0.06	0.02	–	–	1.03	–
Burmese												
2005,1-15.22 spot 4	96.6	1.09	0.84	0.25	0.71	–	0.11	0.02	–	0.11	0.24	–
2005,1-15.21 spot 1	92.4	5.45	0.19	0.52	0.29	1.06	0.10	–	0.02	–	–	–
2005,1-15.21 spot 2	96.2	1.07	0.42	0.13	1.26	–	0.10	–	–	0.32	0.34	0.18
2005,1-15.21 spot 5	96.1	0.94	0.51	0.09	0.28	–	0.04	–	–	0.20	–	1.86
2005,1-15.17 spot 8	96.0	1.63	0.07	0.16	0.55	0.13	0.13	0.02	–	0.09	0.39	0.84
2005,1-15.16 spot 5	94.7	3.56	0.84	0.03	0.49	–	0.06	0.06	–	–	0.11	0.17
2005,1-15.13 spot 5	91.1	5.65	–	0.15	2.69	0.11	0.10	–	–	–	0.25	–
2005,1-15.13 spot 4	90.5	5.63	–	0.21	3.34	–	0.04	–	–	0.12	0.19	–
2005,1-15.13 spot 2	91.2	5.21	–	0.17	2.92	–	0.09	–	–	–	0.44	–
2005,1-15.12 spot 3	95.7	1.55	1.40	0.06	0.46	0.13	0.14	0.04	–	0.22	–	0.34

The metals alloy compositions were analysed using a Bruker AXS ARTAX micro-X-ray spectrometer (X-ray tube settings: 50 kV / 800 µA, collimator 0.2 mm with a collection time of 300 s). Detection limits for the technique are of the order of 0.01% for Cr, Fe, Co, Ni, Cu, Zn, Pb and Sb; 0.02% for Ag and Sn; and 0.05% for As. Precision for copper is ~1–2%, decreasing down to ~50% for elements at or near the detection limit. Analysis was performed on uncleaned patinated surfaces.

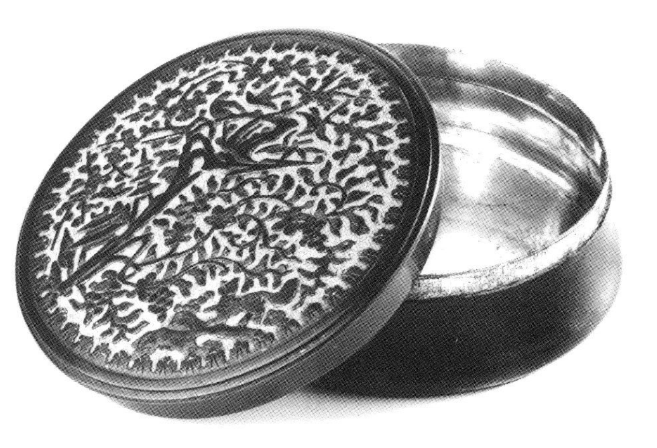

Figure 2 Small circular lidded box of *sawasa* ware (OA Reg. 1983 6-16 1). (Photo: A. Milton/British Museum.)

The various items in the 1998 S*awasa* catalogue were described as being assembled from many components soldered together, using a true copper-silver hard silver solder, otherwise known as a brazing alloy. This would be necessary as the items were usually fire gilded at about 400 °C, which would have melted a soft solder. A variety of techniques was described as being used to shape and decorate the wide range of pieces exhibited, including casting and repoussé work, decorated by engraving. Careful examination of the British Museum's single piece, a circular lidded box, showed that the body and the lid had both been assembled from several pieces of hammered sheet metal of approximately 0.5 mm thickness. Casting the body or the lid of a box of such thinness from an alloy made predominantly of copper would have been well nigh impossible. The recessed decoration on the lid would seem to have been partially carved out and decorated by engraving, chasing and punching. The latter was done with a small ring punch to produce a matting effect that is also seen on the illustrations of the decorated detail on many of the pieces in the 1998 exhibition. The *wu tong* boxes are made from a composite sheet of the gold-silver alloy brazed to a substrate copper sheet, and the brazing alloy is clearly visible on radiographs (Fig. 4). The interior of the *sawasa* box and the underside of the lid are both gilt and thus it was not possible to determine directly whether the sheet metal of the box was also a composite. However, unlike the *wu tong* pieces, no brazing alloy was visible on the radiograph of the base or lid, suggesting that the sheets were single pieces of metal. Following decoration, all that remained was gilding and patination of the metal.

Hallebeek's (1998: app. 1) examination of the *sawasa* pieces led him to the conclusion that although they had been made of the *shakudo* alloy they were not in fact patinated, but rather had been lacquered. Unfortunately his Fourier transform infrared (FTIR) analyses were unable to distinguish the actual lacquer used. Analysis of the black surface of the British Museum's piece failed to detect any lacquer, but XRD analysis revealed the presence of cuprite, identical to that found in the *shakudo* patinas. Another problem for Hallebeek was that the lacquer had apparently been applied *before* the fire gilding where the temperatures of 400 °C would inevitably have damaged it (elsewhere in the 1998 catalogue old accounts are recorded of merchants placing the *sawasa* coffee pots on the fire to demonstrate that the black surface was unaffected). Thus it would seem that the black surface is usually an inorganic patination identical to that found on the *shakudo* metalwork, and perhaps subsequently these were given a protective coat of lacquer.

The gilding and patination could have been accomplished in two ways: either the whole surface was patinated and then the design was carved into the surface and gilded or alternatively the design was carved in the unpatinated metal, gilded and then the remainder was polished and pati-

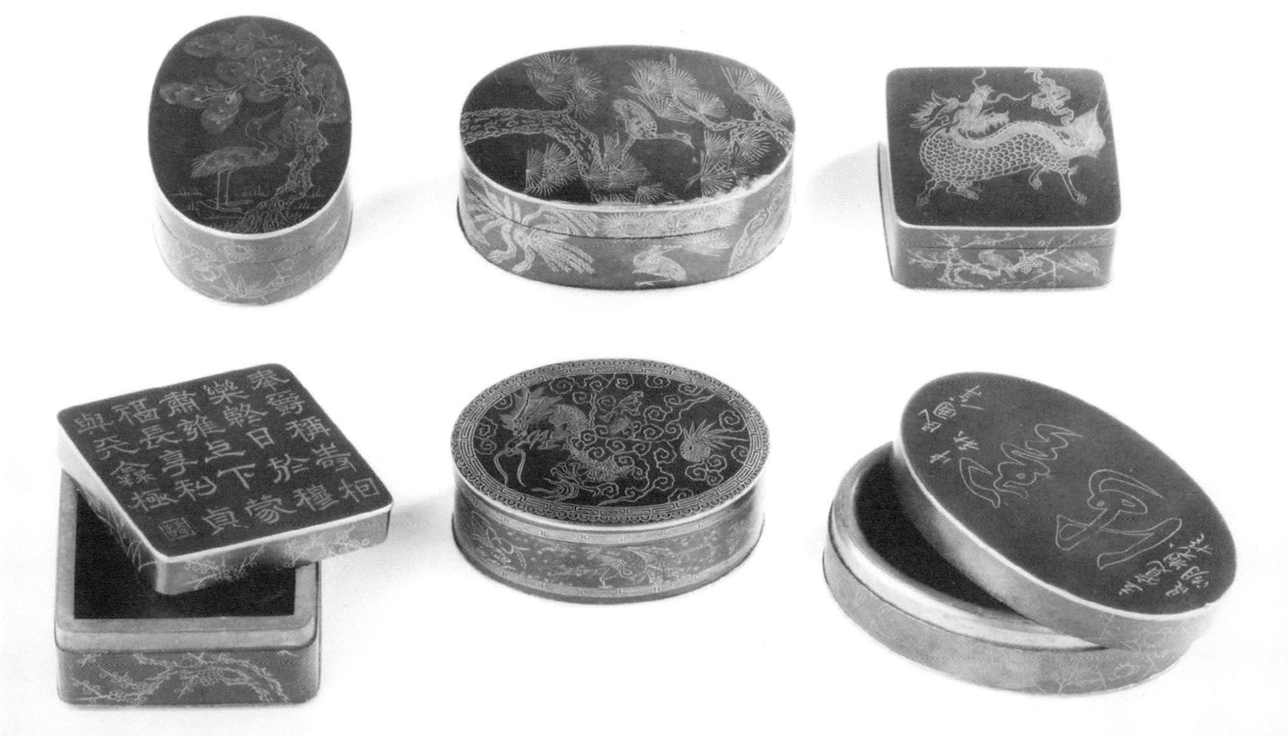

Figure 3 A selection of *wu tong* ink boxes formerly in the Collier collection, now in the British Museum (OA Reg. 1992 11-9 1-6). (Photo: A. Milton/British Museum.)

nated. Inspection of the British Museum's box shows that in many areas the gilding extends as streaks onto the metal that is otherwise patinated. Mercury gilding could not take to a patinated surface and thus the gilding must have been carried out before the box was patinated.

Wu tong

This black-patinated alloy was made in Yunnan in the south-west of China (Wayman and Craddock 1993; Craddock 1996). Production concentrated on small items such as tobacco and ink boxes (Fig. 3) and tobacco pipes. The alloy is usually of copper with about 1% each of gold and silver (see Table 1). This was inlaid with a silver brazing alloy, polished and patinated.

Although the basic components of *wu tong* are the same as *shakudo* and *sawasa*, there are significant differences in construction and inlay techniques. A metallographic examination of a section through the metal revealed it to be a composite. A thin sheet of the alloy containing the precious metal had been brazed with silver brazing alloy (containing about 50% silver, 40% copper, 4% zinc and 1% lead) to a thicker sheet of red brass (copper containing about 5% zinc). The two component sheets of the composite were first hammered, or rolled, down to a thickness of about 0.3 mm for the brass sheet and 0.1 mm for the *wu tong* sheet. It was previously believed that the composite sheet was reduced in thickness *after* brazing, but radiography (Fig. 4) has shown there is no distortion to the brazing joining the composite such as would have occurred if the composite sheet had been worked.

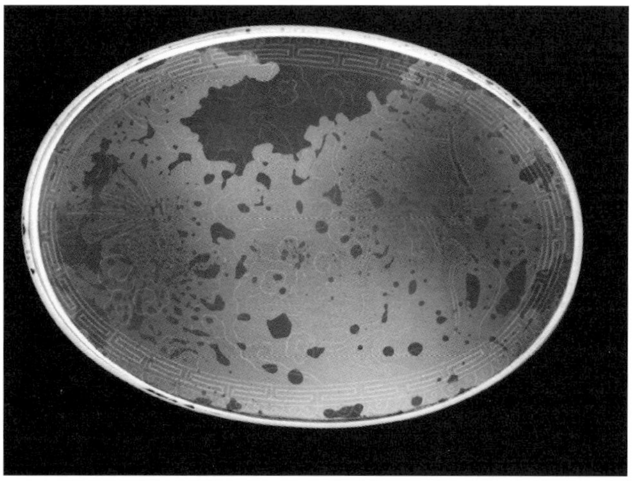

Figure 4 X-radiograph of the lid of a *wu tong* box. Note that the darker areas are cavities in the braze in between the sheets of *wu tong* and copper alloy and that there is no deformation of these cavities (OA Reg. 1992 11-9 4, length 7.1 cm). (X-radiograph: J. Ambers/British Museum.)

The design or inscription was engraved into the surface of the *wu tong*, and then inlaid with wires of a silver brazing alloy. The piece was heated to just above 820 °C, causing the silver brazing alloy to melt into the groove (Fig. 5) and after cooling the whole surface was filed and polished flat. The construction can be exemplified by the oval box illustrated in Figure 6. The boxes comprise a base and sides with a rim or lip all of copper. The *wu tong* composite was soldered to the sides. The lid is a sheet with side walls to both of which the *wu tong* composite was soldered. Raised strips of silver were added at the edges to conceal the soldered joins on the outside, and to protect the *wu tong* surfaces. The patinated surfaces on all the boxes are covered with fine parallel lines, suggesting the boxes were deliberately given a sanded finish. It is still not certain how the patination was achieved. It is claimed both in old information sheets on the process and by the current operatives at the Kunming Arts and Craft Mill (Kunming, Yunnan where patinated bonzes but not *wu tong* is produced) that the patination was achieved by repeated handling, the ammonium and sodium salts in the average sweaty palm being sufficient to produce the black patina over the course of several hours. Experiments by Giumlia-Mair and Lehr (2003) showed the method worked, but this still seems an inherently unlikely method for a regular production process. Mang and Han (1989) experimented with tea water and mixtures of organic acids (acetic, lactic and tannic). They obtained a good black patina except that XRD apparently showed that it was principally of silver oxide with only a little copper oxide whereas the original contains only copper oxide. It seems likely that treatment with solutions similar to those used to patinate the *shakudo* was employed.

Alongside the true *wu tong* there was also a cheaper version, known as *hong tong* (red copper), which has a deep chocolate brown patina. Analysis of an example purchased in Kunming, and now in the British Museum (OA Reg. 1996 01 5), showed the metal to contain a few percent of arsenic rather than precious metal, analogous to the Japanese *kuromido* alloy (Craddock 1996). The alloy was also encountered on Burmese *mylar* lime boxes (see below), used in conjunction with the precious metal alloy in the decorative schemes.

Production of *wu tong* seems to have been centred on the village of Yuejiawan near the town of Shiping on Lake Yinlong, about 200 km south of Kunming (Fig. 1). The process is recorded as being started by the Yue family which had then just arrived in the area in the mid-17th century (Craddock 1996). Production continued in the village until the mid-20th century when the remaining craftsmen moved to Kunming, before ceasing in about 1960.

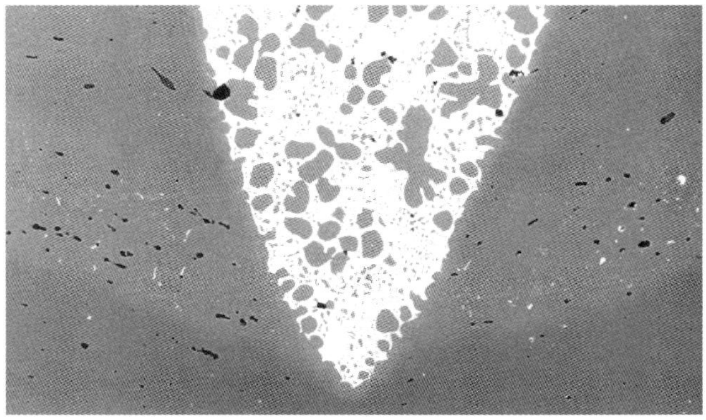

Figure 5 Cross-section through the inlay of silver brazing alloy clearly showing it had been melted into place. The dark regions in the inlay are dendrites of the copper-rich phase (83% copper, 10% silver, 6% zinc) which solidified first. The bright speckled regions are the silver-rich copper-silver eutectic (72% silver, 24% copper, 3% zinc). Scanning electron micrograph, backscattered electron image. Inlay depth 0.18 mm. (Micrograph: M. Wayman/British Museum.)

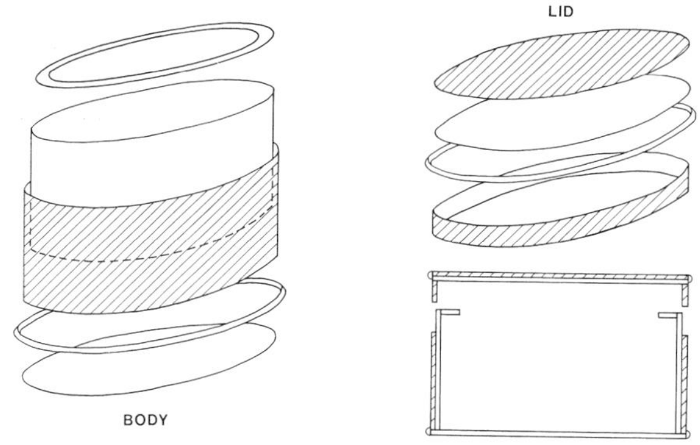

Figure 6 Construction of an oval *wu tong* box, using the central box in Figure 3 as an example. Left: exploded view of the body. Right: (above) exploded view of the lid; (below) cross-section through the body and lid. The hatched components are of the *wu tong* composite sheets and carry the inlay, brazed to the underlying thicker sheets of copper. (Diagram: B. Craddock.)

The black copper metalwork of Korea

There has been a tradition of using black-patinated copper alloys since at least the 17th century in Korea. It was recorded in the *Imwon Simyukji* written by Seo Yu-gu that tobacco pipes of white brass (cupro-nickel) were adorned with 'black copper' which itself was inlaid with silver or gold. Similar tobacco pipes are still made in the Imsil and Anseong regions of Korea (Fig. 1) (Anon. 2001: 74–7) and appear very similar to the 20th-century *wu tong* pipes, except that silver alloys are the usual material for the *wu tong* pipes, not cupro-nickel.

In addition, wires made of black copper are also used to inlay (*sanggam ipsa*) silver (Fig. 7) (Anon. 2001: 43–5). These black-patinated copper alloys do not seem to have been analysed although the white brass is described as containing 58% copper, 37% nickel and 5% zinc.

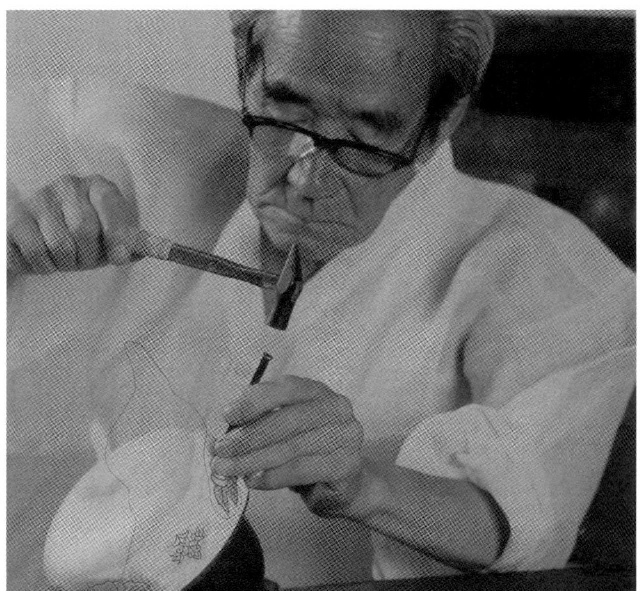

Figure 7 Korean craftsman inlaying a wire of black-patinated copper into a silver bowl. Note the wire is apparently already patinated (from Anon. 2001: 46).

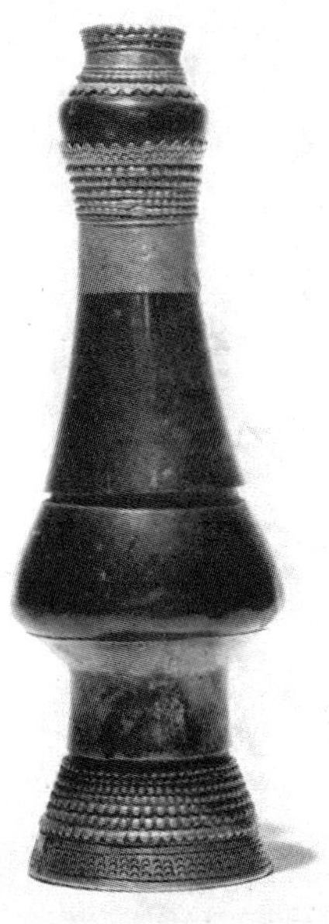

Figure 8 Lime box of *pan ca* from the Chieng Mai district of Thailand, described as having a silver and black colour scheme, now in the Musée de l'Homme, Paris (from Thierry 1969: fig. 227).

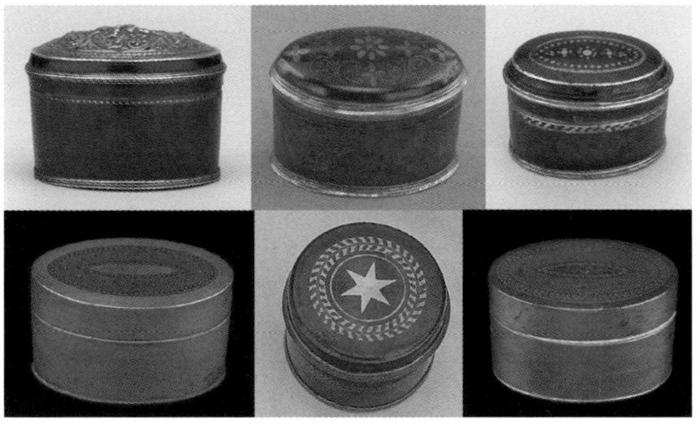

Figure 9 Selection of lime boxes with *mylar* from Myanmar that form part of the Ralph and Ruth Isaacs gift to the British Museum. Top left to bottom right, respectively OA Reg. 2005 01-15 10, 12, 14, 16, 15 and 17. (Photos: T. Springett, M. van Bellegem and J. Williams/British Museum.)

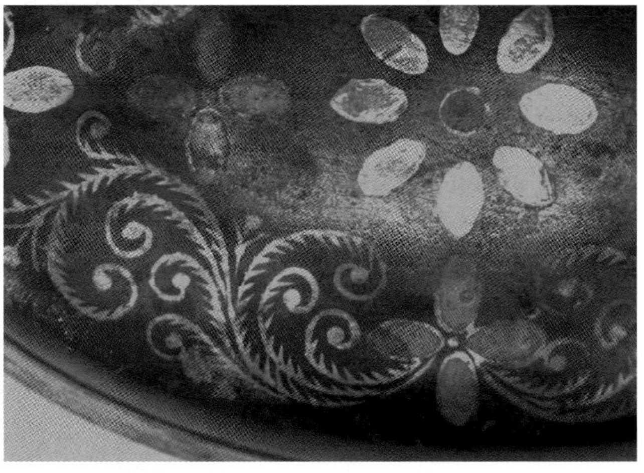

Figure 10 Detail of the lid of box OA Reg. 2005 01-15 12 showing fern-shaped leaves filled with silver-braze alloy and inlaid with silver and copper petals set into chased recesses. (Photo: M. van Bellegem/British Museum.)

Thailand

In recent centuries a black-patinated and inlaid metal was used in the Chieng-Mai region of north-west Thailand to make betel boxes etc. (Thierry 1969: cat. nos. 209, 242, 243) (Fig. 8). The alloy was said to contain five metals: silver, gold, brass, red-copper (ordinary copper) and red-gold (a copper-gold alloy, or more possibly the red-patinated gold, see below). The name of this alloy was *pan ca* which is a Hindi rather than Thai word. Also, Chieng-Mai is adjacent to the Shan states of Myanmar, where a similar range of metalwork, known as *mylar*, was produced.

Mylar and the Myanmar lime boxes

The British Museum has been fortunate to receive a collection of metal lime containers used in the preparation of the betel quid (Isaacs and Blurton 2000: 65–9), presented by Ralph and Ruth Isaacs, who collected them in Myanmar in the early 1990s (OA Reg. 2005 01-15 1-25) (Bellegem *et al.* 2007). Stylistically the boxes are purely local and were certainly made in Myanmar, probably mainly in the Shan States in the north-east bordering Yunnan (Fig. 1). The collection comprises boxes manufactured using various techniques and materials, including silver, various copper alloys, either embossed or engraved with inlays of niello or lacquer. Thirteen boxes in the collection use a black-patinated bronze, locally known as *mylar* (Fig. 9). From the outward appearance of the boxes it was immediately apparent that there were parallels with the *wu tong* boxes and this has been reinforced by their scientific study. Some of the boxes display a greater range of techniques and materials, and many are polychrome with reds and yellows of copper and brass inlays used together with the black-patinated bronze and silvery metals.

The boxes have been built up from sheet components brazed together to form an outer and inner jacket. The outer sheets are usually of the *mylar* metal or of cupro-nickel with an inner jacket of copper. Raised strips of silver or cupro-nickel were added at the edges as decoration and to conceal the brazed joins on the outside, very similar to the silver strips on the *wu tong* boxes. The decoration in the *mylar* sheets on two boxes in the collection are engraved or chased and inlaid with the molten silver brazing alloy in exactly the same technique as found on the *wu tong* boxes. Additionally, sheets of silver and copper were brazed into recesses made by embossing (OA Reg. 2005 01-15 12, Fig. 10). Alternatively chevron patterns were made by twisting together combinations of wires of different metals, such as silver, copper, *mylar* or a copper-arsenic alloy (Fig. 11), to make a cable with a helical seam.

The construction of the lids varies considerably, possibly reflecting different workshops. On some the *mylar* sheet has been punched, chased or engraved to create the channels and cells to hold the inlays (OA Reg. 2005 01-15 12, 13 and 14, Figs 9 and 10), but on other flat lids complex annular polychrome decorative schemes have been built up linearly (OA Reg. 2005 01-15 16, 17 and 20, Figs 9, 11 and 12). On these

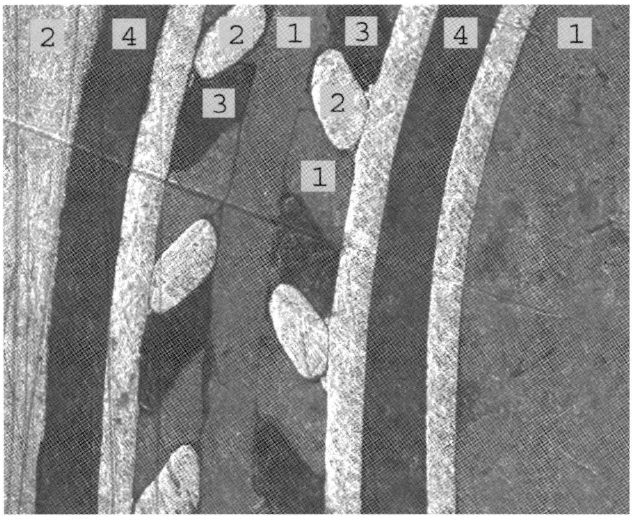

Figure 11 Detail of chevron pattern on the lid of OA Reg. 2005 01-15 16 made of copper (1), white metal (2) and *mylar* (3) wire, respectively red, white and black in colour. Parallel to the chevron pattern there are wires of copper-arsenic (4) alloy and white metal (2) which are respectively brown and white in colour. (Photo: M. van Bellegem/British Museum.)

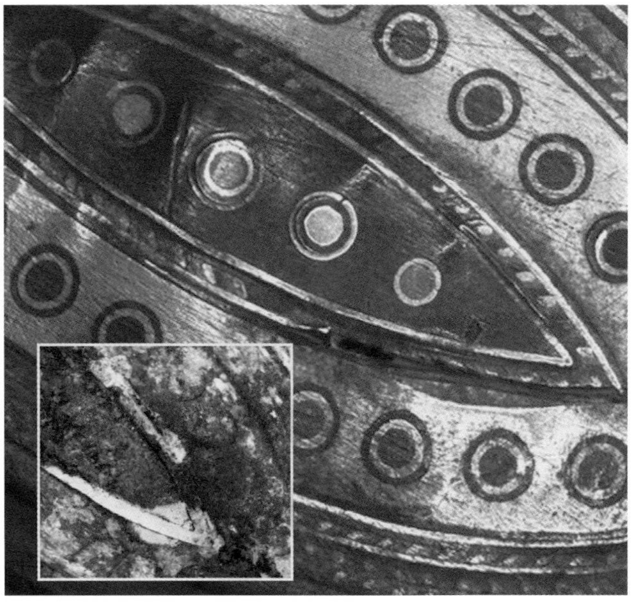

Figure 12 Detail of the lid of box OA Reg. 2005 01-15 20, made up of separate annular components soldered/ brazed to each other, not to a base beneath. The inset is a detail of the inside of the lid where the fracture corresponding to the outside of the lid can be seen. (Photo: M. van Bellegem/British Museum.)

lids there is no underlying metal base plate into which the elements had been set – the components are simply brazed to each other. The use of the various alloys such as copper, cupro-nickel, *mylar* and the copper-arsenic alloy in and next to the chevron patterns indicates that these alloys were intentionally contrasted to create the colour scheme.

Some of the earlier accounts of Burmese metal crafts made in Colonial times (Scott and Hardiman 1906; Tilly 1903) suggest that the various metals were imported, implying that the metal crafts were dependent on European maritime trade. However much this may have been true in the late 19th and early 20th century, it certainly was not true at other times (Ball 1881: copper 280; silver 234; arsenic minerals 162 and gold 230). The Bawdwin mines in the Shan States, for example, were a major source of silver when worked by the Chinese up to the mid-19th century and were an important source of zinc, lead and copper when worked by the British in the 20th century (Coggin Brown and Dey 1955: silver 138; copper 150; zinc 164–70; tin 168–75). Just over the border, Yunnan was a major producer of copper, tin, gold, silver, zinc and nickel (Coggin Brown 1920). Thus all of the metals found in the *wu tong* and *mylar* metals could be obtained locally.

Some analyses have been carried out on the unprepared surfaces using XRF (see Table 1). The *mylar* components are usually of copper with between about 1.7 and 2.2% (average 1.8%) silver; 0.8–1.8% (average 1.2%) gold, and with only traces of other metals, including arsenic. Some of the components were of copper with about 10% silver and 0.4% arsenic but no gold, recalling some of the Japanese *shibuichi* alloys. The inlays of silver brazing alloy that had been melted into place contained approximately 60% silver, 30% copper and 10% zinc. The cupro-nickel components are of copper with variable quantities of nickel and zinc, but both average about 10%. The alloy also contains between about 0.5 and 1% iron, significantly higher than in the other copper-based components. Cupro-nickel alloys were produced in Yunnan and formed a major item of trade to Europe in the 18th century (Mei 1995). It was known as *paktong* and usually contained variable quantities of nickel, typically about 10% and considerable quantities of zinc, typically between 20 and 40%. The *paktong* was made by co-smelting the nickel sulphide ore pentlandite with copper ores to produce an impure copper-nickel *matte* that had to be further purified and zinc added. Even so, the resulting *paktong* almost invariably contained 1% or more of iron. Cupro-nickel began to be made in Europe from the 1820s by mixing copper, nickel and zinc metals, and therefore the alloys only contain traces of iron.

Little is known of the history of the *mylar* process, and the few European reports are both short and not very reliable. Tilly (1903) states that 'copper is blackened by melting it with 5% of gold and a little sulphur'. The gold percentage seems very high and the sulphur is suspect – it was probably added to the list of ingredients by Tilly because it was believed erroneously that it should be there. Scott and Hardiman (1906: II, pt. ii, 723) describe the production of the inlaid lime boxes noting it was a speciality of the brass workers at the town of Pakokku in the Pakokku district of Myanmar (Fig. 1). They clearly had some inkling of the patination process and report that the finished boxes were placed in a clay pot, heated and then buried for some time. Recently experiments have been carried out on *mylar* and related alloys using techniques including handling and chemical treatments (Bellegem *et al.* forthcoming).

Several of the pieces are dated, usually to the first part of the 20th century, together with the owner's name, although one is dated to 1959, but this need not necessarily be the date of manufacture. Production of the patinated boxes had ceased by the time the Isaacs made their collection and Fraser-Lu made her study of Burmese crafts (1994).

Origins of black bronzes in the Orient

Black-patinated copper alloyed and inlaid with precious metals has a long history over a wide cultural area (Craddock and Giumlia-Mair 1993), first appearing at the end of the 3rd millennium BC in the Middle East (Giumlia-Mair and Quirke 1997).

The process seems to have died out in the West in the post-Roman period at much the same time as it probably first appeared in the Far East. Needham (1974: 257–66) considered that the Japanese *shakudo* must have had Chinese precursors. He came across references to a series of coloured golds that apparently existed in a variety of shades from scarlet *chhih chin* through purple, *tzu chin* to *tzu mo chin*, or purple sheen golds. References to these alloys occur in the Chinese literature from about the 5th century AD, together with statements that they were not indigenous but had come to China from the West.

However, when other references to coloured golds are considered, it is by no means clear that these Chinese scarlet and purple sheen golds should be interpreted thus. In the Indian classics, the *Ramayana* and the *Mahabharata*, there are references to *jambunada* gold (Dube 1996), which has been interpreted both as gold coming from the River Jambu, or that it resembled the *jamba*, a red-purple fruit, similar to a damson. Support for the latter view is given by contemporary descriptions that the *jambunada* gold was indeed scarlet. In later centuries, up to the present day, gold of high purity has been treated in India to produce a range of colours from red through purple to blue by treatment with a poultice of tamarind pulp containing nitre, common salt and elemental sulphur (Rây 1956: 218–22). Similar treatments are still carried out on gold to produce a red patination in the Malay peninsula and Indonesia (Anon. 1988). In the absence of other evidence these would seem to be more obvious candidates for *tzu mo chin* than *shakudo*.

References to metals related etymologically to *tzu mo chin* do occur outside China, notably in Tibet, and there the evidence does seem to suggest a patinated copper alloy containing, among other things, precious metals. There was a special alloy, *zi-khyim*, described in Tibetan records from the 10th century AD that was found either as a natural alloy in the ground or had to be synthesised (Lo Boe 1981: 41–4). The artificial alloy contained eight materials: copper with gold, silver, white iron, black and white lead (lead and tin?), rock crystal and mercury. This, when cast, was treated with 'poisonous water' (acid?) which gave it a purple-red iridescence.

It remains to consider the origins of the various indigenous traditions described above and their relationship to the European-inspired and distributed *sawasa* wares. Were there already long-established local schools of metalwork incorporating black-patinated bronzes in their repertoire of techniques or was an essentially Japanese technique spread around East and South-East Asia, inspired by the *sawasa* wares and carried by Europeans traders? Alternatively, if it is accepted that there is some continuity running through these black-patinated and inlaid alloys through the millennia and across the Old World, then possibly the technique did come with the exiled philosophers and alchemists across Central Asia to China in the mid-1st millennium AD. From there, one strand went to Korea and Japan and another to the mountains of south-west China and beyond.

References

Anon. (1988) *Gold Jewelry and Ornaments of Malaysia*. Kuala Lumpur: Muzium Negara.

Anon. (2001) *Korean Intangible Cultural Properties*. Seoul: Hollym.

Ball, V. (1881) *A Manual of the Geology of India III: Economic Geology*. Calcutta: Geological Survey of India.

Bellegem, M. van, Fletcher, P., Craddock, P.T., La Niece, S. and Blurton, R. (2007) 'The black bronzes of Burma', *British Museum Technical Research Bulletin* 1: 55–63.

Bellegem, M. van, Fletcher, P., Craddock, P.T. and Blurton, R. (forthcoming) '*Mylar*: reproductions of a traditional Burmese patinated alloy', in *Metals 07: Proceedings of the International Conference on Metals Conservation, 17–20 September, Amsterdam*.

Coggin Brown, J. (1920) 'The mines and mineral resources of Yunnan', *Memoirs of the Geological Survey of India* 47(1): 1–201.

Coggin Brown, J. and Dey, A.K. 1955. *India's Mineral Wealth*. Oxford: Oxford University Press.

Craddock, P.T. (1995) *Early Metal Mining and Production*. Edinburgh: Edinburgh University Press.

Craddock, P.T. (1996) 'The Ancestors of Japanese *irogane*', in *Proceedings of the Forum for the Fourth International Conference on the Beginning of the Use of Metals and Alloys (BUMA IV)*, 79–84. Sendai: Japan Institute of Metals.

Craddock, P.T. and Giumlia-Mair, A. (1993) '*Hsmn km*, Corinthian bronze, *shakudo*: black-patinated bronze in the ancient world', in S. La Niece and P.T. Craddock (eds), *Metal Plating and Patination*, 101–27. Oxford: Butterworth-Heinemann.

Craddock, P.T. and Hook, D.R. (1997) 'The British Museum collection of metal ingots from dated wrecks', in M. Redknap (ed.), *Artefacts from Wrecks*, 143–54. Oxford: Oxbow Monograph 84.

de Bruijn, M. and Kist, B. (eds) (1998) *Sawasa: Japanese Export Art in Black and Gold 1650–1800*. Amsterdam: Rijksmuseum.

Dube, R K. (1996) 'Some literary and documentary evidences for coloured gold in ancient India', *Bulletin of the Metals Museum of the Japan Institute of Metals* 26(2): 27–32.

Fraser-Lu, S. (1994) *Burmese Crafts: Past and Present*. Oxford: Oxford University Press.

Giumlia-Mair, A. and Lehr, M. (2003) 'Experimental reproduction of artificially patinated alloys identified in ancient Egyptian, Palestinian, Mycenaean and Roman objects', in P. Bellintani and L. Moser (eds), *Archeologie sperimentali*, 291–310. Trento: Provincia Autonoma di Trento.

Giumlia-Mair, A. and Quirke, S. (1997) 'Black copper in Bronze Age Egypt', *Revue d'Egyptologie* 48: 95–108.

Gowland, W. (1910) 'The art of working metals in Japan', *Journal of the Institute of Metals* 2: 4–41.

Hallebeek, P. (1998) 'Scientific analysis of metal alloys and surface layers of Sawasa objects by the Netherlands Institute for Cultural Heritage, Amsterdam', in M. de Bruijn and B. Kist (eds), *Sawasa: Japanese Export Art in Black and Gold 1650–1800*, 32–5 and app.1 (p. 116) and app. 2 (p. 117). Amsterdam: Rijksmuseum.

Harris, V. (1993) 'Why *shakudo*?', in S. La Niece and P.T. Craddock (eds), *Metal Plating and Patination*, 95–100. Oxford: Butterworth-Heinemann.

Isaacs, R. and Blurton, R. (2000) *Visions from a Golden Land: Burma and the Art of Lacquer*. London: British Museum Press.

La Niece, S. (1990) 'Japanese polychrome metalwork', in E. Pernicka, G. Wagner and A. Hauptmann (eds), *Archaeometry '90*, 87–94. Basel: Birkhauser.

Lo Boe, E. (1981) 'Statuary metals in Tibet and the Himalayas', in W.A. Oddy and W. Zwalf (eds), *Aspects of Tibetan Metalwork*, 33–67. British Museum Occasional Paper 15. London: British Museum.

Mang, Z. and Han, R. (1989) 'Studies of the blackening on copper-silver-gold alloy', *Studies in the History of Natural Sciences* 6(4) [in Chinese].

Mei, J. (1995) 'The history, metallurgy and spread of Paktong', *Bulletin of the Metals Museum of the Japan Institute of Metals* 24: 43–55.

Murakami, R. (1993) 'Japanese traditional alloys', in S. La Niece and P.T. Craddock (eds), *Metal Plating and Patination*, 85–94. Oxford: Butterworth-Heinemann.

Needham, J. (1974) *Science and Civilisation in China* 5(2). Cambridge: Cambridge University Press.

Notis, M.R. (1988) 'The Japanese alloy *shakudo*: its history and its patination', in R. Maddin (ed.), *The Beginning of the Use of Metal and Alloys*, 315–27. Cambridge, MA: MIT Press.

Rây, A.C. (1956) *History of Chemistry in Ancient and Medieval India*. Calcutta: Indian Chemical Society.

Roberts-Austen, W.C. (1892) *Report on the Analysis of Various Examples of Oriental Metalwork in the South Kensington Museum*. London: HMSO.

Savage, E.I. and Smith, C.S. (1979) 'The techniques of the Japanese *tsuba*-maker', *Ars Orientalis* 11: 291–328.

Scott, J.G. and Hardiman, J.P. (1906) *Gazetteer of Upper Burma and the Shan States*. Rangoon: Government Printing Office.

Thierry, S. (1969) *Le Bétel I Indie et Asie du Sud-Est*. Supplément au tome IX, 3, d'Objets et Mondes, Revue du Musée de l'Homme. Paris: Catalogues du Musée de l'Homme.

Tilly, H.L.(1903) 'The silverwork of Burma', *Journal of Indian Art and Industry* 10(84): 31–5.

Uno, D. (1929) 'Künstliche Korrosion von japanischen Speziallegierungen', *Korrosion und Metallschutz unter Einbeziehung des allgmeinen Metallschutzes* 5(6): 121–30, 147–56.

Wayman, M. and Craddock, P.T. (1993) '*Wu tong*: a neglected Chinese decorative technology', in S. La Niece and P.T. Craddock (eds), *Metal Plating and Patination*, 128–34. Oxford: Butterworth-Heinemann.

Bronze casting technologies in ancient China

Origins and evolution of the casting technology of Anyang bronze ritual vessels: an exploratory survey

Yu Liu

ABSTRACT This paper presents the origins and evolution of the casting technology of Yinxu bronze ritual vessels. By using new methods to categorise the assembly types of *ding* vessels, the evolution of the section-mould division is examined. Different separate casting techniques and the use of metal spacers are discussed. By looking at the casting techniques of bronze vessels from the Panlongcheng site, the origin of the casting techniques used for Yinxu bronze ritual vessels is explored.

Introduction

Bronze ritual vessels are one of the most characteristic features of ancient Chinese civilisation. Anyang occupies a central place in the history of Chinese archaeology and the casting technology of its bronzes has been discussed by numerous authors such as Karlbeck (1935), Shi Zhangru (1947, 1955), Chen Mengjia (1954), Li Ji and Wan Jiabao (1964, 1966, 1968, 1970, 1972), Gettens (1969), Barnard (1961), Barnard and Tamotsu (1975), Hua Jueming *et al.* (1981), and Bagley (1987, 1999) to name just a few. The intricate design and shape of ancient Chinese bronzes led early Western scholars to argue that Shang bronzes were cast using the same *cire perdue* method known for Greek and Roman bronze sculptures. The view that ancient Chinese metalworkers used section-mould rather than lost-wax casting was firmly established only in the 1960s, although several scholars had already realised during the 1930s–40s that Yinxu bronze vessels were made using section-mould casting technology. Karlbeck (1935) carefully studied the mould fragments and came to the conclusion that the moulds were not used as models but as casting patterns. Shi Zhangru (1947, 1955) described the materials and structures of the moulds, models and other foundry remains and discussed the manufacturing procedure. Chen Meng-jia (1954) suggested that section-mould casting technology was the main method used by Anyang casters. Li Ji and Wan Jiabao (1964, 1966, 1968, 1970, 1972) carried out important work on the Yinxu bronze casting technology; they studied the assembly of different types of vessels and were the first to use changes of the typical vessel casting assembly to deduce their chronology. They undertook a series of reproduction experiments in order to explore the details of the manufacturing process. Based on their studies of the bronzes from the Fuhao tomb, one of the most important finds at the Yinxu site, Hua Jueming *et al.* (1981) discussed the different casting methods of various types of Shang bronzes, especially the use of the separate casting technique.

Casting technology of Anyang ritual bronze vessels

Section-mould division: a round ding *vessel*

In the late Shang-Yinxu period, section-mould casting technology was widely used in the production of bronzes. In the assembly design, the section mould is divided not only along the vertical line, but also along the horizontal line. The vertical casting traces are easily discerned on the surface of bronzes and have therefore been the subject of many previous discussions, however the horizontal division traces are difficult to observe directly, so we have only come to know about them recently from excavated foundry remains (Anyang Archaeological Team 2006; Xiaomintun Archaeological Team 2007). Li Yung-ti (2003) was the first scholar to identify and describe the use of this method.

More than 2,000 Anyang ritual vessels have been unearthed (IA 1980, 1985, 1998, 2007; Anyang City Archaeological Team 1993; IA and Anyang City Institute of Cutural Relics and Archaeology 2008) with complex and different types of section-mould divisions. The section-mould divisions of a round *ding* (鼎) vessel can be clas-

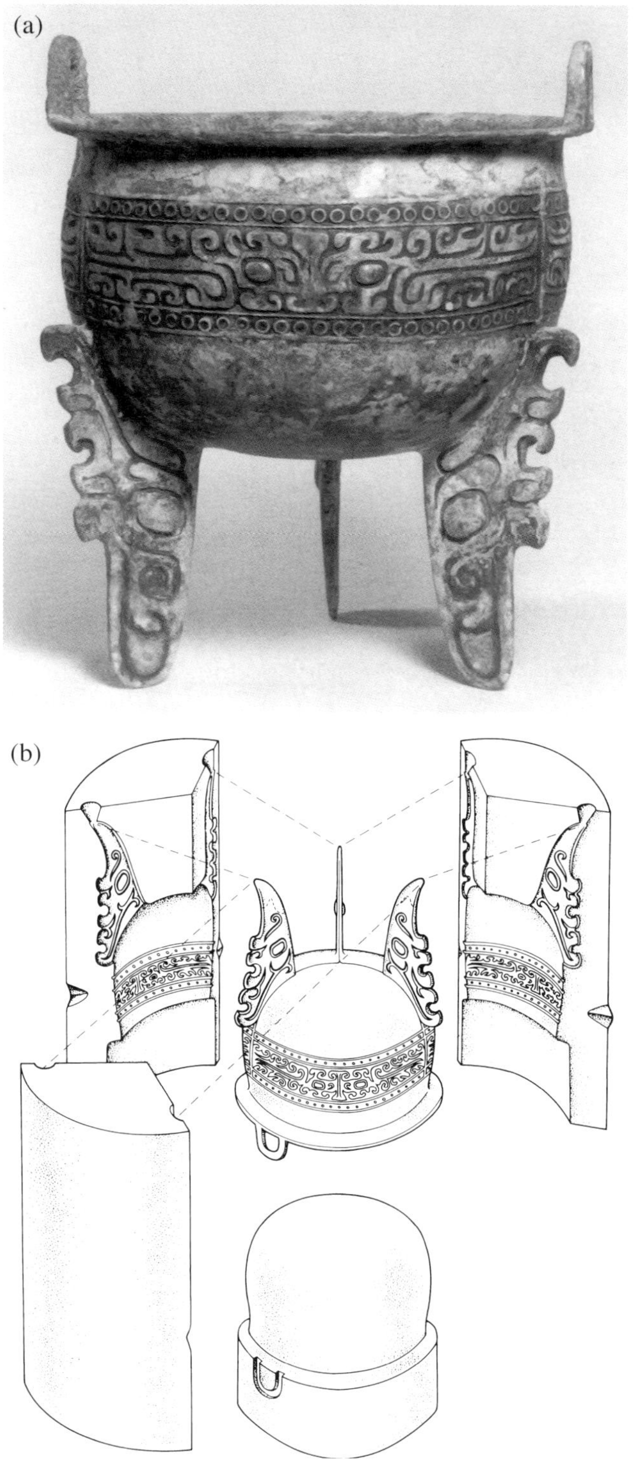

Figure 1 3fz(Y) type: (a) the round *ding* with flat legs (after Bagley 1987: fig 99) and (b) the *ding*'s mould assembly (after Bagley 1987: fig. 100).

sified into seven types. The simplest and earliest assembly includes three vertical mould divisions with legs and one core[1] (Fig. 1). A 'Y' style casting trace can be seen on the outer bottom so this can be designated as the 3fz(Y) type. This type was already in use in the early Shang period. In Anyang this type is still found in the Sanjiazhuang period, i.e. within the late phase of the middle Shang period (1435–1260 BC), slightly earlier than the Yinxu period (1260–1046 BC). However, it had largely disappeared after the 2nd phase of the Yinxu period.

The 3fz+1(△) type probably appeared in the Sanjiazhuang period and was widely used throughout the Yinxu period. It includes three vertical mould divisions and one triangle bottom mould (Fig. 2). The '△' style casting trace can be seen on the outer bottom and usually also on three flanges on the walls of vessels.[2] The 6fz+1(△) type is similar to the 3fz+1(△) type but includes six vertical mould divisions.

The 6j+3fz+1(△) type is more complicated as its assembly includes a horizontal division along the neck. In Chinese, the neck part is called *jing*, so it can be simplified into 'j'. The whole assembly needs six vertical moulds from the ear to the neck, and three vertical moulds from the shoulder to the leg (Fig. 3). The three types are divided along the abdomen, as shown by the section mould and its correspondent part of the *ding* vessel.

The 6jf+3z+1(△) type includes a horizontal division along the lower part of the abdomen, so the whole assembly needs six vertical moulds from the ear to the lower abdomen, and three vertical moulds from the lower abdomen to the foot (Fig. 4). The 3jf+3z+1(△) type is similar but includes only three vertical moulds from the ear to the lower abdomen, and three vertical moulds from the lower abdomen to the foot. The 6j+6fz+1(△) type needs six vertical moulds from the ear to the higher abdomen, and six vertical moulds from the higher abdomen to the foot.

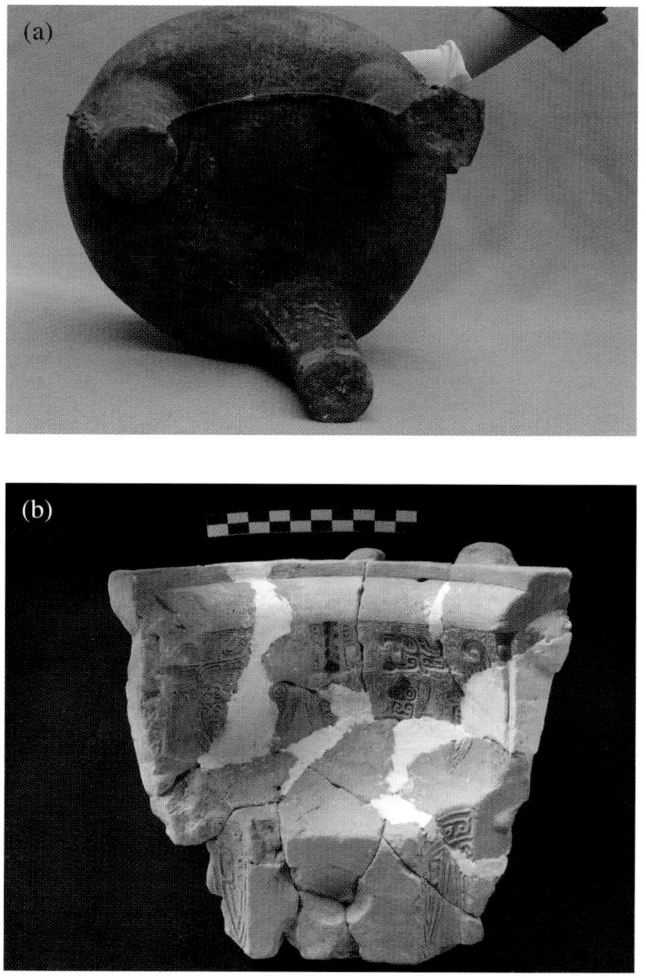

Figure 2 3fz+1(△) type: (a) a round *ding* from the Anyang City Museum and (b) a piece of *ding* mould from the Xiaomintun site, Anyang.

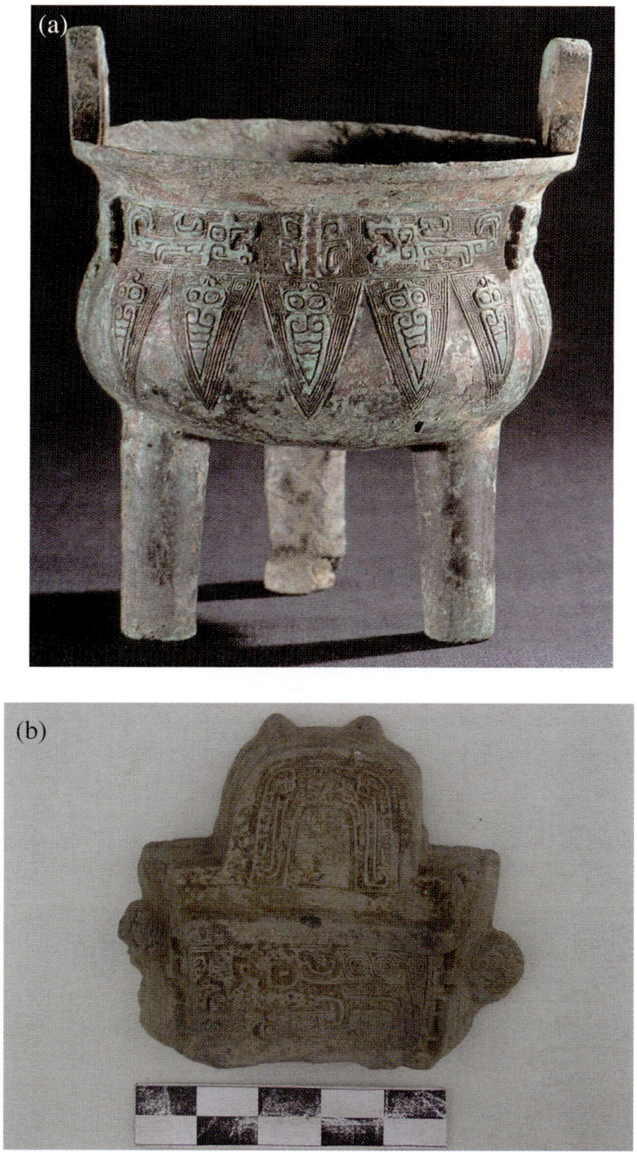

Figure 3 6j+3fz+1(△) type: (a) the round *ding* from GM874, Anyang and (b) the *ding*'s mould piece of the ear–shoulder part from the Xiaomintun site, Anyang.

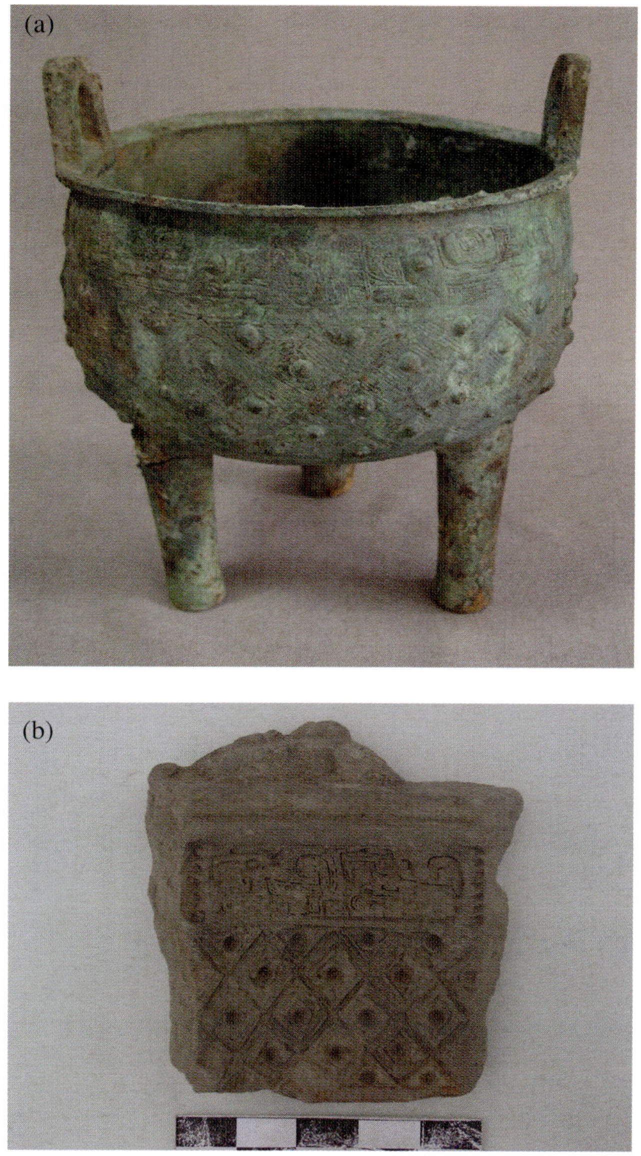

Figure 4 6jf+3z+1(△) type: (a) the round *ding* from M1046, Anyang and (b) the *ding*'s mould piece of the ear–higher abdomen part from the Xiaomintun site, Anyang.

The types of assembly patterns of round *ding* vessels in Anyang are shown in Table 1. From this table, it can be seen that there are some changes of types of assembly patterns of round *ding* in Anyang which vary according to the time period. In the third phase of the middle Shang period, the main pattern found is 3fzY, but 3fz+1(△) had already appeared. In the first phase of the Yinxu culture, the latter came into regular use; after the second phase, several kinds of assemblies with horizontal divisions were commonly used and in the third phase 6jf+3z+1(△) occupied the central position. In the fourth phase, the assembly with the simple pattern reappeared in regular use. Generally, most round *ding* vessels in Anyang were made by the section-mould casting technique, pouring just once, and the design of assembly patterns showed a tendency to standardisation. Compared with previous round *ding* vessels, the assembly pattern of Yinxu bronzes seems to have become more complex, with an increasing dependence on the use of bottom moulds and horizontal division.

Table 1 Types of assembly patterns of round *ding* vessels in Anyang.

Time	3fzY	3fz+1(△)	6fz+1（△）	6j+3fz+1（△）	6jf+3z+1（△）	3jf+3z+1（△）	6j+6fz+1(△)
Middle Shang: 3rd phase	11	2					
Yinxu: 1st phase	4	2					
Yinxu: 2nd phase		15	5	18	12	3	5
Yinxu: 3rd phase		4		2	15		1
Yinxu: 4th phase		11	3	1	14	4	1
Total	15	34	8	21	41	7	7

Separate casting

In the Yinxu period, the separate casting technique became more popular than during the middle Shang dynasty, probably because it was very difficult to attach appendages in the shape of big animal sculptures to the main body of bronze vessels in just one casting operation. Separate casting can be categorised into three types: casting on, casting in and casting joint (Hua *et al.* 1981).

The term 'casting on' implies a metallurgical join, by which the second pour of liquid metal has sufficient heat and fluidity to cause the terminal metal of the previously cast member to melt and fuse with it (Barnard and Tamotsu 1975). In general, handles and ears of big animal sculptures were made using a casting-on process.

Appendages such as core-cast knobs were joined to the body by a casting-in process, in which the precast appendage was put into the clay mould assembly to form the connection by the second casting.

A vessel of a complex shape, such as a *you* (卣), would need more than two castings to connect the handle and the chain (链环), which could be made by a cast-joint process.

The knobs of a *jia* (斝) vessel from M54, a large tomb contemporary with the Fuhao tomb, are joined by a casting-in process (IA 2007). The joint structure is different from other examples from the same period (Fig. 5). The handle is joined by a casting-on process, which was widely used in all vessels with three-dimensional animal-shaped handles. A *you* vessel with a swing handle and projecting beams decorated with long-tailed bird motifs illustrates the use of the cast-joint process to form the complex structure. We only had access to the mould fragments from the Xiaomintun foundry site in Anyang, rather than the vessel themselves (Anyang Archaeological Team 2006) but our investigations revealed 181 vessels show the use of the separate casting technique, accounting for about 17% of the 1150 ritual vessels found. These data suggest that the separate casting technique was commonly used in Anyang during the late Shang dynasty.

Metal spacers

In Anyang, metal spacers were used in the assembly of many ritual vessels, usually placed between the moulds and the core in order to maintain the wall thickness and hold the moulds in position. The earliest bronzes so far known which used the metal spacers technique are the bronze vessels from Panlongcheng, a middle Shang site in Hubei province. After the second phase of the Yinxu period, the use of metal spacers in assembly became standard. They were usually placed between the moulds and the abdomen core in three-legged vessels, and between the abdomen core and the round-foot core in round-foot vessels. Three metal spacers were used between the bottom mould and the abdomen core in a round *ding* vessel, and one metal spacer was placed between the bottom mould centre and the abdomen core in a *jia* or *jue* (爵) vessel. Three or four metal spacers were placed between the abdomen core and the round-foot core in round-foot vessels such as *gui* (簋) (Fig. 6).

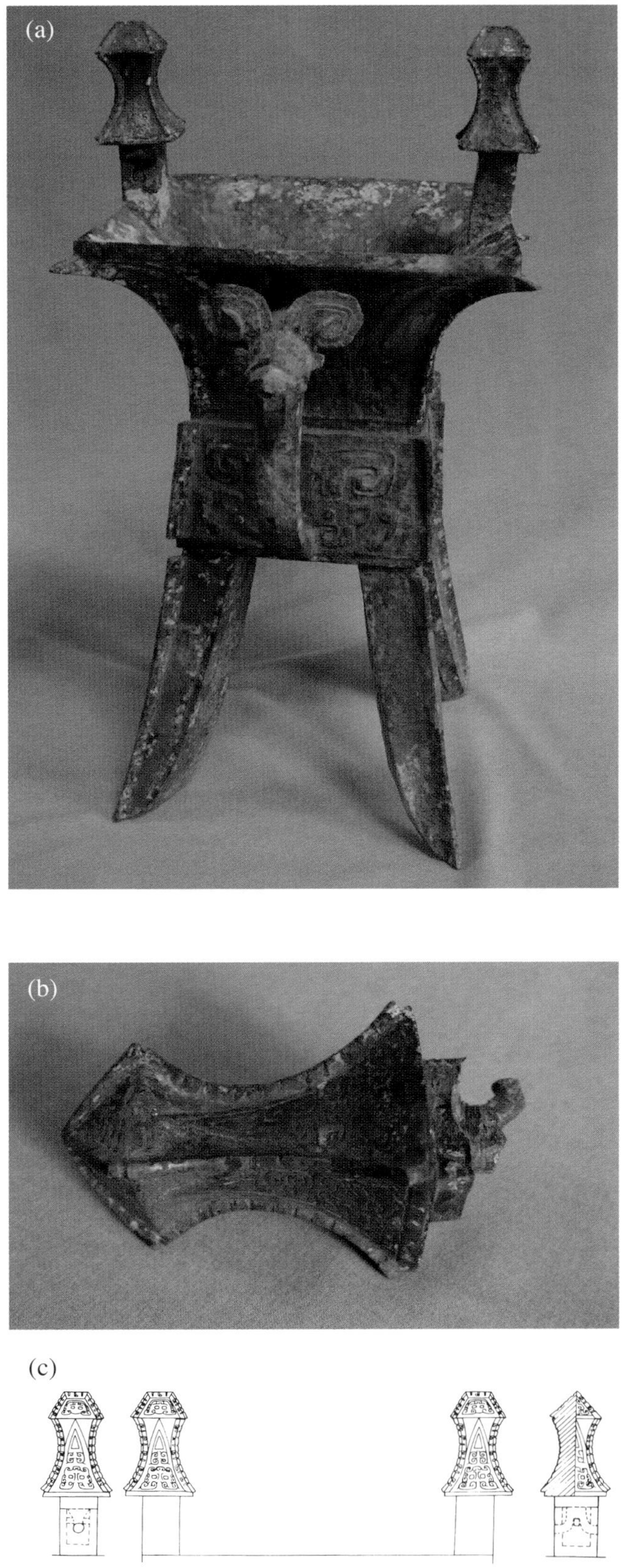

Figure 5 (a) The square *jia* vessel from M54, Anyang; (b) the vessel's knob; (c) the knob's connection structure.

It was previously thought that the ritual bronze vessels from Fuhao tomb were made without the use of metal spacers because no traces could be seen on the surface, however X-ray photography has now shown that a considerable number of metal spacers were used between the moulds and the core. The statistical data for those vessels displaying the

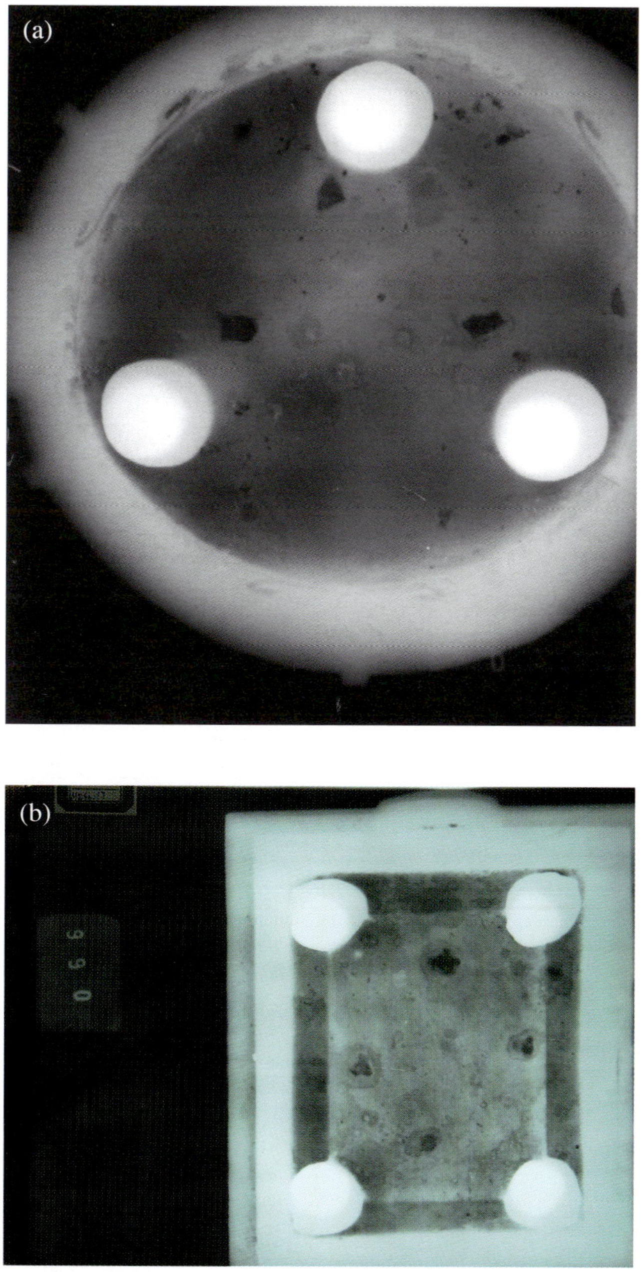

Figure 6 (a) Distribution of metal spacers in the round *ding* from M663, Anyang and (b) distribution of metal spacers in the square *ding* from M160, Anyang.

Table 2 Vessels made using metal spacers.

Period	Middle Shang: 3rd phase	Yinxu: 1st phase	Yinxu: 2nd phase	Yinxu: 3rd phase	Yinxu: 4th phase	Total
Vessels with metal spacers	12	9	49	30	16	116
Investigated vessels	32	24	110	66	40	272

use of metal spacers are given in Table 2. Of the 272 vessels investigated, 116 vessels exhibit the use of metal spacers, indicating that this technique of using metal spacers was prevalent in Anyang during the late Shang dynasty.

Origins of the casting technology of Yinxu bronze ritual vessels

The earliest bronze ritual vessels excavated to date are from the Erlitou site, Yanshi city (1880–1529 BC). In that period, all the vessels were cast using a section-mould process, pouring just once. The casting method was fundamentally established in the early Shang period (1600–1415 BC); this traditional method was inherited by Anyang casters, although the shape of the objects – especially the handles, the bottom and the knobs – looked very different. After the 2nd phase of Yinxu period, the separate casting technique was used regularly.

The separate casting method was employed widely by Anyang casters to form all kinds of bronze vessels with complex designs. The appearance of this important technique can be traced back at least to the early/middle Shang period at the Panlongcheng site in Hubei (Hubei Provincial Institute of Cultural Relics and Archaeology 2001). A *gui* vessel and a *you* vessel from the site were made using separate casting technology and represent the earliest intentional cases known so far. Alongside them are some cases indicating unintentional use of this technology, perhaps

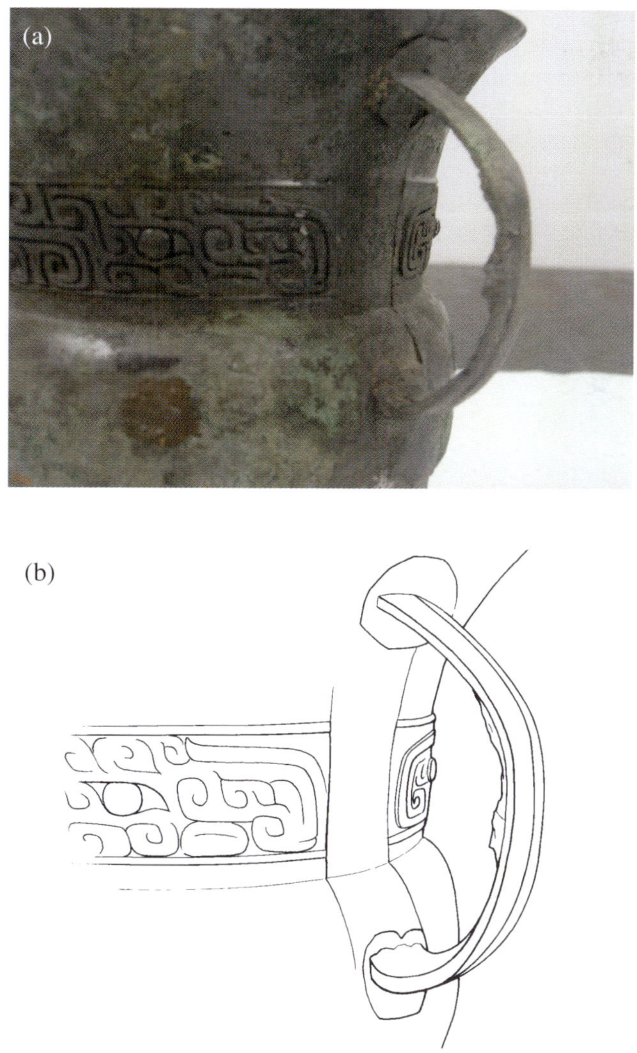

Figure 7 (a) PLZM1: 12 *jia* vessel from the Panlongcheng site. (b) the vessel's cast-on handle.

representing the initial stages of development. An entire precast bottom was put into the assembly of the *zun* to form the shape. Such a method would look like a casting-in process.

The handles of two *jia* vessels were both shaped by the casting-on process – in one case, a repair patch is seen (Fig. 7), and the other is an intentional design, although both joints seem to be a little crude (Fig. 8). Under the handle of first *jia* vessel (Fig. 7) are two casting lines suggesting that the section-mould assembly included the handle, but for some unknown reason this did not work and so it had to be added using a second casting. No casting line is apparent under the handle of the other *jia* vessel (Fig. 8), implying that the handle was added intentionally using a second casting. The development of the casting technique can be seen clearly from the former *jia* vessel, via the latter *jia* vessel, to the *gui* vessel.

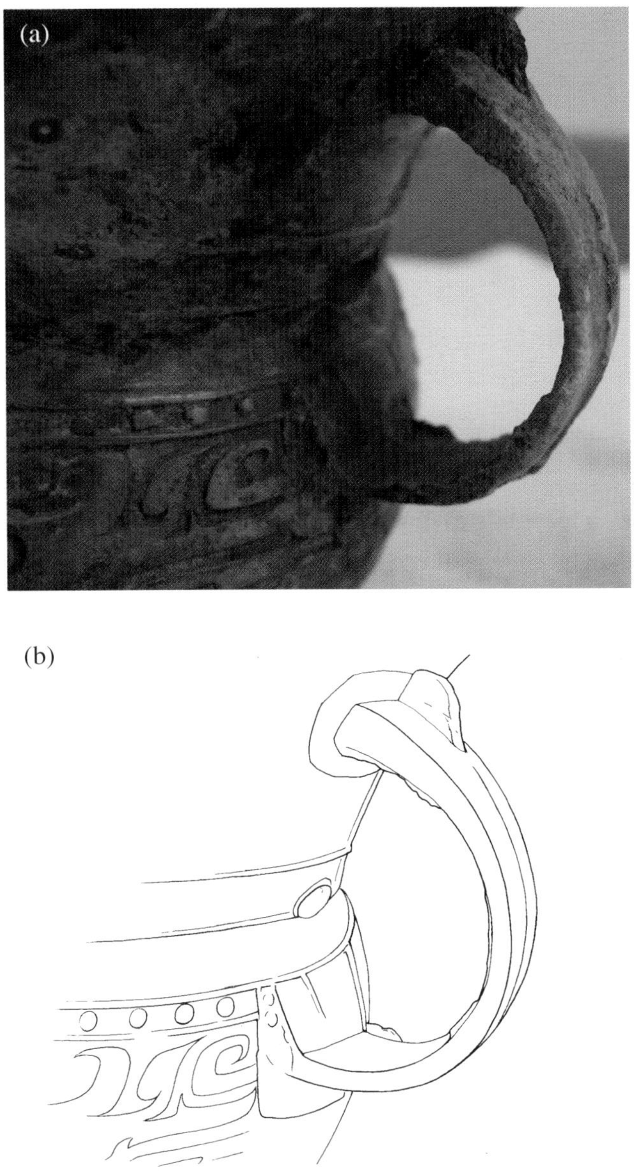

Figure 8 (a) PLZM1: 13 *jia vessel* from the Panlongcheng site and (b) the vessel's cast-on handle.

The method used to shape the Anyang square *ding* is the process of whole casting, which is different from the method used to shape the square *ding* from Zhengzhou and Xingan; they are all made by a cast-joint method with different parts.

Discussion

Barnard suggested that casting-on technology could have originated from the practice of repairing (Barnard and Tamotsu 1975). Many bronze vessels from the Panlongcheng site have repair patches on their bodies, especially to the legs of *dings*, which were all created using a second or even a third casting. Some bronze vessels from Panlongcheng used metal spacers, which are hitherto the earliest such cases found. Because the metal spacers were easily dislodged, small holes on the body were common, some of which have been repaired by small patches. Bronze vessels from Zhengzhou also have small patches, but their number is considerably smaller than of those from Panlongcheng.

In the Yinxu period, the shape, the motif and the appendages became more complicated, especially when shaping animal sculpture appendages, which required the use of the separate casting technique. The design and casting technique of such bronzes seemed to have influences from south China (Bagley 1977; Rawson 1992–93). For example, the jade tiger from Anyang is very similar in shape to the bronze tiger from Xingan, Jiangxi province. The *gu* vessels with hollow motifs are found only in Panlongcheng and Anyang, a fact which probably implies that there existed some technological relation between Anyang and other areas, especially south China.

Conclusions

The technology of Anyang ritual bronze vessels is characterised by the employment of the section-mould casting method, which was already established in the early and middle Shang period in the Central Plains of China. The design of mould assembly tended to develop towards a standard of both vertical and horizontal division. Separate casting techniques, which seem to have originated from Panlongcheng, were used widely in Yinxu period in order to cast the three-dimensional appendages.

Acknowledgement

I would like to thank John Moffett of the Needham Research Institute for useful suggestions on my use of English in this paper. I would also like to thank Mr Yue Zhanwei of the Institute of Archaeology, CASS, and Professor Zhang Changping of the School of History, Wuhan University, for permission to use some of their photos.

Notes

1. As all types have one core, this is not mentioned in the following text.
2. As the one triangle bottom mould exists in all the following kinds of assembly it is not mentioned further in the text.

References

Anyang Archaeological Team，Institute of Archaeology, Chinese Academy of Social Sciences (中国社会科学院考古研究所安阳考古队) (2006) 2000–2001安阳孝民屯东南地商代铸铜遗址发掘报告 ('The 2000-2001 excavation of the Shang bronze foundry site at Xiaomintun Southeast, Anyang'), 考古学报（*Acta Archaeologica Sinica*）3: 351–84.

Anyang City Archaeological Team（安阳市文物工作队）(1993) 安阳殷墟青铜器 (*Yinxu Bronzes in Anyang*). Zhengzhou: Zhongzhou Guji Press.

Bagley, R.W. (1977) 'P'an-lung-ch'eng: a Shang city in Hupei', *Artibus Asiae* 1977 (XXXIX, 3/4): 165–219.

Bagley, R.W. (1987) *Shang Ritual Bronzes in the Arthur M. Sackler Collections*. Washington, DC: Arthur M. Sackler Foundation.

Bagley, R.W. (1999) 'Shang archaeology', in M. Loewe and E.L. Shaughnessy (eds), *The Cambridge History of Ancient China: From the Origins of Civilization to 221 B.C.*, 124–231. Cambridge: Cambridge University Press.

Barnard, N. (1961) *Bronze Casting and Bronze Alloys in Ancient China*. Monumenta Serica Monograph 14. Canberra: Australian National University.

Barnard, N. and Satō, Tamotsu (1975) *Metallurgical Remains of Ancient China*. Tokyo: Nichiōsha.

Chen, Fang-mei (陈芳妹) (1998) 商代青銅艺术「区域风格」之探索- 研究课题与方法之省思 ('Inquiry into the "regional styles" of the art of Shang bronzes: some thoughts on research topics and methodology'), 故宫学术季刊 (*Gugong Academic Journal Quarterly*) 15(4): 109–50.

Chen, Mengjia (陈梦家) (1954) 殷代铜器 ('Bronze vessels of the Shang dynasty'), 考古学报 (*Acta Archaeologica Sinica*) 7: 15–60.

Gettens, R.J. (1969) *The Freer Chinese Bronzes*. Volume 2: *Technical Studies*. Washington, DC: Smithsonian Institution.

Hua, Jueming (华觉明) (1999) 中国古代金属技术---铜和铁造就的文明 (*Metallurgy in Ancient China: A Civilization Based on Copper and Iron*). Zhengzhou: Daxiang Press.

Hua, Jueming, Feng, Fugen, Wang, Zhenjiang, and Bai, Rongjin (华觉明, 冯富根, 王振江, 白荣金) (1981) 妇好墓青铜器群铸造技术的研究 ('Study on the casting technology of bronzes from the Fuhao Tomb'), 考古学集刊 (*Collected Papers in Archaeology*) 1: 244–72.

Hubei Provincial Institute of Cultural Relics and Archaeology (湖北省文物考古研究所) (2001) 盘龙城一九六三年～一九九四年考古发掘报告 (*The 1963–1994 Excavation of Panlongcheng Site*). Beijing: Wenwu Press.

IA (Institute of Archaeology, Chinese Academy of Social Sciences, 中国社会科学院考古研究所) (1980) 殷墟妇好墓 (*The Fuhao Tomb at Yinxu*). Beijing: Wenwu Press.

IA (Institute of Archaeology, Chinese Academy of Social Sciences, 中国社会科学院考古研究所) (1985) 殷墟青銅器 (*Yinxu Bronzes*). Beijing: Wenwu Press.

IA (Institute of Archaeology, Chinese Academy of Social Sciences, 中国社会科学院考古研究所) (1998) 安阳殷墟郭家庄商代墓葬: 1982年—1992年考古发掘报告 (*Shang Burials at Guojiazhuang, Yinxu, Anyang: Excavations between 1982–1992*). Beijing: China Encyclopedia Publishing House.

IA (Institute of Archaeology, Chinese Academy of Social Sciences, 中国社会科学院考古研究所) (2007) 安阳殷墟花园庄东地商代墓葬 (*Report on the Excavations at Huayuanzhuang Locus East in Anyang*). Beijing: Science Press.

IA (Institute of Archaeology, Chinese Academy of Social Sciences) and Anyang City Institute of Cutural Relics and Archaeology (中国社会科学院考古研究所，安阳市文物考古研究所) (2008) 殷墟新出土青铜器 (*Ritual Bronzes Recently Excavated in Yinxu*). Kunming: Yunnan People Press.

Karlbeck, O. (1935) 'Anyang moulds', *Bulletin of the Museum for Far Eastern Antiquities* 7: 39–60.

Li, Ji and Wan, Jiabao (李济, 万家保) (1964) 殷墟出土觚形器之研究 (*Studies of the Ku Beaker*). Taipei: Institute of History and Philology, Academia Sinica.

Li, Ji and Wan, Jiabao (李济, 万家保) (1966) 殷墟出土爵形器之研究 (*Studies of the Chüeh-cup*). Taipei: Institute of History and Philology, Academia Sinica.

Li, Ji and Wan, Jiabao (李济, 万家保) (1968) 殷墟出土斝形器之研究 (*Studies of the Jia-cup*). Taipei: Institute of History and Philology, Academia Sinica.

Li, Ji and Wan, Jiabao (李济, 万家保) (1970) 殷墟出土鼎形器之研究 (*Studies of the Ding Cauldron*). Taipei: Institute of History and Philology, Academia Sinica.

Li, Ji and Wan, Jiabao (李济, 万家保) (1972) 殷墟出土五十三件青铜容器之研究 (*Studies of Fifty-three Ritual Bronzes*). Taipei: Institute of History and Philology, Academia Sinica.

Li, Yung-ti（李永迪）(2003) *The Anyang Bronze Foundries: Archaeological Remains, Casting Technology, and Production Organization*, unpublished doctoral thesis, Harvard University.

Rawson, J. (1992–93) 'Contact between southern China and Henan during the Shang period', *Transactions of the Oriental Ceramic Society* 57: 1–24.

Shi, Zhangru (石璋如) (1947) 殷墟最近之重要发现：附论小屯地层 ('Recent important discoveries at Yinxu and a discussion of the stratigraphy at Xiaotun'), 田野考古报告 (*Field Excavation Report*) 2: 1–82.

Shi, Zhangru (石璋如) (1955) 殷代的铸铜工艺 ('Bronze casting techniques of the Shang dynasty'), 历史语言研究所集刊 (*Bulletin of the Institute of History and Philology*) 26: 95–130.

Xiaomintun Archaeological Team, Institute of Archaeology, Chinese Academy of Social Sciences (中国社会科学院考古研究所孝民屯考古队) (2007) 2003－2004安阳孝民屯商代铸铜遗址发掘简报 ('The 2003–2004 excavation of the Shang bronze foundry site at Xiaomintun, Anyang'), 考古（*Archaeology*）1: 14–25.

Three Western Zhou bronze foundry sites in the Zhouyuan area, Shaanxi province, China

Wenli Zhou, Jianli Chen, Xingshan Lei, Tianjin Xu, Jianrong Chong and Zhankui Wang

ABSTRACT A large amount of Western Zhou bronzes and some Proto-Zhou bronzes have been discovered in the Zhouyuan area, the foremost region for the origin and development of Zhou culture. Since 2003, three bronze foundry sites have been located and excavated, providing new materials for investigating the bronze-making technology in this area. The Zhougongmiao site dates to the early Western Zhou period and even the Proto-Zhou period, while the other two sites have been dated to the middle and late Western Zhou period. This paper concentrates on the melting, refining and alloying aspects of metallurgical technology by examining small bronze artefacts, fragments and slag using a metallographic microscope and analytical techniques such as scanning electron microscopy with energy-dispersive spectroscopy (SEM–EDS) and energy-dispersive electron probe microanalysis (ED-EPMA). The results suggest that copper, tin and lead ingots and bronze scraps were probably brought to the foundry sites as raw materials. The Zhougongmiao site exhibits quite different characteristics from the other two sites with regard to the sources of copper ores, alloying and mould-making technology.

Introduction

The region of Zhouyuan, located in the Guanzhong Plain in Shaanxi province, north-west China, was the most important area for the origin and development of Zhou culture. Zhou people had already inhabited this region for many centuries before the establishment of the Western Zhou dynasty (mid-11th century BC–771 BC), an epoch called the Proto-Zhou period. The toponym 'Zhouyuan' is used to cover two geographical areas (Shi 1981): in a narrow sense, it refers to the main site, the Zhouyuan site, covering approximately 30 square kilometres and bordering a number of villages in Qishan and Fufeng counties; in a broad sense, it refers to the larger region, covering hundreds of square kilometres spanning from Wugong county in the east to Fengxiang county in the west, and from the Qishan Mountains in the north to the Wei River in the south. In this paper, the term Zhouyuan area refers to this latter region.

Since ancient times, numerous Western Zhou bronzes have been found, mostly from burials and hoards, in the Zhouyuan area, especially at the Zhouyuan site (Beida 2002). The huge number of these bronzes, together with the delicate craftsmanship and informative inscriptions, are testament to the important status of the Zhouyuan area during the Western Zhou period. Most Chinese archaeologists have focused on the dating, inscription interpretation and artistic dimensions of the bronzes (Cao 1994); only a few unsystematic technical studies of bronzes have been carried out (Zhang 1997; Yang and Han 2007).

Before 2003, a limited number of moulds, slag and ingots had been found in the Zhouyuan area (Luo 1993: 22), implying that these bronzes were likely to have been manufactured locally. Since 2003, however, three important bronze foundry sites – Lijia, Zhougongmiao and Kongtougou (Fig. 1) – have been found and excavated in the region, yielding a wealth and variety of new materials that allow a more detailed study of the bronze-making technology. A joint team from Peking University and Shaanxi Provincial Institute of Archaeology carried out the field investigations and excavations. These are the only Western Zhou foundry

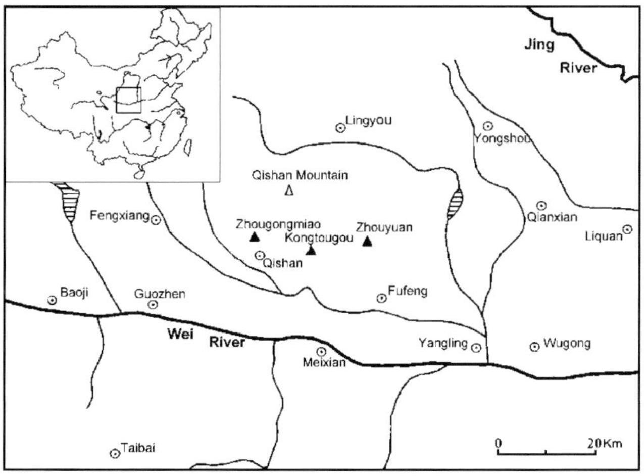

Figure 1 Map showing the distribution of the three bronze foundry sites in the Zhouyuan area.

sites excavated apart from the Beiyao foundry site of the early Western Zhou period in Luoyang, Henan province. Based on the typology of coexisting pottery, together with the styles and motifs of moulds (which can be related to those of extant bronze vessels), the Zhougongmiao foundry site can be dated to the early Western Zhou period and may even have been established as early as the Proto-Zhou period, while the Lijia and Kongtougou foundry sites date to the middle and late Western Zhou periods.

The Zhougongmiao site, located in the Fenghuang Mountains, Qishan county, yielded a number of bronze vessels of the Western Zhou and Proto-Zhou period. The discovery of high-status burial sites, a large number of inscribed animal bones and a bronze foundry in 2004 (Xu 2006) suggested that this site was an important arena of Zhou people. Field surveys revealed a small kiln, on the bottom of which was a piece of a *luanling* (銮铃) mould showing typical early Western Zhou features. The kiln, apparently different from ordinary pottery kilns, is likely to have been used to bake the moulds. A small-scale excavation of the foundry site revealed hundreds of broken mould fragments with few motifs remaining. These moulds were identified to relate to vessels (*ding*鼎 and *gui*簋), weapons (knives and arrowheads), and chariot accessories (horse bit, *pao*泡, *luanling*), etc. In 2006, a waste pit (H27) of the late Proto-Zhou period was discovered, which contained a bronze knife, mould and crucible fragments, and slag (Chong and Lei 2007).

The Lijia bronze foundry site, located within the Zhouyuan site, west of Lijia village, Fufeng county, was first excavated in spring 2003 and then from September 2003 to January 2004 by the same team (Zhouyuan Archaeology Team 2004a,b). Thousands of moulds were excavated which had been used to cast a variety of bronzes, including vessels (*ding*, *li*鬲, *gui*, *hu*壶), musical instruments (*zhong*钟), weapons (*ge*戈), chariot accessories (*luanling*, *wei*軎, *xia*辖, *biao*镳, horse bits, *jieyue*节约), etc. Some of the well-preserved moulds had clear motifs on the inner surfaces (Fig. 2).

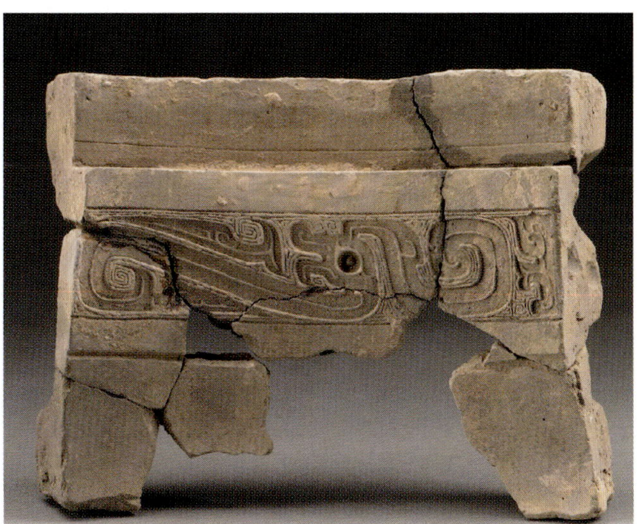

Figure 2 Mould fragments for a *ding* vessel (H66:95) from the Lijia site.

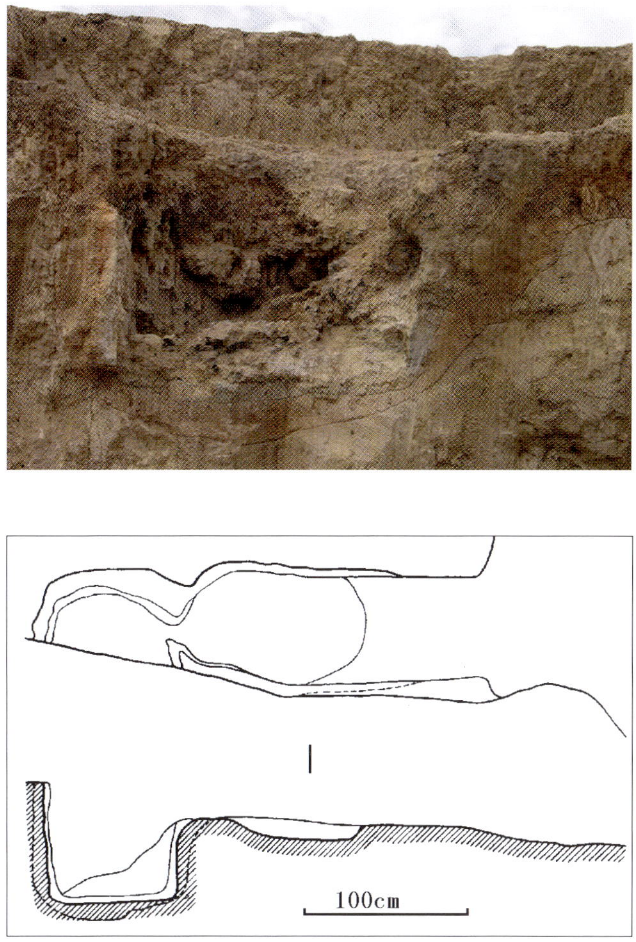

Figure 3 The furnace unearthed from Qizhen (above: section; below: drawing).

In 2006, a third foundry site was excavated at the Kongtougou site, located between Zhougongmiao and Zhouyuan (Chong *et al.* 2007). Moulds were identified to cast vessels (*ding* and *gui*), musical instruments (*zhong*), chariot accessories (*biao* and *xia*), etc. The motifs on the inner surfaces of moulds were less diverse than those of contemporary moulds from the Lijia site.

Many furnace fragments were found in the three foundry sites, typically mixed with mould fragments in large waste pits, particularly abundant in context H11 at the Kongtougou foundry site; all were made of coils of clay mixed with straw. From the largest fragments, diameters of furnaces are estimated to have ranged from 40 to 100 cm. Only one furnace was identified *in situ* (Fig. 3) (Wei and Li 2007), discovered when sectioned by a bulldozer digging clay at Qizhen village, about 1 km west of Lijia foundry at the Zhouyuan site. The preserved part was a pit dug into the soil, about 80 cm in diameter, 85 cm deep and with a wall thickness of 3–4 cm exhibiting clear evidence of exposure to high temperatures. Several pieces of moulds were discovered inside the furnace, while some slag and bronze fragments were found in nearby waste pits of the section. Local villagers reported the presence of two similar furnaces in the vicinity. Initially, it was expected that these furnaces had been used for bronze melting and casting, given their context and their overall structure. However, inductively

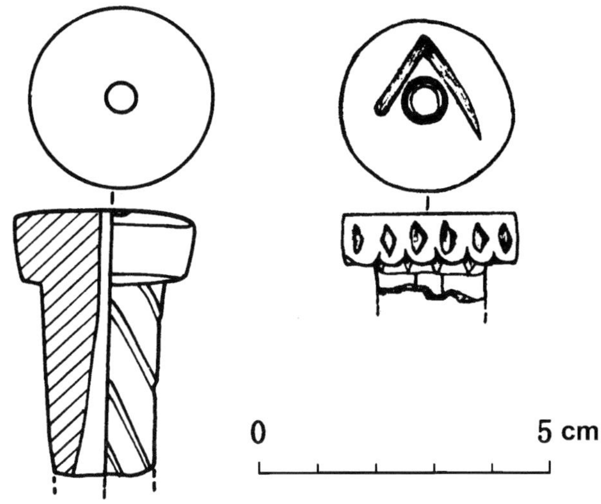

Figure 4 Ceramic tubes from the Lijia site (left: M2:19; right: H97: 10).

coupled plasma atomic emission spectrometry (ICP-AES) analyses of the slag found inside the furnace showed no obvious enrichment of copper, tin or lead, suggesting that the slag constituted only molten furnace clay. In addition, no clear evidence of crucibles or tuyères was found at any of the foundry sites. On the contrary, some small columnar or mushroom-like ceramic tubes, whose functions remain unknown, were found at all sites (Fig. 4).

The most impressive finds from the three bronze workshops are the thousands of Western Zhou ceramic moulds employed for bronze casting. Chen Yang (2005) studied the mould-making technology of the Lijia and Zhougongmiao sites, reconstructed the mould-making process and discussed the technological changes from the early to the middle and late Western Zhou period. Some analytical studies were carried out on Lijia moulds and furnaces from the first excavation in the spring of 2003, providing some preliminary information on the choice of raw materials and mould-making process in the Zhouyuan area (Liu *et al.* 2007). There is no doubt these moulds deserve more detailed technical studies, and they will yield more information for our reconstruction of the piece-mould casting technology during the Western Zhou period (Nickel 2006).

This paper, however, will concentrate on other metallurgical remains, less abundant and perhaps less impressive at first glance, but also informative of the Western Zhou metallurgical technology. Based on analytical studies carried out on metal artefacts, metal fragments and slag, this paper examines the melting, refining and alloying technology of the three foundry workshops in the Zhouyuan area. Given that the samples from Zhougongmiao date to the early Western Zhou period while those from Lijia and Kongtougou date to the middle and late Western Zhou periods, an attempt will be made here to consider chronological changes by comparing the evidence from the different sites. Samples from Lijia, Zhougongmiao and Kongtougou sites are labelled SZY, ZGM and ZJT, respectively.

Methodology

Samples of metal and slag were mounted in phenolic resin and ground and polished following established procedures. Metallographic examinations were performed with a LEICA DM4000M optical microscope at the Archaeological Science Laboratory in the Department of Archaeology and Museology, Peking University. Chemical data were obtained according to two methods, depending on the availability of equipment during different stages of this research. Some samples had previously been analysed with a FEI QUANTA600F SEM and an Oxford Instruments EDS at the Chinese National Museum. Further samples were analysed by a JXA-8100 EPMA with an INCA-400 Oxford EDS at the School of Earth and Space Sciences, Peking University. The accelerating voltage was 20 kV in both cases. Although the chemical results from these different instruments can be considered comparable, it should be noted that ED-EPMA analyses quantified the oxygen content, unlike the other technique (see tables below). In all cases, results are presented as weight percentages (wt%) normalised to 100%, and as stoichiometric oxides where relevant. Heavily corroded samples tend to have a higher content of oxygen and tin. Based on the chemical compositions, metals and alloys were classified as copper, bronze, etc., using conventional terminology. It should be noted, however, that only alloys containing lead in excess of 2% were identified as 'leaded'.

For metal artefacts and metal fragments, cross-sections were first investigated to identify inclusions and assess the degree of corrosion. They were subsequently carbon coated and studied in the backscattered electron mode of the SEM/EPMA. Several EDS area measurements were performed to obtain average bulk compositions and spot analyses were carried out on inclusions. After microanalysis, the carbon coating on the surface of the sections was removed, and samples were etched in alcoholic ferric chloride and observed under the metallographic microscope for microstructural studies. Similar analytical procedures were followed for the analysis of slag, although the phase structure was more complicated and thus required more detailed consideration.

Results

Metal artefacts

Only from the site of Lijia were artefacts obtained of recognisable shapes – all the other metals analysed for this study constituted smaller fragments of possible broken objects or casting spillage. Thirteen small metal artefacts from the Lijia site were collected for examination, including two awls, two chisels, five knives and three arrowheads.

It can be seen from the results of SEM–EDS/ED-EPMA (Table 1) that most of these artefacts were made of tin

Table 1 Chemical composition of bronze artefacts from the Lijia site (wt%).

Lab no	Type	Cu	Sn	Pb	Fe	S	O	Other	Alloy	Technology
SZY37	awl	62.4	22.2	3.7	–	0.5	10.1	Cl:1.0	Cu-Sn-Pb	cast
SZY38*	awl	84.4	15.0	0.3	0.3	0.1	na		Cu-Sn	cast, cold-worked
SZY53	awl	60.6	28.0	1.4	–	–	10.0		Cu-Sn	hot-worked
SZY50*	chisel	83.0	15.0	2.0	0.0	0.1	na		Cu-Sn-Pb	cast, cold-worked
SZY51*	chisel	77.6	21.9	0.3	0.1	0.1	na		Cu-Sn	cast, cold-worked
SZY39	knife blade	84.5	14.6	–	0.5	0.3	–		Cu-Sn	hot-worked, cold-worked
SZY42	knife blade	83.2	14.9	–	–	–	1.9		Cu-Sn	hot-worked, cold-worked
SZY43	knife handle	75.8	18.7	1.2	–	–	4.4		Cu-Sn	cast
SZY44*	knife blade	81.5	17.8	0.1	0.2	0.4	na		Cu-Sn	hot-worked
SZY45	knife blade	84.9	13.5	–	–	–	1.6		Cu-Sn	cast, annealed
SZY30	arrowhead	87.6	9.2	1.6	–	–	1.7		Cu-Sn	cast
SZY31	arrowhead	82.7	14.7	1.1	0.8	–	0.8		Cu-Sn	cast
SZY34	arrowhead	79.8	16.1	2.5	–	–	1.0	As:0.6	Cu-Sn-Pb	cast

Note: * = analysed by SEM–EDS; na = oxygen not analysed. The other samples were analysed by ED-EPMA including oxygen. Tin or lead exceeding 2% is regarded as an alloying component.

Table 2 Chemical composition of metal fragments from Lijia, Zhougongmiao and Kongtougou sites (wt%).

Lab no	Cu	Sn	Pb	Fe	S	O	Alloy	Technology
SZY22-1	84.6	15.0	–	–	0.4	–	Cu-Sn	cast
SZY22-2	73.3	26.7	–	–	–	–	Cu-Sn	cast
SZY55	81.3	13.1	3.9	–	–	1.0	Cu-Sn-Pb As: 0.6%	cast
SZY56*	84.9	13.7	0.6	0.6	0.2	na	Cu-Sn	cast
SZY62	80.0	15.6	1.3	–	–	3.2	Cu-Sn	cast
SZY63	88.2	9.0	1.2	0.4	–	1.2	Cu-Sn	cast
SZY64	88.7	9.2	1.0	–	–	1.1	Cu-Sn	cast
SZY65-6	82.2	13.5	2.0	–	–	2.2	Cu-Sn-Pb	cast
SZY65-17	72.2	17.3	–	1.0	0.9	8.6	Cu-Sn	cast, annealed
SZY66-2*	82.3	14.8	2.4	0.2	0.3	na	Cu-Sn-Pb	cast, annealed
SZY66-4	67.7	21.3	–	1.3	0.7	8.9	Cu-Sn	cast, annealed
SZY67	79.5	20.5	–	–	–	–	Cu-Sn	quenched
SZY99	85.7	14.3	–	–	–	–	Cu-Sn	hot-worked
SZY100	84.9	14.6	–	0.4	–	–	Cu-Sn	cast
ZGM26	69.2	12.0	11.7	1.1	–	6.1	Cu-Sn-Pb	cast, annealed
ZGM29-1	83.6	15.1	–	–	0.2	1.1	Cu-Sn	cast
ZGM29-2	84.8	11.2	1.3	–	–	2.6	Cu-Sn	cast
ZGM30	84.5	15.2	–	–	–	0.3	Cu-Sn	cast
ZGM33-2	80.1	12.3	3.8	–	0.2	3.6	Cu-Sn-Pb	cast, annealed
ZGM36-1	54.8	29.2	0.6	0.5	0.2	14.7	Cu-Sn	cast, annealed
ZGM36-2	79.4	–	16.3	1.1	0.6	2.7	Cu-Pb	hot-worked
ZGM36-3	86.1	12.6	–	–	–	1.3	Cu-Sn	cast, hot-worked
ZGM37	84.3	11.5	–	2.1	0.4	1.8	Cu-Sn	hot-worked, then cold-worked
ZGM38-1	82.6	14.0	–	–	–	3.4	Cu-Sn	cast
ZGM39	79.3	12.7	5.8	–	–	2.3	Cu-Sn-Pb	cast
ZGM40	96.8	0.8	–	0.8	–	1.6	Cu	cast
	62.8	21.5	–	1.7	0.5	12.4	Cu-Sn	cast
ZGM41	75.1	17.0	4.4	–	–	3.5	Cu-Sn-Pb	cast
ZGM44	73.9	18.5	3.1	–	–	4.5	Cu-Sn-Pb	cast
ZJT10	89.8	0.8	–	6.8	0.8	1.7	Cu	cast
ZJT11	80.7	12.9	–	0.8	0.2	5.3	Cu-Sn	cast, annealed
ZJT14-1	82.2	8.2	–	0.7	0.6	8.3	Cu-Sn	cast
ZJT14-2	71.9	16.6	–	0.5	0.4	10.6	Cu-Sn	cast
ZJT15	82.0	10.9	1.6	1.0	–	4.5	Cu-Sn	cast, annealed
ZJT17	96.4	–	–	1.5	1.0	1.1	Cu	cast
ZJT19-1	82.6	13.9	0.4	–	0.2	2.9	Cu-Sn	cast
ZJT21-1	96.5	–	1.3	–	–	2.1	Cu	cast
ZJT21-2	85.5	9.2	–	0.6	0.6	4.0	Cu-Sn	cast
ZJT23-1	88.0	10.4	–	–	–	1.6	Cu-Sn	cast, annealed

Note: * = analysed by SEM–EDS; na = oxygen not analysed. The other samples were analysed by ED-EPMA including oxygen. Tin or lead exceeding 2% is regarded as an alloying component.

Table 3 Alloy types of metal fragments.

Site	Cu	Cu-Pb	Cu-Sn	Cu-Sn-Pb	Total
Lijia			11	3	14
Zhougongmiao	1	1	7	5	14
Kongtougou	3		7		10
Total	4	1	25	8	38

bronzes, seven of which contained a few percent of lead. Tin levels were around 15% in most cases. Only three objects showed higher tin levels, but these were heavily corroded bronzes whose remaining microstructures were not indicative of such high tin concentrations. Lead contents were typically low, rarely exceeding 2%. Most samples had many small copper iron sulphide inclusions; in two cases (SZY44 and 34), sulphide inclusions were found to contain nearly 1% selenium.

The metallographic study confirmed that all 13 artefacts were formed by casting. Nine samples showed normal casting dendrites and (α+δ) eutectoids. Among them, one awl and two chisels (SZY38, 50 and 51) were found to have some strain lines at the edges, which were probably caused by usage, while the handle of a knife (SZY45), with unobvious dendrites, seemed to have been slightly annealed after casting. The remaining four objects displayed clear evidence of working. The microstructures of a chisel (SZY53) and a knife (SZY44) showed equi-axed hexagonal grains, some of which were twinned. The middle part of the cross-section of the knife (SZY44) still preserved some dendrites, indicating that the annealing process was not extensive. Two other knives (SZY39 and 42) were cold-worked following hot-working. Their equi-axed grains showed bent twins and strain lines, while the sulphide inclusions and remaining (α+δ) eutectoids had become elongated along the orientation of the cold-working.

In summary, all 13 artefacts were made of quite similar copper alloys with around 15% tin but little lead. Such bronzes are of high tensile strength, hardness and some degree of extensibility, and some tools were worked to improve their mechanical properties.

Metal fragments

Many small metal fragments were excavated from the foundry sites, but it was difficult to identify their original functions or technological origins. They could either be metal raw materials (ingots or scraps for recycling), fragments of the products, or simply foundry spillage. Their analysis, however, yielded information about the types of alloys employed in the different foundries (Tables 2 and 3).

Lijia

Fourteen metal fragments from the Lijia foundry site were examined; most are tin bronzes containing 10–18% tin and little lead. Only samples SZY22-2 and SZY67 contained tin above 20%, while SZY55 was singled out as a leaded tin bronze with 0.6% arsenic. Overall, these compositions were consistent with those of the recognisable artefacts found at the site. Most of the fragments appeared as cast structures, although some seemed to have been annealed after casting. Only one sample (SZY99) was hot-worked, with sulphide inclusions aggregating and small iron-rich phases scattering. Sample SZY67, with a high tin content, showed acicular crystals indicative of quenching (Figs 5 and 6).

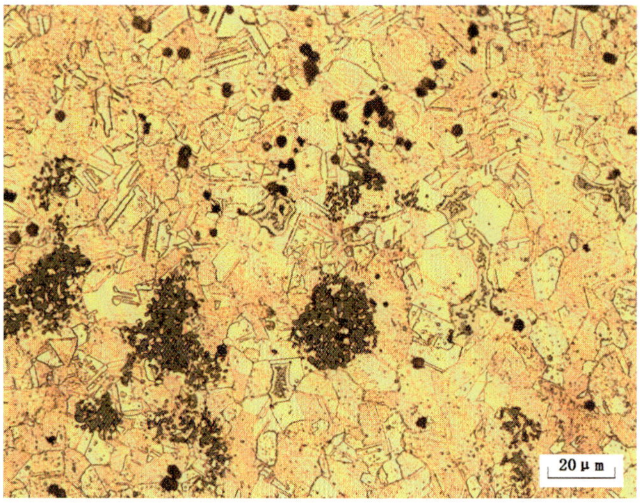

Figure 5 Sample SZY99 microstructure, etched, equi-axed and twinned grain with remnant (α+δ) eutectoids, aggregated sulphide inclusion and scattering.

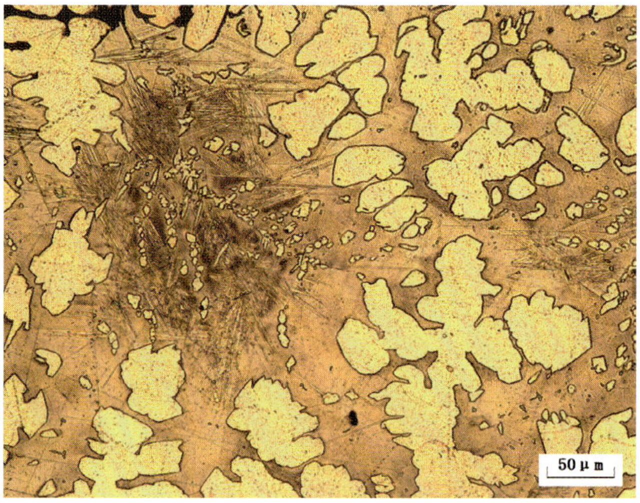

Figure 6 Sample SZY67 microstructure, etched, needle-like phase in (α+δ) eutectoids.

Zhougongmiao

There are seven tin bronzes, five leaded tin bronzes, one leaded copper and one mixed metal (see below) among the 14 metal fragments from the Zhougongmiao foundry site. Four out of the five leaded tin bronzes had less than 6% lead, while the fifth contained as much as 11.7%. The only leaded copper sample (ZGM36-2) had a high lead content of 16.3% (Fig. 7). Overall, the presence of leaded alloys was much more frequent than at Lijia. Most of these objects exhibited cast or annealed structures, with few showing evidence of working.

Of particular interest is sample ZGM40, showing the mixture of two types of metal: pure copper and tin bronze (Fig. 8). The pure copper was embedded in a tin bronze matrix, with distinct boundaries. Although both metals appeared as

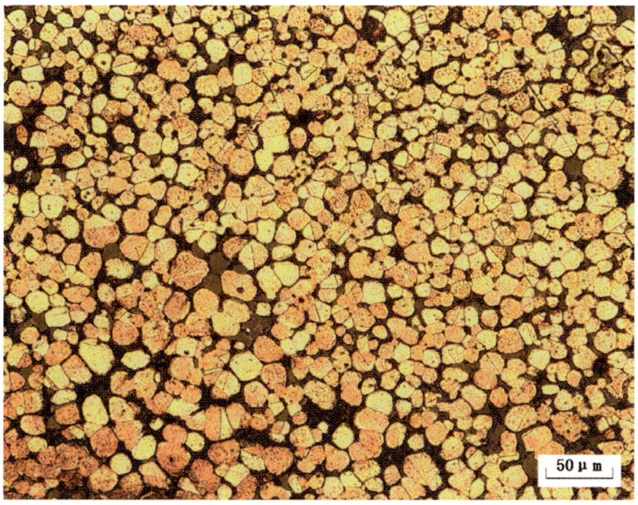

Figure 7 Sample ZGM36-2 microstructure, etched, equi-axed and twinned grain with intergranular lead prills and sulphide inclusions.

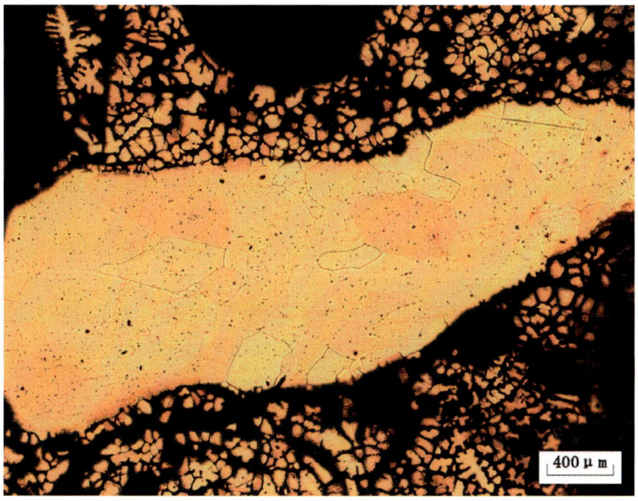

Figure 8 Sample ZGM40 microstructure, etched, two metals.

cast microstructures, the sulphide inclusions had different shapes and compositions: those in the pure copper were small prills of copper iron sulphides containing selenium, whereas those found in the intergranular (α+δ) eutectoids of the tin bronze were copper sulphides without detectable selenium. This piece of metal fragment probably formed when copper ingots and bronze scraps were mixed together and heated at a temperature too low to melt the copper completely.

Finally, it is notable that nine samples of these metal fragments have sulphide inclusions with selenium. Selenium is sometimes found as a trace element of copper ores, coexisting with sulphides, and it tends to remain with the sulphides when copper ores are smelted and refined (Rehren 1991). Therefore, the selenium in these sulphide inclusions of copper and bronze probably derived from the copper ores.

Kongtougou

Among the 10 samples analysed from Kongtougou are seven tin bronzes which were cast or annealed after casting. There are, however, two samples of unalloyed copper (ZJT17 and ZJT21-1). Interestingly, ZJT10 is a mixture of raw copper and slag (described and discussed in detail in the next section). Only one sample (ZJT17) was found to contain selenium-rich sulphide inclusions.

Slag

Some slag was found adhering to furnace wall fragments, while other samples were recovered as small lumps. Based on microanalytical results, most of the samples appear as typical copper-alloy melting slag, but a few are particularly rich in iron, making the interpretation more complicated. These two slag types will be discussed in turn.

Melting slag

Melting slag forms as the molten metal reacts with the furnace or crucible wall and fuel ash during the melting and alloying process. When exposed to air, the metal partly oxidises, thus forming diverse oxidic phases containing the alloying components – copper, tin and/or lead. Typical melting slag can be expected to contain silicates derived from the ceramic material, oxides of the alloying components, and a glassy matrix possibly enriched in alkali or alkali earth oxides from the fuel ash, but they rarely contain significant amounts of iron or other gangue-related elements. These are precisely the features noted in most of the Zhouyuan slag.

Sample ZJT7-3 from Kongtougou is presented here as a typical tin bronze melting slag, similar to those identified at other sites. It contains primary crystals of tin oxide, together with copper prills (sometimes with a rim of copper oxide) in a glassy matrix characterised by a high lime content. Exsolving from this matrix were also fine dendrites (probably of cuprite), together with very small globules, which were too small to be identified by EDS. Sample ZGM33-1 from Zhougongmiao was slag produced during the melting of leaded tin bronze. In this sample, the dominant phases were tin oxide, copper oxide and lead oxide, in a lead silicate matrix. Pure copper prills and heavily corroded leaded tin bronze prills were also found in this sample (Figs 9 and 10).

The presence/absence of compounds of copper, tin and lead allows us to infer the alloys melted in each case (Table

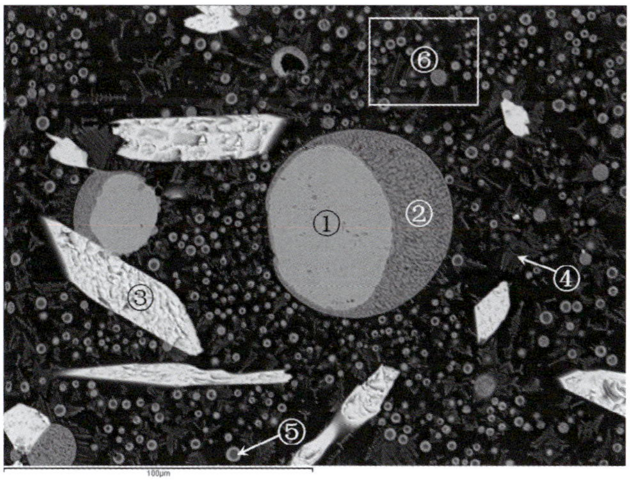

Figure 9 Backscattered electron (BSE) image of sample ZJT7-3, some copper prills (1) with a rim of copper oxide (2), tin oxide crystals (3), fine dendrites (4) and small globules (5) in a glassy matrix (6).

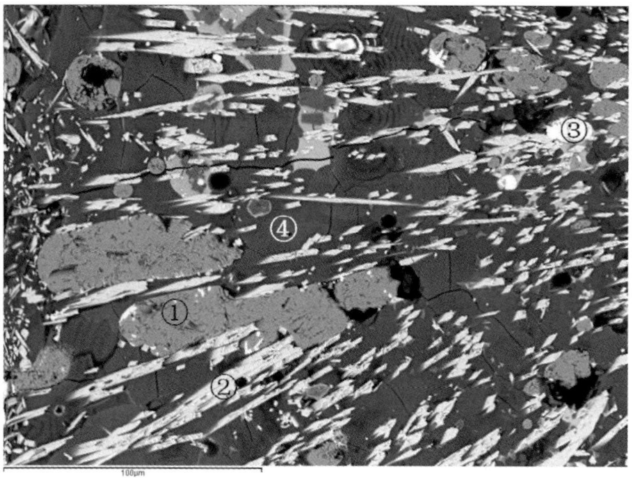

Figure 10 BSE image of sample ZGM33-1, some blocks of copper oxides (1), acicular tin oxide crystals (2) and lead oxides (3) in lead-rich silicate glassy matrix (4).

Table 4 Alloy types of melting slag.

Site	Cu-Pb	Cu-Sn	Cu-Sn-Pb	Total
Lijia		3	2	5
Zhougongmiao	1		1	2
Kongtougou		5	1	6
Total	1	8	4	13

4). In the case of Zhougongmiao, although only two slag samples have been analysed, they show particularly high lead concentrations, a feature in good agreement with the generally higher lead contents of the metal fragments from the site. It is more difficult, however, to determine whether these constituents entered the alloy as pure metals or scrap alloys, or whether tin entered the reaction as relatively pure mineral cassiterite that would be reduced and alloyed to the copper by cementation. A number of scholars (Dungworth 2000; Rostoker *et al.* 1983; Rovira 2005) have attempted to define analytical criteria for discerning these variable technological histories; however, given the lack of experimental data that can be used as a reference, these criteria must remain tentative. In our view, the abundance of tin oxide crystals in the slag cannot be taken as an indication of the addition of cassiterite to the melt; none of these phases seem residual minerals, but rather they are primary euhedral or acicular crystals formed from the slag melt. The free energy of the formation of tin oxide is much lower than that of copper and lead (cf. Ellingham diagram). Thus, all the tin in the alloy can oxidise, while part of the copper and the lead remain in the metallic state. As tin oxide is insoluble in other metals and it shows no reactivity with silica, it will tend to crystallise in a pure form. In some cases, however, relatively large islands of metallic copper are found within the slag, suggesting that this metal could have entered the charge in a relatively pure state, i.e. we would be dealing with bronze alloying rather than simply melting. Finally, where present, lead oxide will react heavily with the siliceous ceramic, resulting in the lead silicate glass identified in some slag. The proportions among the different alloying constituents, the temperatures reached, the duration of the process and the redox environment, will altogether add further variability to the crystallography and composition of the resulting slag, making technological interpretations even more difficult.

Altogether, at present we can only identify the Zhouyuan slag as products of melting (rather than smelting) and alloying of a range of alloys, without more precise technological identifications. In particular, there is no evidence of the use of mineral cassiterite; it seems that tin could have entered the reaction as a metal (see further discussion below).

Iron-rich slag

A few small black slag pieces from the site of Kongtougou (ZJT6 and 20) were more unusual, most notably in their high iron content. Samples ZJT6 and ZJT20 showed well-developed blocks of fayalite and globules of wüstite, with small prills of iron-rich copper (in ZJT6) or matte (in ZJT20), and no detectable levels of tin or lead (Figs 11 and 12). Sample ZJT10 was especially peculiar, as it exhibited a mixture of raw copper and iron-rich slag: the metallic part was composed of raw copper containing copper-iron-sulphide inclusions and globules of metallic iron, while the slag part was dominated by fayalite and wüstite crystals which were not fully developed; between them was a layer of copper-iron-sulphide (Figs 13 and 14).

Some of these features could appear more indicative of copper smelting, rather than melting, slag. However, considering the very low incidence of this type of slag at the site, we are reluctant to take these as indications that copper smelting took place in Kongtougou. If they truly were smelting slag, we would be inclined to regard them as acci-

Table 5 Chemical composition of typical melting slag and iron-rich slag (wt%).

Lab no	Na_2O	MgO	SiO_2	Al_2O_3	SO_3	K_2O	CaO	FeO	CuO	SnO_2	PbO
ZJT7-3	1.0	3.8	49.2	11.1	–	1.2	20.1	6.6	5.0	–	2.1
	1.4	2.3	63.6	13.8	–	3.1	10.7	5.1	–	–	–
ZGM33-1	–	–	56.3	10.4	–	–	–	–	1.8	4.6	26.9
	0.8	1.4	32.3	6.4	–	1.5	7.5	6.2	–	–	43.9
ZJT6	–	1.0	20.9	2.5	0.7	0.6	5.4	68.9	–	–	–
	–	1.0	21.7	3.1	–	0.6	5.3	68.3	–	–	–
ZJT20	–	1.0	25.7	3.0	7.2	0.5	3.6	56.1	2.9	–	–
	–	0.9	28.3	2.5	5.0	0.7	0.9	56.8	4.8	–	–
ZJT10	–	0.7	23.0	3.5	4.8	–	2.2	43.9	21.8	–	–

Note: The chemical compositions of ZGM33-1 were spot compositions of the glassy matrix, while the others were area compositions of the samples. All the S, Fe, Cu, Sn and Pb contents are reported in their oxidic forms.

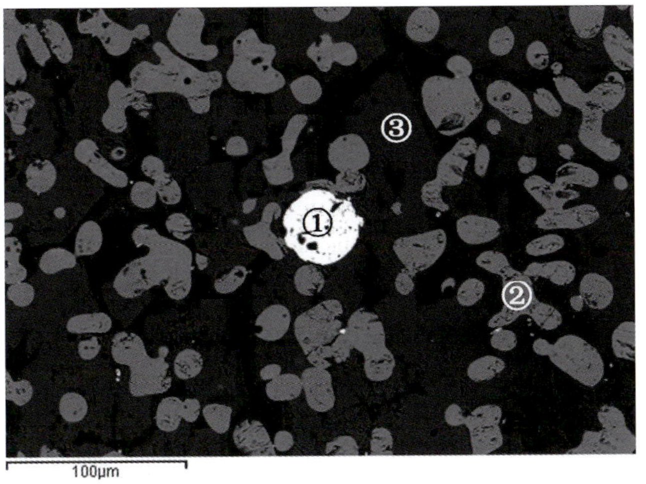

Figure 11 BSE image of sample ZJT6, a copper prill (1), wüstite globules (2) and blocky fayalite (3) in a glassy matrix.

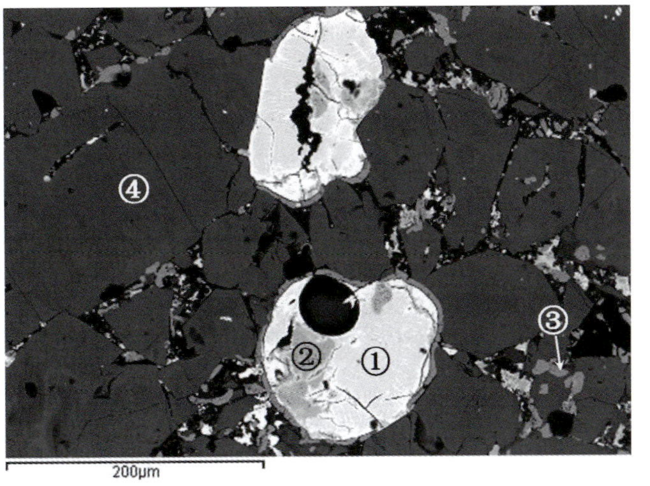

Figure 12 BSE image of sample ZJT20, some whitish copper iron sulphide prills (1) with some iron sulphide part (2), small wüstite globules (3) and blocky fayalite (4) in a glassy matrix.

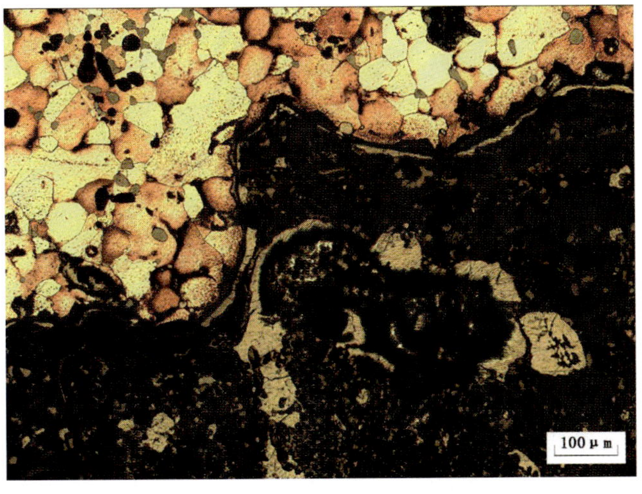

Figure 13 Sample ZGM40 microstructure, etched, a layer of copper-iron-sulphide between the raw copper and the slag.

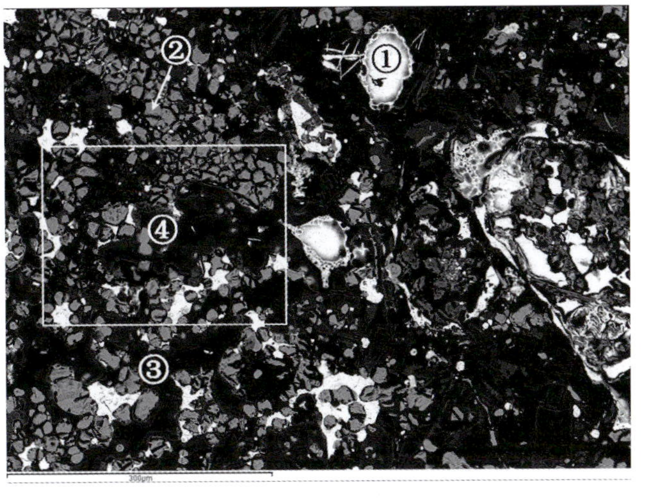

Figure 14 BSE image of sample ZJT10, some copper prills with a few percent of iron (1), wüstite globules (2) and fayalite (3).

dental lumps that reached the foundry with the raw copper imported for the casting activities. However, it is also possible to interpret these as copper refining slag. If raw copper containing unduly high iron and/or sulphur reached the site, this could be refined by melting it in relatively oxidising conditions, so that these elements and any other slag inclusions present in the metal could float on the surface and be skimmed off. The result would be a slag or slag/matte layer similar to those documented here. In this sense, the presence of wüstite and fayalite would not be the result of a smelting reducing environment (where Fe^{3+} from the ore is reduced to Fe^{2+} or metallic iron), but rather the products of an oxidising atmosphere (where metallic iron contained in the copper could be oxidised to Fe^{2+}). Sample ZJT10 may be documenting exactly this process, preserving part of the raw iron-rich copper undergoing refining and forming slag. Arguing against the smelting scenario is also the predominance of Fe^{2+} in the form of wüstite, which would denote conditions unnecessarily reducing for the smelting of copper – iron oxide in copper smelting slag is typically present in the more oxidised forms, such as magnetite.

Discussion

Copper was an important commodity in the Western Zhou period. In the 1970s and 80s, two bun ingots were discovered at the Shaochen and Shaoli villages of the Zhouyuan site (Zhou 1972; Luo 1985). The one from Shaochen is 24 cm in diameter, 4.65 kg in weight and contains 97.7% copper and 2.3% lead; the other, from Shaoli, is 32.5 cm in diameter, 5 kg in weight and contains 97.6% copper, 0.6% iron and 0.6% sulphur (Fig. 15). These two ingots found not far from the Lijia foundry site provided strong evidence that copper ingots were used as raw materials. In addition, some of the pure copper and raw copper fragments described above were probably fragments of copper ingots. Together with the evidence provided by the iron-rich slag, it can be inferred that refined pure copper ingots, unrefined raw copper ingots and also leaded copper ingots (such as the Shaochen ingot) were introduced to foundry workshops for bronze-making. Although the precise provenance of the copper ingots has yet to be determined, it seems likely that a number of sources were supplying copper to the Zhouyuan workshops (see further discussion below). Different copper ingots have been discovered in other areas dated to the Western Zhou period. For example, several kilograms of diamond-shaped ingots were found in Muyu Hill, Tongling, Anhui province. Two of the three analysed ingots contain iron (2% and 9.5%). These iron-rich copper ingots were cast directly with relatively unrefined raw copper, and are thus similar to some of the copper materials in the Zhouyuan area.

Another important source of copper could be bronze fragments, such as broken artefacts, and by-products generated during the melting and casting process. At the Zhouyuan site, fragments of large bronze vessels, sprues and even slag were found in bronze hoards, some of which were kept in bronze or pottery containers. These scraps might have been regarded as valuable metal materials by ancient people and deliberately preserved for recycling. More relevant evidence is provided by the newly excavated foundry sites. One piece of evidence is the large number of metal fragments found, some of which might have been collected from other places to be recycled; some generated during the melting and casting processes could have been used for local re-melting. Sample ZGM40 proves the use of bronze scraps as one kind of raw material. Another important piece of evidence is presented by some vitrified clay cores found in association with some slag at the three foundry sites. Clay cores are inner moulds of ears or legs of bronze vessels, so they usually remain inside bronze vessels after casting. If some broken bronze vessels were re-melted, the low-density clay cores would float to the surface of the molten metal and then be discarded, ending up as sintered lumps with some slag such as those recorded here (Fig. 16).

Tin was a more precious metal component for bronzes since tin ores were not as widespread as copper ores. In addition, ancient metallic tin rarely survived and few pure tin objects have been discovered to date. In ancient China, the earliest tin objects with reliable scientific analysis are one tin *ding*, one tin *gui* and eight tin fishes from Yu state cemeteries, dating back to the early and middle Western Zhou period (Su *et al.* 1988). One of the tin fishes, with 99% tin, indicates that contemporary workers were able to produce pure tin. More tin objects are known from the Eastern Zhou period (771–220 BC). Although neither metallic tin objects or fragments nor tin ores (usually cassiterite) were found in the Zhouyuan area, the availability of metallic tin in the period and the absence of residual cassiterite in the melting slag together suggest that bronzes were made with metallic tin.

Lead was easier to smelt and lead ores were more common. Many lead objects and fragments, dating back to the Western Zhou period and even earlier, have been found (Li 1984). No direct archaeological evidence for lead materials was found in the Zhouyuan area. However, metallic lead ingots or scraps were probably used as the lead source for the leaded copper and bronze, while in bronze objects with lower lead concentrations it might have been introduced as an impurity from copper ingots.

From the analyses of metallurgical debris in the Zhouyuan area, coupled with other relevant archaeological finds of the Shang and Zhou periods, it seems that generally similar metal materials, including ingots and scraps, were used at the Zhouyuan foundries. However, the metallurgical technology of the Zhougongmiao site is quite different from that of the other two sites. First, metal fragments and melting slag from Zhougongmiao generally displayed higher levels of lead than those from the other two foundries. Secondly, nearly two-thirds of the sulphide inclusions in the Zhougongmiao metal fragments contained selenium, whereas it was only found in a few cases in Lijia and Kongtougou metal artefacts and fragments. This strongly indicates that the copper materials at the Zhougongmiao foundry came from lead- and selenium-rich copper ores, which were no longer the main copper sources for the Lijia and Kongtougou foundries in the later period. A third difference is the relatively primi-

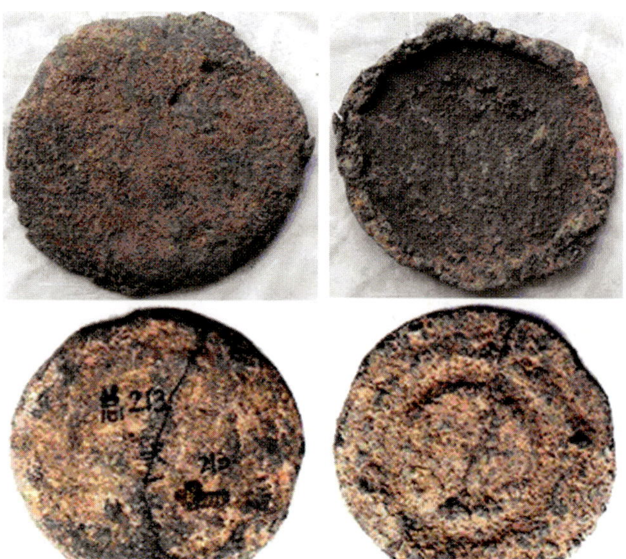

Figure 15 Copper ingots from Shaochen (top: 24 cm in diameter) and Shaoli (bottom: 32.5 cm in diameter). (Photo: Yang Junchang.)

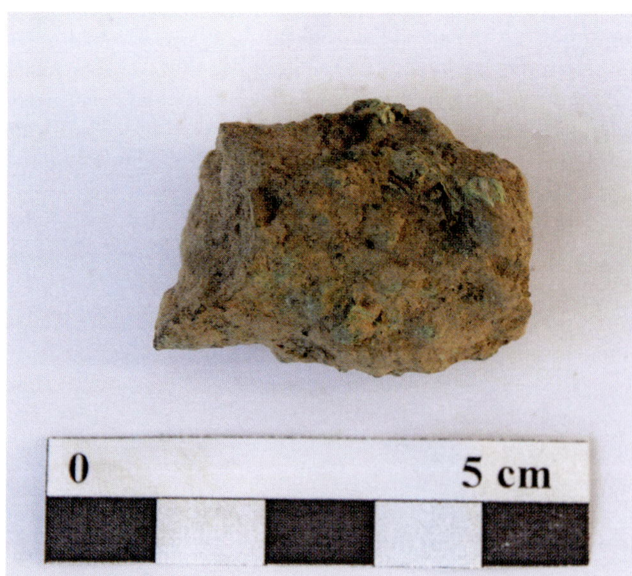

Figure 16 Vitrified ceramic casting cores from Zhaojiatai.

tive mould-making technology in Zhougongmiao, which has been discussed elsewhere (Chen 2005; Zhou 2008). These technological differences might be explained partly by chronological changes, but the different characteristics of the settlements where these foundry sites were located should also be taken into consideration.

The Zhougongmiao site was identified by archaeologists to be a settlement of Zhou people from the Proto-Zhou to the Western Zhou period. The discovery of late Proto-Zhou casting remains in 2006 demonstrated that Zhou people could cast bronzes before the demise of the Shang dynasty. Their more rudimentary bronze-making technology was reflected in the restricted variety and number of Proto-Zhou bronzes in the Zhouyuan area, typically showing a relative rough craftsmanship. The bronze-making technology at the Zhougongmiao foundry in the early Western Zhou period can probably be considered an indigenous development from the Proto-Zhou period in this area, with little influence from Shang people. In the later period, however, Shang descendants in Lijia and Zhou people in Kongtougou might have inherited the more advanced casting technology from Shang people and also developed it from that of the Zhou people, thereby producing bronzes of higher quality.

Conclusions

These three bronze foundry sites, spanning the early, middle and late Western Zhou periods, are another important discovery following that of the Beiyao foundry site (dated to the early Western Zhou period) in Luoyang, Henan province. They give clues pertaining to the study of the technology of bronze casting, the sources of metal materials and the distribution of settlements in the Zhouyuan area during the Western Zhou period.

This paper has revealed aspects of bronze-casting technology at three Zhouyuan foundry sites by analysing bronze artefacts, bronze fragments and slag. It was found that the bronze tools and weapons were made primarily of tin bronze. Some of the bronze tools were hot-worked after casting and were of a good quality. Bronze fragments from the Zhougongmiao site were mainly made of tin bronze and leaded tin bronze, while those from the other two sites were mostly tin bronzes. A majority of the slag samples were typical melting slag or alloying slag, with phases containing oxidised alloying components. A few pieces of iron-rich slag from the Kongtougou site were interpreted as possible refining slag.

From the results of the analytical studies, it is estimated that copper, tin and lead ingots were brought to the foundry sites as raw ingredients, as well as scrap bronze fragments. The Zhougongmiao foundry (early Western Zhou period) is distinctive in its sources of copper ores, alloying and mould-making technology, which were different from those present at the later Lijia and Kongtougou foundries.

This paper focused primarily on the metallurgical activities of melting, refining and alloying by analysing metal artefacts, metal fragments and slag. It should be noted, however, that more work needs to be done. More bronze artefacts excavated from foundry sites and surroundings, especially vessels and chariot accessories, need to be examined further, particularly with a view to address the provenance of the different metals. In addition, furnace fragments call for future careful work and scientific examinations. The conspicuous lack of crucibles should also be considered.

Acknowledgements

The research was supported by NSFC (No.10405003) and the State Administration of Cultural Heritage of China. We are especially grateful to Professor Chen Tiemei, Professor Wu Xiaohong and Dr Cui Jianfeng of Peking University; Professor Han Rubin, Professor Sun Shuyun, Professor Li Yanxiang and Professor Mei Jianjun of the University of Science and Technology Beijing; Dr Yang Junchang of Shaanxi Provincial Institute of Archaeology; Professor Thilo Rehren and Dr Marcos Martinon-Torres of University College London for their useful supervision, advice and help with the research.

References

Beida (Research Centre for Ancient Civilisation, Peking University 北京大学考古文博学院 北京大学古代文明研究中心) (ed.) (2002) 吉金铸国史 (*A National History as Seen in Bronze Casting*). Beijing: Wenwu Press.

Cao, Wei (曹玮) (1994) 周原西周铜器的分期 ('The periodisation of Zhouyuan Western Zhou bronzes'), 考古学研究 (*Archaeology Studies*) 2: 24–58.

Chen, Yang (陈阳) (2005) 周原铸铜遗址所出陶范的初步研究 (*Preliminary Study of Moulds from Zhouyuan Foundry Sites*). 北京大学硕士学位论文 (unpublished MA thesis, Peking University).

Chong, Jianrong (种建荣) and Lei, Xingshan (雷兴山) (2007) 先周文化铸铜遗存的确认及其意义 ('The identification of bronze casting remains of Proto-Zhou culture and its significance'), 中国文物报 (*China Cultural Heritage*), 30 November.

Chong, Jianrong (种建荣), Zhang, Min (张敏) and Lei, Xingshan (雷兴山) (2007) 岐山孔头沟遗址商周时期聚落性质初探 ('Preliminary discussions of characteristics of Shang-Zhou settlements at the Kongtougou site in Qishan'), 文博 (*Antiquity and Museum*) 5: 40–45.

Dungworth, D. (2000) 'Serendipity in the foundry? Tin oxide inclusions in copper and copper alloys as an indicator of production process', *Bulletin of the Metals Museum* 32: 1–5.

Li, Minsheng (李敏生) (1984) 先秦用铅的历史概况 ('The historical summary of lead in the pre-Qing period'), 文物 (*Relics*) 10: 84–8.

Liu, Yu (刘煜), Song, Jiangning (宋江宁) and Liu, Xinyi (刘歆益) (2007) 周原出土铸铜遗物的分析检测 ('Analytical examination of casting remains from the Zhouyuan site'), 考古与文物 (*Archaeology and Relics*) 4: 94–100.

Luo, Xizhang (罗西章) (1985) 从西周出土文物试论西周货币 ('Discussion on Western Zhou currency from excavated Western Zhou antiquities'), 中国钱币 (*Chinese Coins*) 2: 14–21.

Luo, Xizhang (罗西章) (1993) 扶风县文物志 (*Antiquity Chorography of Fufeng County*). Shaanxi: People Education Press.

Nickel, L. (2006) 'Imperfect symmetry: rethinking bronze casting technologies in ancient China', *Artibus Asiae* 66(1): 5–39.

Rehren, Th. (1991) 'Selenium and tellurium in Mediterranean copper ingots', in N.H. Gale (ed.), *Bronze Age Trade in the Mediterranean*, 240–48. Sävedalen: Paul Åströms Förlag.

Rostoker, W., McNallan, M. and Gebhard, E.R. (1983) 'Melting/smelting of bronze at Isthmia', *Historical Metallurgy* 17(1): 23–6.

Rovira, S. (2005) 'La producción de bronces en la Prehistoria', in J. Molera, J. Farjas, P. Roura and T. Pradell (eds), *Avances en Arqueometría 2005. Actas del Congreso Ibérico de Arqueometría*, 21–35. Cádiz: Universidad de Cádiz.

Shi, Nianhai (史念海) (1981) 周原的变迁 ('The transition of Zhouyuan'), in Shi, Nianhai (史念海) (ed.), 河山集 (*Rivers and Mountains*) 2, 214–31. Beijing: SDX Joint Publishing Company.

Su, Rongyu (苏荣誉), Lu, Liancheng (卢连成) and Hu, Zhisheng (胡智生) (1988) 弓鱼国墓地青铜器铸造工艺考察和金属器物检测 ('Bronze casting technology and examination of metal artefacts from Yu state cemeteries'), in Lu, Liancheng (卢连成) and Hu, Zhisheng (胡智生) (eds), 宝鸡弓鱼国墓地 (*Baoji Yu State Cemeteries*), 530–89. Beijing: Cultural Relics Press.

Wei, Xingxing (魏兴兴) and Li, Yalong (李亚龙) (2007) 陕西扶风齐镇发现西周炼炉('Western Zhou furnace found in Fufeng, Shaanxi'), 考古与文物 (*Archaeology and Relics*) 1: 22–3, 61.

Xu, Tianjin (徐天进) (2006) 周公庙遗址的考古所获及所思 ('Archaeological finds at the Zhougongmiao site and some thoughts'), 文物 (*Relics*) 8: 55–62.

Yang, Junchang and Han, Rubin (2007) 'Technical studies of Proto-Zhou and early Western Zhou bronze vessels excavated in Guanzhong, Shaanxi province, China', in J.G. Douglas, P. Jett and J. Winter (eds), *Scientific Research of the Sculptural Arts of Asia: Proceedings of the Third Forbes Symposium at the Freer Gallery of Art*, 39–47. London: Archetype.

Zhang, Xiaomei (张晓梅) (1997) 周原遗址及弓鱼国墓地出土青铜器锈蚀研究 (*Studies on the Corrosion of Bronzes from the Zhouyuan Site and Yu State Cemeteries*). 北京大学硕士学位论文 (unpublished MA thesis, Peking University).

Zhou Wen (周文) (1972) 新出土的几件西周铜器 ('Some new excavated Western Zhou bronzes'), 文物 (*Relics*) 7: 9–12.

Zhou Wenli (周文丽) (2008) 周原地区西周时期铸铜遗物初步研究 (*Preliminary Studies on Bronze Casting Ruins of Western Zhou Dynasty in the Zhouyuan Area*). 北京大学硕士学位论文 (unpublished MA thesis, Peking University).

Zhouyuan Archaeological Team (周原考古队) (2004a) 陕西周原遗址发现西周墓葬与铸铜遗址 ('Western Zhou burials and foundry sites found at the Zhouyuan site, Shaanxi'), 考古(*Archaeology*) 1: 3–6.

Zhouyuan Archaeological Team (周原考古队) (2004b) 2003年秋周原遗址（ⅣB2区与ⅣB3区）的发掘 ('The excavation of area IVB2 and IVB3 of the Zhouyuan site in autumn 2003'), 古代文明 (*Ancient Civilization*) 3: 436–90.

New research on lost-wax casting in ancient China

Weirong Zhou, Yawei Dong, Quanwen Wan and Changsui Wang

ABSTRACT Since 1979, the *zun* and *pan* vessels unearthed at the tomb of Marquis Yi of Zeng in Sui county, Hubei province, and the bronze *jin* unearthed at the Chu tomb at Xiasi, Xichuan, Henan province, have been universally regarded as typical examples of bronzes cast by the lost-wax method in the pre-Qin dynasty. In recent years, we carefully researched these bronze utensils and related decorations and found that there are mould lines, pouring spots and other mould casting marks on them. Hence, neither the Marquis Yi of Zeng's *zun* and *pan* vessels nor the bronze *jin* unearthed at Xiasi were cast by the lost-wax casting method. Our further research revealed that highly systematic and developed bronze mould casting techniques were used in the Chinese Bronze Age, techniques which could make possible the production of any complicated bronze vessels and utensils from Chinese culture.

Introduction

In June 1979, a meeting on traditional precision casting was held by the Casting Institute of the China Mechanical Engineering Society. It was agreed at the meeting that the hollow decorative attachments to *zun* and *pan* vessels unearthed from the Warring States period tomb of Marquis Yi of Zeng (曾侯乙), situated in present-day Sui county, Hubei province, were cast using the lost-wax method (Hua 1983). This announcement immediately aroused great interest among experts in the fields of ancient metallurgy, archaeology and history, who set about finding out whether lost-wax technology had been used to make other unearthed bronzes (Fig. 1).

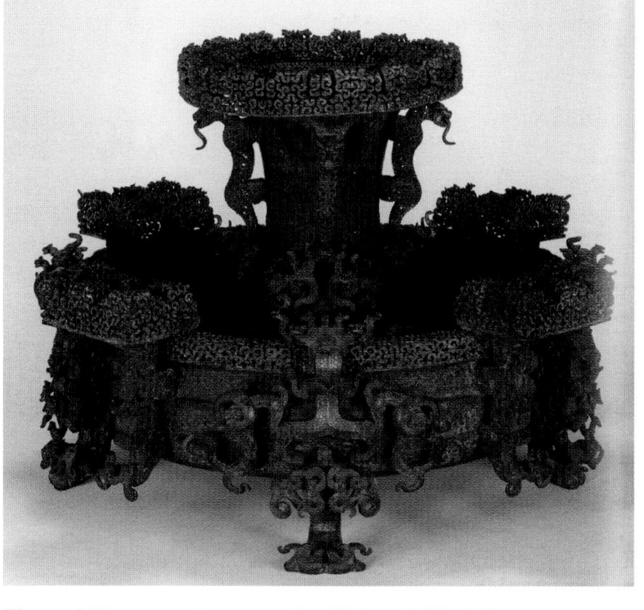

Figure 1 The *zun* and *pan* vessels of Marquis Yi of Zeng.

Not long after, several Warring States period bronzes from the state of Chu were unearthed from tombs no. 1 and no. 2 at Xiasi, Xichuan county, Henan province (Henan Institute of Cultural Relics and Archaeology 1991). These were also identified as having been cast using the lost-wax method, in particular the accessories attached to small bronze cups unearthed from tomb no. 1, the main body and zoomorphic decorations attached to a *jin* (禁) bronze vessel, as well as the zoomorphic decorations attached to the *ding* of Prince Wu (王子午鼎) unearthed from tomb no. 2, etc. Following this discovery, a number of other finds were considered to be products of lost-wax technology, including the Warring States period gold and silver inlaid 'pot of Chen Zhang' (陈璋壶) unearthed at Xuyi county, Jiangsu province; the Western Han dynasty gilded Boshan burner from the tomb of Liu Sheng, King Jing of the Zhongshan State, found in Mancheng county, Hebei province; the Eastern Han dynasty 'horse treading on a flying swallow' unearthed in Wuwei, Gansu province, etc. At that time, it was believed that during the Spring and Autumn period, craftsmen in ancient China had grasped the technology of the lost-wax method of casting (Figs 2–4).

The Research Group on the Technology of Bronze Casting with Piece-Moulds, of which the authors are members, has been engaging for a number of years in a systematic exploration of the development of China's Bronze Age casting technology. Since October 2005, Weirong Zhou, Yawei Dong and other members of the Research Group visited a number of museums in Beijing, Hangzhou, Shanghai, Anqing, Fanchang, Jingzhou, Taiyuan, Houma,

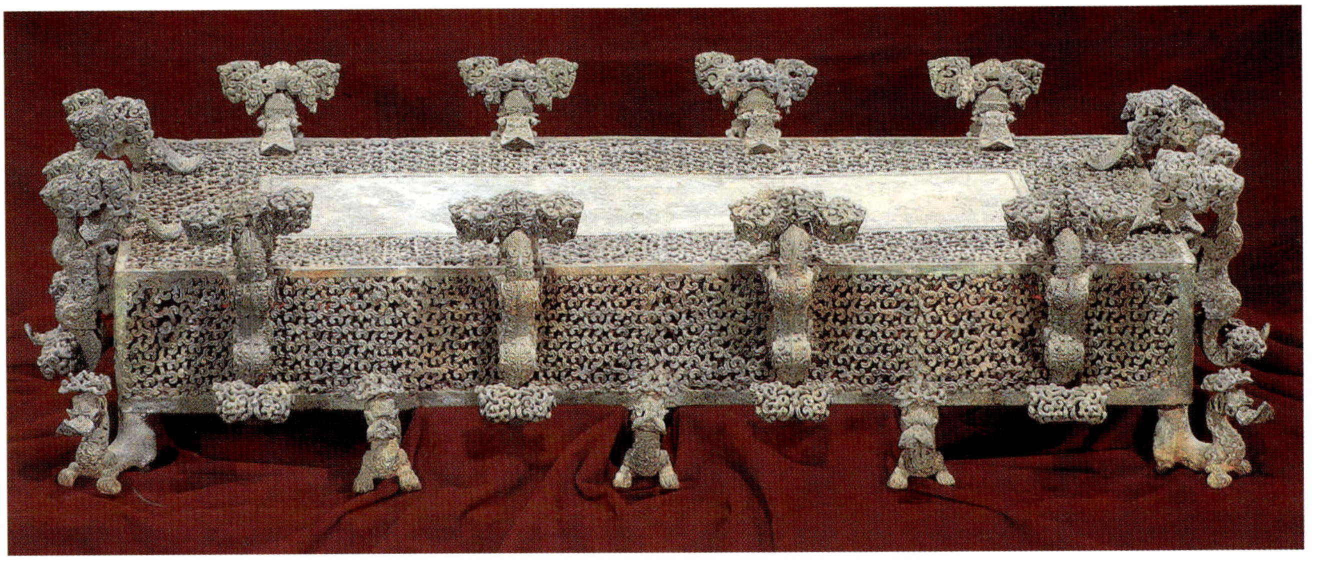

Figure 2 Bronze *jin* vessel from Xiasi (下寺铜禁).

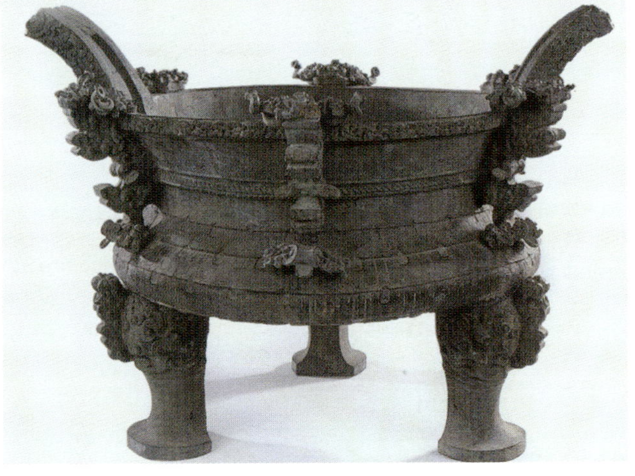

Figure 3 *Ding* of Prince Wu (王子午鼎).

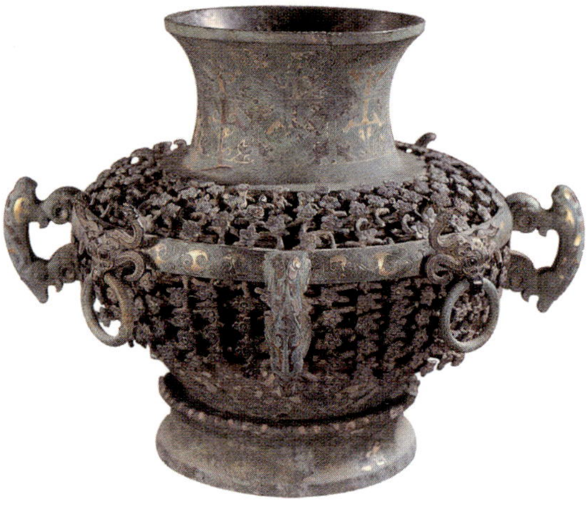

Figure 4 Pot of Chen Zhang (陈璋壶).

Xinzheng, Luoyang, Zhengzhou, Wuhan, Xi'an, Baoji, Qishan, Fengxiang, Zhouyuan etc., performing detailed observations and meticulous analyses of the rich collections of bronze objects in these museums. The Research Group held many discussions during this period and came to the unanimous conclusion that the technology of bronze casting with piece-moulds in ancient China had reached a superb level of skill, one that had evolved over a long period of time. Prior to the Spring and Autumn period, the overwhelming majority of bronze objects were cast in an integrated way, using techniques such as piece-mould casting, modelling with removable blocks, utilisation of clay cores, etc. However, from the beginning of this period, the prevailing casting technique changed to one of joining pre-cast components onto an integrated body by casting-on, welding or riveting. In the later part of this period, the technology of first breaking up the whole into parts and then reassembling the parts after casting reached a level of exceptional finesse. It could be used to cast almost any object, no matter how complicated its design (Dong 2006).

Re-examination of the *zun* and *pan* vessels of Marquis Yi of Zeng

In view of the extraordinary skill level in the use of piece-moulds of the early bronze casters of the Shang and Zhou dynasties, we considered it possible that the highly intricate *zun* and *pan* (wine vessel and tray assembly) found in the tomb of Marquis Yi of Zeng were also the products of this traditional casting technology. Consequently, we travelled to Hubei Provincial Museum, meticulously examined these bronzes and repeatedly studied many high resolution digital photographs. As a result, mould lines and casting gates were identified on sub-unit decorations, and traces of welding between the vessel brim and the decorations were clearly visible. These phenomena show irrefutably that the technology developed for making these highly complicated vessels was not the lost-wax method, but rather one of separately casting small parts and then joining these parts together by casting-on and welding. These methods have been used universally since the Spring and Autumn period (Guo 1981) (Figs 5–7).

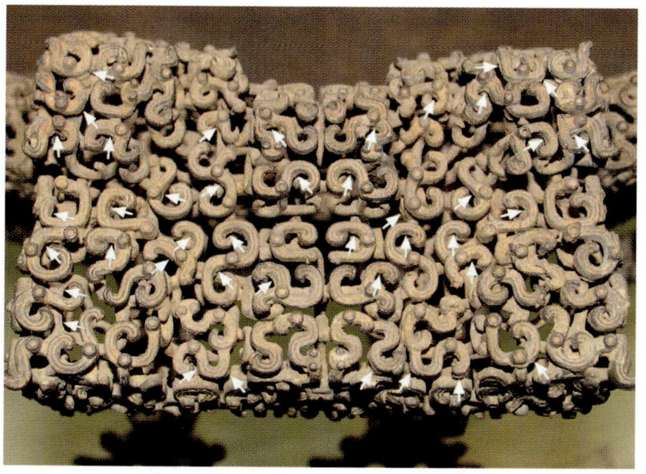

Figure 5 Mould lines.

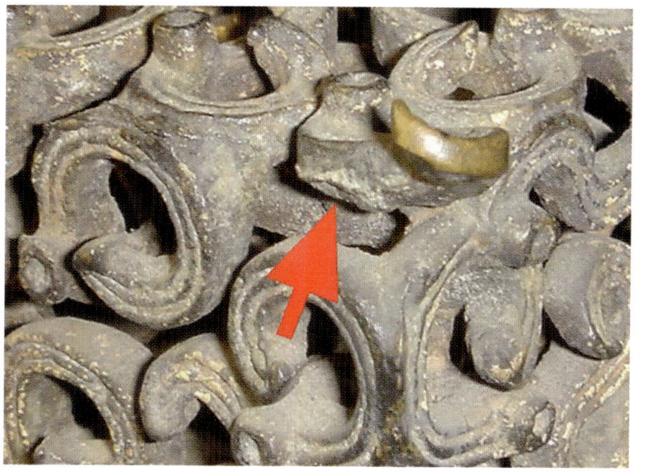

Figure 6 Casting gate (浇口).

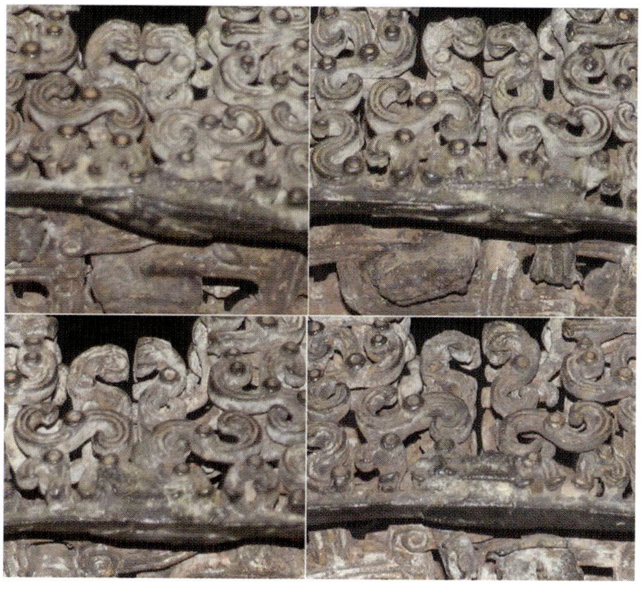

Figure 7 Welding traces.

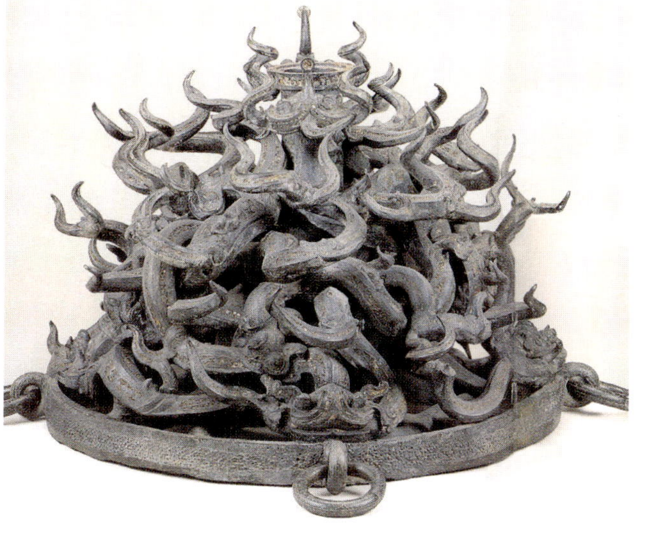

Figure 8 The drum stand.

The next step was to conduct observations and technical analyses of the drum stand, bell-stand base and decorative attachments on the ends of the bell-stand crossbeam, all unearthed from the same tomb. The results were also clear-cut, i.e. small components were made by piece-mould casting and connected to the main parts either by casting-on and then welding (in the case of the drum stand, Fig. 8), or by direct casting-on with combination moulds for the decorative attachments (for the bell-stand base and crossbeam). None was the product of lost-wax technology. It is clear from this research that the initial determination that the *zun* and *pan* of Marquis Yi of Zeng were products of lost-wax technology was premature and incorrect.

Shortly thereafter, principal members of the Research Group meticulously examined the above-mentioned *jin* vessel and other items from Xiasi, Xichuan county at the Henan Provincial Museum. The results of this survey concluded that they were also made by first casting separate sub-units and then connecting them together by welding or casting-on.

Differences between Chinese and Western casting technologies

It is well known that there were many differences between Chinese and Western ancient bronze casting technologies (Tylecote 1992). However, there are several misinterpretations on this issue. One of these is the misunderstanding of the use of the lost-wax method in Bronze Age China. We shall endeavour to analyse and summarise the differences between Chinese and Western bronze casting technologies as follows:

- Along with the casting technology, the use of hammering for making bronze objects was traditional in the West. In ancient China, casting was the traditional method used over a long period of time, and some hammered bronzes occurred in limited numbers after the middle of the Warring States period.
- The technology of casting-in moulds appeared in the West quite early, but it changed very little over time.

The highest level reached was merely the insertion of a clay core between two piece-moulds, falling into the category of casting technology with simple block moulds. However, the lost-wax method was developed to a very high degree and was used to cast objects with extraordinarily complex patterns. In comparison to this, the technology of casting with piece-moulds in ancient China developed over time to a considerably more complicated degree. It would appear at present that no objects made by the lost-wax method ever existed in China, at least not before the Warring States period.

- The use of stone moulds for the fabrication of bronzes has a long history in the West, and it was developed roughly 1,000 years later in China (Tylecote 1992). Ceramic moulds were not extensively developed in the West, whereas in China, the principal moulds used were ceramic, although stone moulds were also used. Bronze moulds began to appear after the middle of the Warring States period (Kaoguyanjiusuo1988; Jiangxi Museum 1975; Zhou 2008). The use of the ceramic mould reached its zenith in Bronze Age China, as casters learned that they could adapt it to suit all their needs, regardless of how complicated the shape. Even delicate decorations with high relief were made with ceramic moulds. Such an example is the square *zun* with four goats unearthed in Ningxiang county, Hunan province, and the *zun* with a dragon and a tiger found in Funan county, Anhui province.
- The majority of bronzes made in the Bronze Age in the West were tools for production and weapons (Tylecote 1992), while those in China were predominantly sacrificial vessels. Tools and weapons constituted only a small percentage of the bronze objects. This aspect is important because it reflects an essential distinction in the social functions of bronze objects between China and the West and exerted considerable influence on the path of development of casting technologies selected. The Bronze Age in the West led primarily to the progress and improvement of production forces, while the Bronze Age in China appears to have been tied to the spiritual life, ideology and authority of the ruling elite, embodied in the bronze sacrificial vessels.

In addition, apart from the social factors, the distinction between Chinese and Western bronzes is closely related to the different climatic and geographical environments of the two regions.

The relationship between the technology of casting with moulds and climatic-geographical environments

The key to the technology of casting with piece-moulds is the fabrication of the moulds. Stone and bronze moulds, limited by the quality of their raw materials, can only be used to cast small-scale objects with simple shapes and designs. However, the situation is very different with clay moulds. Clay is very soft before being baked to make pottery and may be moulded into various complicated shapes and patterns. This is especially true of the loess soil of north China. This type of soil, after being kneaded repeatedly and then left for a while, can result in moulds with a firm and clean surface, allowing fine decorations to be pressed onto them. However, only the right type of soil, with a certain degree of viscosity, can be used. Soils from other regions, such as the Gobi desert, are not suitable. Climatic factors, especially humidity, also played a very important role in the development of the technology. Dong Yawei, during his investigation of the casting technology of bronze mirrors in ancient China, found that the quality of the bronze mirrors cast in northern China is usually better than those cast in southern China. This suggests that the casting quality of bronze mirrors bears a rather close relationship to the climatic environment (Dong 1999). We know that rainfall in southern China has generally been heavier than that in the north, leading correspondingly to a higher ground water level in the south, as well as a generally wetter south and drier north (Xu 1955). Experience has shown that when exposed to a humid atmosphere, the baked moulds easily reabsorb moisture, thus exerting an adverse influence on the casting quality. The environment of southern China with its moister atmosphere makes it more difficult to cast objects with ceramic moulds.

Therefore, archaeological discoveries of bronze casting mould remains in ancient China are concentrated mainly in northern areas, in locations such as the Shang dynasty sites at Anyang, Henan province, Western Zhou sites at Beiyao, Luoyang, Henan province and Zhouyuan county, Shaanxi province, and Eastern Zhou sites at Houma county, Shanxi province, Xinzheng county, Henan province and the ancient Qi city site in Linzi county, Shandong province, etc. (Han and Ke 2007). No large-scale casting ruins have been found in coastal areas. To date, no important remains of bronze casting with moulds have been discovered in the Gobi and other desert areas in Gansu and Qinghai provinces and the Xinjiang Autonomous Region, and in the western part of Inner Mongolia. This helps confirm that the relatively dry climate conditions in northern China and the type of soil found there are necessary for making ceramic moulds.

The climatic conditions in the West were quite different. From the coastal area around the Mediterranean Sea to the European continent, the principal types of climate were a dry desert climate, a moist oceanic climate, and the Mediterranean climate characterised by tepid, dry summers and warm, humid winters. The first two conditions are clearly disadvantageous for making ceramic moulds and the latter conditions occur in coastal areas, also presenting difficulties for their fabrication. It is possible that climatic conditions may have contributed to the fact that the West did not take piece-mould casting to a higher stage of development. In contrast, the prehistoric environment of the West encouraged the development of the lost-wax technology for casting bronze objects. Besides the other inherent factors, since the development of the technology of casting with piece-moulds was limited, this may have been partly responsible for the further development of lost-wax technology in the West.

A new evaluation of piece-mould casting technology in ancient China

Any large-scale technology that exerts considerable influence on society must have a firm material base and be in steady demand – it is not possible to dispense with either of these conditions. In addition, these technologies must constantly adapt to changes in society in order to progress. They can neither surpass the demands of the time, nor lag behind the development of a society. The technology of casting with piece-moulds in ancient China extended from the pre-Shang, through the Shang and Western Zhou dynasties, and into the Spring and Autumn period. Each of the four periods had a technological system that met its needs.

In the pre-Shang period and the earlier stage of the Shang dynasty, casting with piece-moulds was in its early stage, and the technology adopted for making bronzes was a simple section-mould method, designing models for the entire object and adding or pressing decorations onto the surfaces of the moulds. For any object, its models and moulds were designed. However, the decorations were carefully made onto the moulds' surfaces by hand so not only would all examples of a type of bronze have a different pattern, but the symmetry on either side of the central line of the pattern for each object would rarely be identical. This can be seen in the decorations with zoomorphic faces that appear in this period. These objects exhibit a primitive simplicity, bold and unconstrained. However, the efficiency of production was very limited and would only have satisfied the demands of the rulers and a few high-ranking nobles. Furthermore, at this time the manufactures could only supply a limited quantity of the relevant materials.

From the middle of the Shang dynasty, society's demand for bronze objects increased, encouraging the development of the casting technology and bringing about the emergence of the technique of using decoration models. This involved making decorations on a single model and then replicating directly from it onto multiple moulds. This led to a standardisation of decorations as well as an increase in work efficiency. In the meantime, the invention of the technique of affixing clay slip designs broadened the form of decorations, notably enhancing their aesthetic quality, as seen in the famous three-layered flowers. This resulted directly in complete groups or sets of bronzes with identical decorations and laid the foundations for the inclusion of lengthy inscriptions on bronze vessels during the Zhou dynasty.

After the Spring and Autumn period, bronze vessels became more secular. Bronzes could be owned not only by rulers and nobles, but also by lower-ranking officials and even a segment of the common people. At the same time, the needs of production and warfare gave rise to the large-scale production of bronze farm implements, handicraft tools and weapons (Chen 2004). Qualitative changes thus occurred both in the use of bronzes and the methods of their fabrication. It was during this period that the concept of first dividing the whole object into parts and then reassembling the parts was widely adopted. The main body and the attachments of an object were cast separately and then assembled together. The decoration was made by using a single model and multiple moulds. The model was used to replicate multiple identical moulds, which were then assembled into decorations. For the fabrication of complicated attachments, the attachment was divided into several sub-units that were cast and then combined. During the Spring and Autumn and the Warring States periods almost any object could be made by the use of this fabrication method, regardless of its complexity. This efficient method satisfied the ever-increasing demand for bronze objects.

It can be seen that the technology of bronze casting with moulds followed a step-by-step path of development that evolved alongside changes in society from the early stage of the pre-Shang period up to the Spring and Autumn and the Warring States periods. The technology of casting consistently supported the fabrication of bronze objects, which in turn reflected the technology of each era. The course of this path was determined both by the progress of the technology and the growth in demand from society, each factor complementing the other. On the one hand, the prerequisite for the casting of bronzes was guaranteed by the developments in technology, while on the other hand demand urged the technology to move forward.

In summary, it is clear that the impetus for the development of lost-wax technology in the West was strongly influenced by the environmental, social and cultural conditions in that region. During the Bronze Age in China, however, neither the cultural and technical bases for lost-wax casting technology nor demand from society existed. The lost-wax method was therefore not used to make the *zun* and *pan* of the Marquis Yi of Zeng and other ancient bronze objects previously thought to have been the product of this technology. It was not a technique chosen by metalworkers of ancient China.

When was the lost-wax method transmitted to China? We suggest according to the evidence available so far that it was probably introduced into China to cast statuettes of the Western-style Buddha in the period between the Northern dynasties and the Sui and Tang dynasties (Zhou 2009).

References

Chen, Peifen (陈佩芬) (2004) 夏商周青铜器研究 (*Studies on the Bronzes of the Xia, Shang and Zhou Dynasties*). Shanghai: Ancient Books Press.

Dong, Yawei (董亚巍) (1999) 中国南北气候差异对古代青铜镜铸造质量的影响及其相关问题 ('Influences of distinctions in climate between southern and northern China on casting quality of ancient bronze mirrors and related issues'), 东南文化 (*Southeast Culture*) 2: 126–8.

Dong, Yawei (董亚巍) (2006) 范铸青铜 (*The Chinese Bronze Casting Technologies*). Beijing: Art and Science E-Press.

Guo, Baojun (郭宝钧) (1981) 商周青铜器群综合研究 (*An Integrative Research on the Bronzes of the Shang and Zhou Dynasties*). Beijing: Wenwu Press.

Han, Rubin and Ke, Jun (韩汝玢, 柯俊) (2007) 中国科学技术史·矿冶卷 (*Mining and Metallurgy: A History of Science and Technology in China*). Beijing: Science Press.

Henan Institute of Cultural Relics and Archaeology (河南省文物考古研究所) (1991) 淅川下寺春秋楚墓 (*The Tombs of the Chu State in the Spring and Autumn Period at Xiasi, Xichuan*). Beijing: Wenwu Press.

Hua, Jueming (华觉明) (1983) 曾侯舆尊盘和失蜡工艺的起源 ('The *zun* and *pan* of Marquis Yu of Zeng and the origins of lost-wax technology'), 自然科学史研究 (*Studies in the History of Natural Sciences*) 4: 352–9.

Jiangxi Museum (江西省博物馆) (1975) 江西清江吴城商代遗址发掘简报 ('A preliminary excavation report of the Shang sites at Wucheng, Qingjiang, Jiangxi province'), 文物 (*Cultural Relics*) 7: 51–71.

Kaoguyanjiusuo (中国社会科学院考古研究所等) (1988) 夏县东下冯 (*Dongxiafeng Sites at Xiaxian*). Beijing: Wenwu Press.

Tylecote, R.F. (1992) *A History of Metallurgy*. London: Institute of Metals.

Xu, Jinzhi (徐近之) (1955) 黄淮平原气候历史记载的初步整理 ('A preliminary research on historical records of the climate of Huang and Huai plains'), 地理学报 (*Acta Geographica Sinica*) 2: 189–90.

Zhou, Weirong (周卫荣) (2008) 中国古代铸钱工艺及其对传统铸造业发展的贡献 ('Ancient Chinese coin-casting technology and its contribution to Chinese traditional casting industry'), in China National Institute of Cultural Property (中国文化遗产研究院) (ed.), 文物科技研究第6辑 ((*Scientific and Technological Research on Cultural Relics, No. 6*), 49–63. Beijing: Science Press.

Zhou, Weirong (2009) 失蜡工艺的起源与失蜡铸造的工艺特征 ('The origins of lost-wax techniques and the characteristics of lost-wax casting'), 南方文物(*Southern Cultural Relics*) 4: 101–5.

Incipient metallurgy in Yunnan: new data for old debates

Tzehuey Chiou-Peng

ABSTRACT This work examines issues regarding metal production and the making of artefacts in Yunnan around 1000 BCE. Perceptible signs of incipient copper-based metallurgical activities at late 2nd-millennium sites on the western fringes of the Yunnan plateau bear witness to the earlier stages of a full-blown metal tradition. Metallographic studies and compositional analysis of Yunnan artefacts in conjunction with recent data from upper Jinsha sites suggest impetus from steppe-related cultures to the emergence of Yunnan metallurgy. Knowledge of metal technology, in turn, was disseminated through interconnected plains and river valleys, including those at the Yunnan–Guizhou borders near the course of the eastward-flowing Jinsha River. Rudimentary metal artefacts and slag discovered at sites around the eastern periphery of Yunnan and in the vicinities of Lake Dian also point to the possible existence of a locally based pre-Dian substratum, which could have taken shape at the beginning of the 1st millennium BCE, evolving as part of Jinsha-derived development.

Introduction

Archaeologically excavated metal materials in Yunnan comprise mainly bronze artefacts taken from burial sites near Lake Dian and sporadic finds from different regions, in a mosaic of complex topographic reliefs with diverse ecosystems. They include rare items – some displaying superb technical skill and others embellished with unique surface treatments. These artefacts have expanded our knowledge of various aspects of Yunnan societies during the 1st millennium BCE, as well as of prehistoric communications between the Yunnan plateau and areas beyond its borderlands (Ciarla 1994; Watson 1992: 177–9). The step-by-step development of the metal industry in prehistoric Yunnan has remained a subject for ongoing investigations up to the present, but much less is known about the role played by the rich Yunnan metal deposits in the early metallurgical landscape of Bronze Age Asia. The problems in exploring issues regarding the beginnings of Yunnan metallurgy are compounded by the fact that past studies of Yunnan archaeology had a lopsided focus on burials surrounding the Lake Dian region (Allard 1998: 333) – less effort was directed toward locating habitation sites, which could provide good insights into metallurgical activities in the area. Moreover, the problem of the lack of excavated habitation sites is coupled with a shortage of records on ore distribution and mining activities in ancient Yunnan. However, in light of the expanding archaeological activities in Yunnan and its neighbouring regions, as well as of the increasing popularity of metallurgical studies of archaeologically excavated materials from Yunnan (Li and Han 2006; Cui and Wu 2008), many of the pending issues in the research of Yunnan metallurgy can now be discussed with concrete evidence, as well as with broader perspectives. In order to set up an interpretive framework for studying ancient Yunnan metallurgy, this work examines available data pertaining to metallurgical activities in Yunnan around 1000 BCE.

Current study of the early use of metal in Yunnan

Among the Chinese archaeological finds, the Haimenkou site, one of the most ancient metal-using sites in Yunnan, has been regarded as a good candidate for addressing issues of the beginnings of Yunnan metallurgy. The archaeological site at Haimenkou is located on the bank of a creek that feeds into a small lake situated in a hilly area in the north-western Yunnan plateau. Immediately to its north-west are the highlands intersected by the Jinsha (upper Yangzi), Lancang (upper Mekong), and Nu (upper Salween) rivers, which run parallel to each other (see Fig. 1).

Archaeological excavations at the site were carried out for the first time in 1957 (Preparatory Committee for the Yunnan Provincial Musuem 1958), followed by a second campaign in 1978 (Yunnan Provincial Museum 1995), and finally by a third (and the most extensive) project in 2008 (Yunnan Provincial Institute of Cultural Relics and Archaeology 2008). During the first excavation, carried out to record cultural layers that were being exposed during a local field project, 14 copper-based artefacts were retrieved (Fig. 2), along with more than 1,000 pieces of bone, ceramic and lithic artefacts, grain specimens, wooden posts/beams (thought to be the remains of *ganlan* (pile dwellings) structures) and one half of a stone bivalve mould for axes of

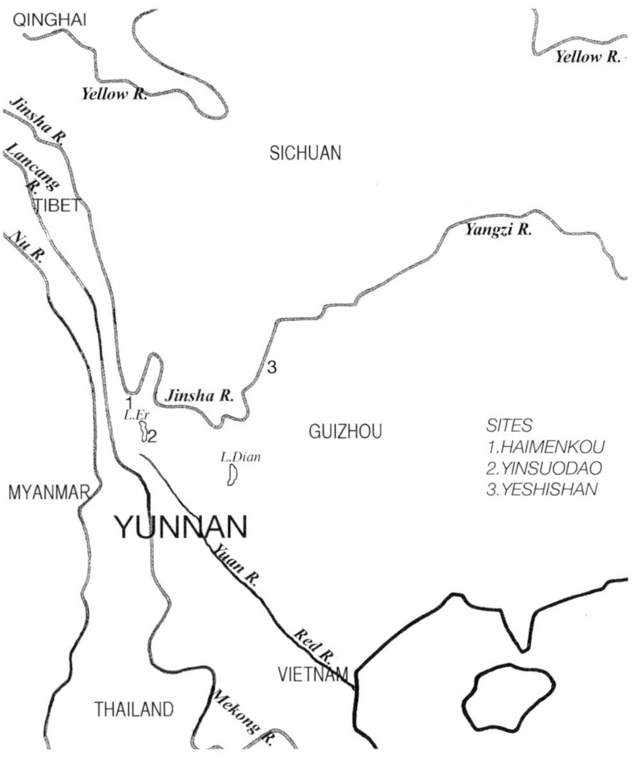

Figure 1 Map of sites mentioned in the text, Yunnan and environs.

a distinct shouldered type (Preparatory Committee for the Yunnan Provincial Museum 1958). According to the site report, all metal artefacts are made of copper, some of which were hammered to shape, while others were cast in single or bivalve moulds. None of these 14 artefacts was examined with scientific instruments at the time, but samples of timber posts were radiocarbon dated subsequently. From data gained from studying the site, the excavated materials, and the resulting radiocarbon age of 1150±90 BCE (Laboratory at the Institute of Archaeology of the Chinese Academy of Social Science 1972: 57), the site is identified as a metallurgical stage of development contemporaneous with the Anyang phase of the Shang civilisation in the Chinese Central Plains.

Discussions of the Haimenkou site and materials have generated considerable interest as well as disputes (Ma 1981). The problems are clearly linked to the fact that the cultural deposits at the excavated area were already dug up prior to the arrival of archaeologists, and that the stratigraphic association between the artefacts and timber samples used for radiocarbon dating was not confirmed. Moreover, approximately 20 years had to pass before fieldwork resumed at Haimenkou. At that time, the surviving documents could not effectively assist in locating the perimeter of the excavated area (Yunnan Provincial Museum 1995). Nevertheless, additional wooden, ceramic, bone and metal artefacts were collected during this second excavation, including some from one of the layers where intrusive objects of more recent dates can be recognised. A piece of metal-bearing ore material from an undisturbed lower layer was taken for radioisotope X-ray fluorescence (XRF) analysis, together with samples of 11 artefacts from the previous excavation. The results of this metallurgical investigation were presented at a 1983 conference and subsequently published (Wang Dadao 1985).

Contrary to previous claims, the study indicates that only two of the 14 items from the 1957 excavation are confirmed to be copper artefacts, while the rest are believed to be made of binary metal, mostly bronzes with tin contents that range from 1% to 10 wt%. In the meantime, data from the XRF analysis of the ore material from the 1978 excavation, together with studies of modern ore materials in nearby areas as well as other regions in Yunnan, suggested to the author that local tin-free copper ores were commonly used at the Haimenkou site (Wang Dadao 1985: 247–8). He concluded that the tin detected in Haimenkou bronzes was 'conscientiously introduced by human intervention' (Wang Dadao 1985: 248). By the same token, according to the author, the fact that pure copper was accessible at the site also proves that lead was added intentionally to the alloy of a fishhook (Fig. 2), a hammered bronze artefact containing 3–5% lead (Wang Dadao 1985: 248). Combining the presumed evidence from studying this single case of ore material and referencing relevant materials, the author is convinced that Haimenkou represented a metallurgical tradition progressing beyond its very first step, which occurred during the 2nd millennium BCE (Wang Dadao 1985: 249).

Additional methodological issues in subsequent interpretations of Haimenkou finds, in conjunction with some seemingly advanced features of metal artefacts from the second excavation, has led to these claims being challenged (Murowchick 1989: 84–96). Moreover, the publication of a markedly more recent radiocarbon date, obtained from samples of timber materials from the second excavation (Laboratory at the Institute of Archaeology of the Chinese Academy of Social Science 1990: 667), further underscores the existing problems in interpreting the data. None of the more convincing evidence, such as stratigraphic sequence, metallurgical data, or typological studies, was taken into account in the final appraisal of the Haimenkou site (Yunnan Provincial Museum 1995: 785-7). Its argument that artefacts in the Haimenkou assemblage delineated a progressive metallurgical development of considerable chronological spread appears problematic, and thus its conclusion that Haimenkou technology had evolved during the 2nd millennium BCE also becomes highly questionable. Therefore, the old Haimenkou issues continue to be debated (Murowchick 2001: 135–8; Higham 1996: 139; Li and Han 2006).

Metal artefacts known in Yunnan prior to 1000 BCE

Data from past studies of Yunnan metals and mounting evidence from recent archaeological activities in western Yunnan (Wang Xiaoyun 2008; Yunnan Provincial Institute of Cultural Relics and Archaeology 2008) bear witness to the existence in the region of the earlier phases of an evolving metal industry, which led to the subsequent occurrence of a distinct metal tradition in the same region (Zhang Zengqi 1981; Chiou-Peng 1998; 2008: 40–41). Interactions

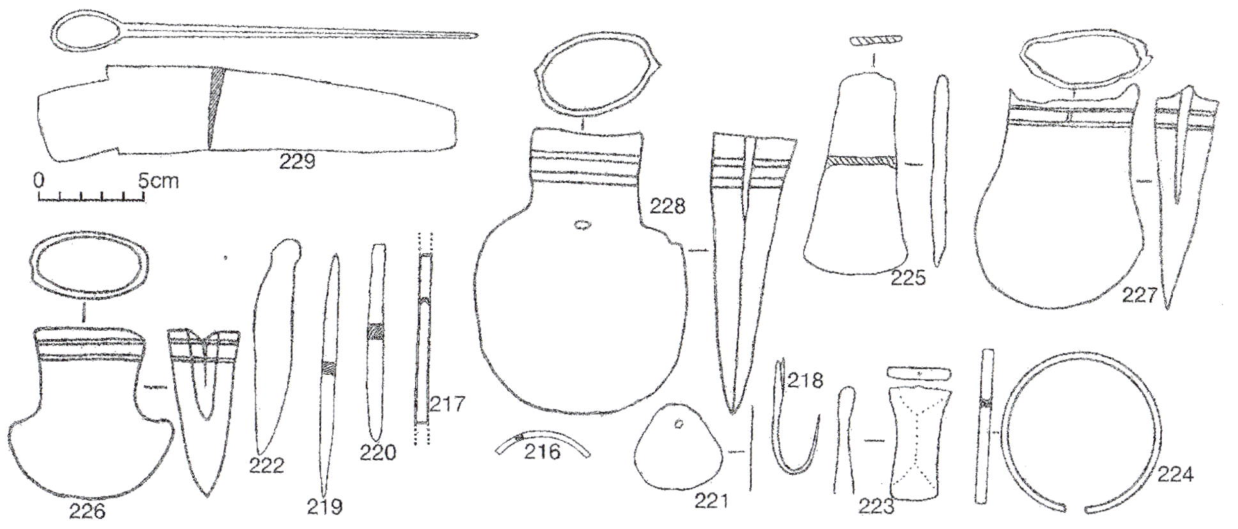

Figure 2 Line drawings of artefacts excavated at Haimenkou in 1957.

among regional centres in and around Yunnan during the first half of the 1st millennium BCE eventually culminated in the emergence of a series of technically advanced sites around Lake Dian (Pirrazoli-t'Serstevens 1974; Jiang 2002) and in areas along the Yuan (upper Red) River (Yin 2001). An analysis of the aforementioned metal artefacts uncovered from the earlier archaeological strata at Haimenkou in 1957 offers a glimpse of the technological aspects of the incipient metal age of Yunnan.

Compositional analysis

Radioisotope XRF studies indicate that small items such as an awl (no. 219) and chisel (no. 220) are made of fine copper (Wang Dadao 1985: 247; Li and Han 2006: 82); a socketed sickle (no. 229) (Fig. 2) and a thin-walled clip (no. 223) containing mostly copper and trace metals can also be regarded as copper artefacts (Li and Han 2006: 82). The bronze alloy of a socketed axe (no. 227) (Fig. 3) shows a mere 1.13–2% of tin (Li and Han 2006: 83; Cui and Wu 2008: 147), which possibly represents trace metal. The rest, including a simple, solid adze (no. 225) (Fig. 3), a socketed axe with pronounced shoulders (no. 228) (Fig. 3), a thin-walled knife (no. 222), a fragmented (no. 216) and a well-preserved bracelet (no. 224), and another fragmented piece (no. 217) are made of bronze with tin ranging from 2.31% to 11.75% (Li and Han 2006: 82; Cui and Wu 2008: 147). The alloy of a fishhook (no. 218) shows 3–5% of lead (Wang Dadao 1985: 248), but no detectable tin (Li and Han 2006: 82).

Manufacturing method

Small, thin-walled bronzes such as a clip (no. 223), knife (no. 222), pierced ornament (no. 221), and fishhook (no. 218) appear to have been hammered to shape. The copper awl (no. 219) and solid bronze adze (no. 225) (Fig. 4) were cast using an open mould (Wang Dadao 1985: 245) and subsequently hot-hammered; others, with sockets or exhibiting evident mould seams, were cast in bivalve moulds, and some may have additionally been finished with hammering. Clay plugs were inserted in the bivalve moulds to create artefacts with sockets; clay pins are also claimed to have been used to suspend such plugs (Wang Dadao 1985:

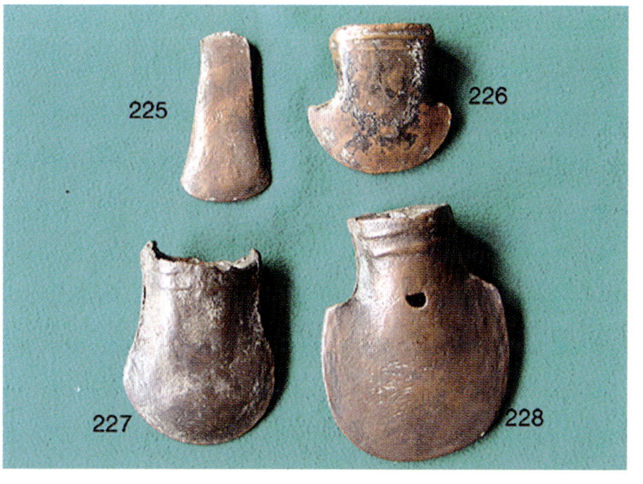

Figure 3 Four metal artefacts from Haimenkou.

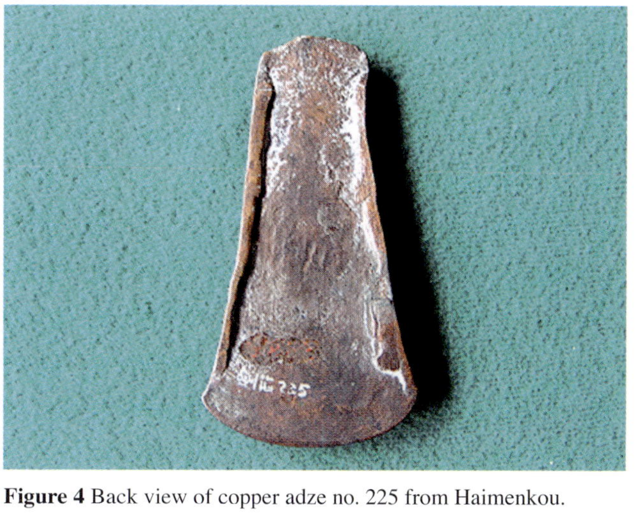

Figure 4 Back view of copper adze no. 225 from Haimenkou.

247–8; Yunnan Provincial Museum 1995: 776), each resulting in a small perforation near the socket of an artefact (Fig. 3). All these conclusions are based on visual observation, except for a socketed adze/axe (no. 227) (Fig. 3), 10.5 cm long and embellished with parallel relief lines near its socket; its test samples, examined under a metallographic microscope, show signs of post-cast hammering (Li and Han 2006: 83).

To sum up, elemental analyses indicate that of these 14 artefacts, five are made of copper, eight are tin bronzes, and one is lead bronze. Of the five copper artefacts, three are small items created using hammering techniques, and two, with sockets, are cast. The mixture of hammered and cast copper artefacts along with copper alloy artefacts at the same site appears to indicate a distinct tradition that continued well into the 7th century BCE in western Yunnan, and therefore the presence of copper artefacts at Haimenkou may not necessarily be indicative of the very beginning of a metal industry. This is consistent with the fact that a number of the copper artefacts from the Haimenkou site were produced using skilled casting processes uncharacteristic of the embryonic stage of a metal industry. For example, the harvest knife (no. 229), 20.2 cm in length, with a transverse socket (Fig. 2), is clearly a mature metal type that also has no precedence in the local lithic inventory. The origin of its prototype may be a subject of future interest.

It is also noted that the 11.75% of tin in the material of a fragmented metal piece (no. 217) (Cui and Wu 2008: 148) (Fig. 2) ranks among the highest of all tin bronzes in this assemblage. This item had been mislabelled but was later identified as part of a bracelet because the curved shape of its cross-section resembles that of an identifiable bracelet (no. 224), an artefact that contains around 10% tin (Li and Han 2006: 82). Comparing these two artefacts with others in the same group, it may be speculated that copper or low tin alloys (3–5%) were reserved for utilitarian objects, while alloys with greater amounts of tin were purposely used to create body ornaments for aesthetic reasons. Except for these, the relationship between artefact type and alloy composition does not show any discernible pattern.

The issue of lead bronze production at the Haimenkou site also demands further investigation. The only example of lead-bronze alloy in this collection is a fishhook (no. 218) produced by a hammering process (Wang Dadao 1985: 247). A microscopic metallographic analysis will be needed to unravel further details in order to address questions of whether lead-containing material was deliberately introduced in the alloy or if lead was present in the natural copper ore, which is indeed accessible in the area (Wang Dadao 1985: 247). Nevertheless, it is clear that complex bronze alloys, such as those combining both lead and tin, are completely absent at this stage.

In comparison with the sophisticated techniques and bronze recipes utilised for metal artefacts representing a more developed phase (645±75 BCE) of the same tradition at the Haimenkou site (Li and Han 2006), the metal assemblage discussed above exhibits technical features consistent with its assigned radiocarbon age of 13th–11th century BCE (Laboratory at the Institute of Archaeology of the Chinese Academy of Social Science 1972: 57). Metal slag and one half of a bivalve stone mould discovered at the same site testify that some of these items were produced locally, probably using metal-bearing ore materials from nearby sources (Wang Dadao 1985: 247; Cui and Wu 2008: 101-4). Given the fact that small personal items and weapon-tools were the only items present in the list of the Haimenkou metal inventory, metallurgical activities at Haimenkou during this stage appear to be small-scale, most likely conducted without the presence of controlling elites.

Early metals along the Jinsha Corridor

A better understanding of the early stages of Haimenkou technology, however, is pending on the publication of archaeological and metallurgical data from its 2008 excavation, during which sequential pre-metal and metal age strata alike were uncovered (Yunnan Provincial Institute of Cultural Relics and Archaeology 2008). The materials collected at the site, as well as ethnological studies of western Yunnan people, point to cultures near the upper Yellow River as the source for inspiring the emergence of Haimenkou metals. Excavated remains of wheat, millet and wooden cultivation tools of northern types at Haimenkou attest to the impact of intrusive elements on the site, whereas the inclusion of double-handled painted pottery also hints at affiliations with Majiayao-derived cultures (Yunnan Provincial Institute of Cultural Relics and Archaeology 2008). Moreover, the typological features of the copper and bronze axes/adzes found at the site clearly are non-local derivatives – they are metal facsimiles of a unique type of flaked shouldered stone adze characteristic at Neolithic sites in deep gorges and valleys along the Lancang-Nu Corridor (Tong 1980: 439–41), which originates in a region near the Qinghai-Tibet borders and comes down to southwestern Yunnan to extend into mainland South-East Asia. In areas around the middle Lancang and Nu rivers, comparable shouldered stone adzes (Fig. 5) have been found side by side with rudimentary copper axes of the same shapes (Geng and Zhang 1992: 448), as well as with a more rudimentary type of solid copper axe/adze (Wang Jinlin 1992). Stone moulds for casting and metal slag thought to be ear-

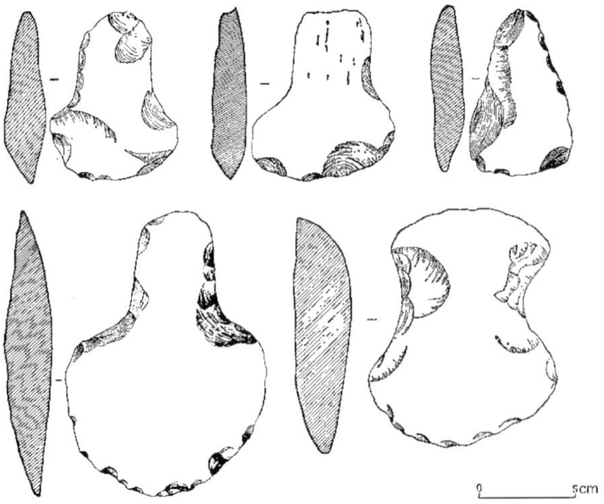

Figure 5 Shouldered stone adzes from areas near the middle Nu River.

lier than the 13th century BCE (Wang Dadao 1994: 270–71) were also discovered in the same region. In this large area inhabited by people who were akin to steppe-related Di/Qiang stock (Tong 1980: 440; Wang Ningsheng 1979: 437), shouldered stone or copper tools, designed for land clearing, were effective implements for vertical transhumance. The distinct tool-making technology used in the Lancang-Nu River corridor was apparently shared with the inhabitants of Haimenkou, where typical polished stone adzes, rectangular or trapezoidal in shape, were produced (Yunnan Provincial Institute of Cultural Relics and Archaeology 2008).

In piecing together available data to address issues regarding the inception of Yunnan metallurgy, it may be argued that northern cultural affinities arrived in Yunnan concurrently with continuous southward shifting population movements out of their homeland near the upper Yellow River beginning in the Neolithic period (van Dreim 2005: 98–102; Huo 2005). While some diverted to the upper Min River of Sichuan at one point (Chen 2005), others trickled down the valleys of the grand rivers and eventually colonised various regions of western Yunnan, where they became intermingled with indigenous inhabitants. Carried along with these peoples were pottery designs, tool-making technologies, cultivated grains, as well as other unique northern-style artefacts and iconographical elements, including some revealed in archaeological finds near the Yunnan–Myanmar borders (Huo 2005: 276–8). A similar storyline possibly accounted for the metallurgical activities at Haimenkou, occurring at the time when metalworking became a new addition to the inventory of items transported by migratory people.

How the people at the Haimenkou site communicated with people in areas of a different ecosystem further east and south-east is not known at this stage; however, it is evident that the interconnected plains along the Jinsha valley were ready to embrace the new technology. Small metal artefacts, slag and ore materials recently excavated at Yinsuodao are concrete evidence of the early use of metal in flat lands, possibly during a period of time that was contemporaneous with the earliest finds of metal at Haimenkou. The Yinsuodao site, a shell midden at the south-eastern edge of Lake Er, had been occupied continuously from as early as the 3rd millennium BCE (Wang Xiaoyun 2008). Additional metal artefacts from sites near the lower Jinsha valley (Zhang and Luo 2006; Sun 2007; Liu and Sun 2008) were possibly also produced as part of the same technological continuum. The bronze adzes and slag discovered at the Yeshishan site near the Yunnan–Quizhou borders, in particular, provide concrete evidence of an embryonic metal industry in eastern Yunnan around 1000 BCE (Sun 2007: 14; Liu and Sun 2008). At this site, metal tools, modelled after indigenous stepped stone prototypes, were used alongside double-handled ceramic jars that appear to be intrusive elements in the ceramic repertoire of the area. It can be argued that the introduction of metal technology in eastern Yunnan and contiguous areas galvanised the local technological substratum into growth, and heralded the advent of a vigorous metal industry (4th century BCE–1st century CE) on the plains near Lake Dian (Pirazzoli-t'Serstevens 1974). Rudimentary copper-based *ge* daggers, stepped adzes and slag from the Wanjiadun site near Kunming city (Li and Wang 1983; Wang Dadao 1981: 82) are also tangible evidence which exemplifies aspects of pre-Dian industry in eastern Yunnan. Chance finds of undecorated copper axes and crudely made solid bronzes from areas along the Yuan (upper Red) River valley further south (Yin 2001: 162) probably came into existence along similar lines. These were modelled after local stone tools, believed to be 'contemporaneous with Haimenkou productions' (Yin 2001: 164). Considering the fact that the Haimenkou-related technology had dispersed beyond the eastern borders of modern Yunnan by 1000 BCE (Zhang and Luo 2006), it should be plausible for metal production to have become accessible at an early date in ore-rich regions outside of Yunnan, in particular areas open to access through the valleys of the lower Nu, Lancang, the Yuan rivers, and their tributaries (Watson 1992).

Conclusions

Perceptible signs of incipient copper-based metallurgical activities at 2nd-millennium sites in western and north-western Yunnan point to steppe-related cultures as the source for inspiring metallurgical activities at the Haimenkou site, which commenced during the latter part of the 2nd millennium BCE. The arrival of metallurgical technology revolutionised the tool-making industry in western regions of Yunnan; knowledge of metal production and the making of artefacts was also disseminated through interconnected valleys and flat lands along the eastward-flowing Jinsha River, and possibly also in the same manner toward the headwater of the upper Yuan River. Around 1000 BCE, metalworking was active in areas near the Yunnan-Guizhou borders, and began to be integrated into the indigenous technological substrata in the lacustrine plains surrounding Lake Dian.

Acknowledgements

I wish to thank the Chiang Ching-Kuo Foundation for funding a research project on Yunnan metallurgy during 2002–04, which led to the discussions included in this work. Special thanks are also extended to Professors Han Rubin, Sun Shuyun, Li Xiaocen, and Li Yanxiang for their assistance in preparing this manuscript, and to Professor Mei Jianjun for inviting me to participate in the BUMA VI conference in Beijing.

References

Allard, F. (1998) 'Stirrings at the periphery: history, archaeology, and the study of Dian', *Antiquity* 2(4): 321–41.

Chen, Jian (陈剑) (2005) 营盘山遗址再现"藏彝走廊"5000年前的区域中心 ('Uncovering the 5000 year-old cultural hub along the "Zangyi Corridor"'), in Shi Shuo (石硕) (ed.), 藏彝走廊: 历史与文化 (*Zangyi Corridor: History and Culture*), 300–14. Chengdu: Sichuan Chuban Jituan and Sichuan Renmin Press.

Chiou-Peng, Tzehuey (邱兹惠) (1998) 'Western Yunnan and its steppe affinities', in V.H. Mair (ed.), *The Bronze Age and Early Iron Age Peoples of Eastern Central Asia Vol. 1*, 280–304. Journal of Indo-European Studies Monograph Series, Monograph No. 2. Washington, DC: Institute for the Study of Man.

Chiou-Peng, Tzehuey (邱兹惠) (2008) 'Dian bronze art: its source and formation', *Bulletin of the Indo-Pacific Prehistory Association* 28: 34–43.

Ciarla, R. (1994) 'A long debated question. Pastoralism in south-west China: a different approach', in B. Genito (ed.), *The Archaeology of the Steppes: Method and Strategies*, 73–86. Napoli: Istituto Universitario Orientale.

Cui, Jianfeng and Wu, Xiaohong (崔剑锋 吴小红) (2008) 铅同位素考古研究 (*The Study of Lead Isotopic Archaeology*). Beijing: Wenwu Press.

Geng, Deming and Zhang, Shaoquan (耿德铭 张绍全) (1992) 昌宁青铜器综说 ('An overview of the bronzes from Changning'), 考古 (*Archaeology*) 5: 443–50.

Higham, C. (1996) *The Bronze Age of Southeast Asia*. Cambridge: Cambridge University Press.

Huo, Wei (霍巍) (2005) 论横断山脉地带先秦两汉时期考古学文化的交流与互动 ('On the pre-Han cultural communications along the Hengduan intermountain chain'), in Shi Shuo (石硕) (ed.), 藏彝走廊: 历史与文化 (*Zangyi Corridor: History and Culture*), 272–99. Chengdu: Sichuan Chuban Jituan and Sichuan Renmin Press.

Jiang, Zhilong (蒋志龙) (2002) 滇国探密 (*Exploring the Mystery of the Dian Kingdom*). Kunming: Jiaoyu Press.

Laboratory at the Institute of Archaeology of the Chinese Academy of Social Science (中国社会科学院考古研究所实验室) (1972) 放射性碳素测定年代报告（二） ('Radiocarbon dating report-2'), 考古 (*Archaeology*) 5: 56–8.

Laboratory at the Institute of Archaeology of the Chinese Academy of Social Science (中国社会科学院考古研究所实验室) (1990) 放射性碳素测定年代报告（十七）('Radiocarbon dating report 17'), 考古 (*Archaeology*) 7: 663–8.

Li, Xiaocen and Han, Rubin (李晓岑 韩汝玢) (2006) 云南剑川海门口出土铜器的技术分析及其年代 ('On the technical aspects and chronology of bronzes from the Haimenkou site in Jianchuan county, Yunnan'), 考古 (*Archaeology*) 7: 80–85.

Li, Yongheng and Wang, Han (李永恒 王涵) (1983) 昆明市西山区王家墩发现青铜器 ('The bronzes discovered at the Wangjiadu in the Xishan district of Kunming'), 考古(*Archaeology*) 5: 479.

Liu, Xu and Sun, Hua (刘旭 孙华) (2008) 试论野石山文化 ('A study of the Yeshishan culture'), paper presented at the *Fifth Conference of the Western Region on Archaeological Cooperation and International Academic Symposium on 'China's Western Region in the Prehistoric Age with the Yunnan-Guizhou Plateau as the Center'. Kunming, China, 3 November 2008.*

Ma, Changzhou (马承源) (1981) 劍川海門口古文化遺址的几个问题 ('Questions about the ancient cultural site at Haimenkou, Jianchuan'), in Yunnan Provincial Museum (ed.), 云南省博物馆建馆三十周年纪念文集 (*Essays Commemorating the Thirtieth Anniversary of Yunnan Provincial Museum*), 68–72. Kunming: Yunnan Provincial Museum.

Murowchick, R.E. (1989) *The Ancient Bronze Metallurgy of Yunnan and Its Environs: Development and Implications*. Unpublished PhD thesis, Harvard University.

Murowchick, R.E. (2001) 'The political and ritual significance of bronze production and use in ancient Yunnan', *Journal of East Asian Archaeology* 3(1–2): 133–92.

Pirazzoli-t'Serstevens, M. (1974) *La civilisation du royaume de Dian à l'époque Han, d'après le matériel exhumé à Shizhai shan (Yunnan)*. Paris: Ecole Française d'Extrême-Orient.

Preparatory Committee for the Yunnan Provincial Museum (云南省博物馆筹备处) (1958) 劍川海門口古遺址清理簡報 ('Preliminary report of the excavation of the Haimenkou site in Jianchuan'), 考古通訊 (*Kaogu Newsletter*) 6: 5–12.

Sun, Hua (孙华) (2007) 滇东黔西青铜文化初论 ('On the bronze cultures in eastern Yunnan and western Guizhou'), 四川文物 (*Sichuan Cultural Relics*) 5: 12–25.

Tong, Enzheng (童恩正) (1980) 近年来中国西南民族地区张国秦汉时代的考古发现及其研究 ('A study of recent archaeological finds of Warring States to Han periods in southwest China'), 考古学报 (*Acta Archaeologica Sinica*) 4: 417–41.

van Dreim, G. (2005) 'Tibeto-Burman vs Indo-Chinese: implications for population geneticists, archaeologists and prehistorians', in L. Sagart, R. Blench and A. Sanchez-Mazas (eds), *The Peopling of East Asia*, 81–106. London/New York: Routledge Curzon.

Wang, Dadao (王大道) (1981) 滇池地区的青铜文化 ('The bronze culture in the Lake Dian region'), in Editorial Board (ed.), 云南青铜器论丛 (*Essays on Yunnan Bronzes*), 77–91. Beijing: Wenwu Press.

Wang, Dadao (王大道) (1985) 云南劍川海門口早期青铜研究 ('A study of the early bronzes from Haimenkou,Yunnan'), in Zhongguo Kaogu Xuehui (ed.), 中国考古学会第四次年会论文集 (*Essays for the 4th Annual Meeting of the Chinese Archaeological Association*), 244–51. Beijing: Wenwu Press.

Wang, Dadao (王大道) (1994) 云南青铜文化与新石器晚期文化的关系 ('Links between the bronze and late Neolithic cultures in Yunnan'), in Chung Tang (ed.), 南中国及邻近地区古文化的研究 (*Ancient Cultures of South China and Neighbouring Regions*), 269–79. Hong Kong: Chinese University of Hong Kong.

Wang, Jinlin (王锦麟) (1992) 龙陵梅子寨发现铜石器和小田新石器遺址 ('Bronze and lithic artefacts discovered at Meizizhai, Longling and the Neolithic site at Xiaotian'), 云南文物 (*Yunnan Cultural Relics*) 31: 83.

Wang, Ningsheng (汪宁生) (1979) 云南晋宁石寨山青铜器图像所见古代民族考 ('Ancient tribes in Yunnan as represented on bronzes unearthed at Shizhaishan in Jinning county'), 考古学报 (*Acta Archaeologica Sinica*) 4: 423–39.

Wang, Xiaoyun (王晓云) (2008) 海门口考古发掘特别报道 ('Exclusive report on the excavation at Haimenkou'), 大理日报 (*Dali Daily*), 15 May, B4.

Watson, W. (1992) 'Pre-Han communication from west China to Thailand', in I. Glover, P. Suchitta and J. Villars (eds), *Early Metallurgy, Trade and Urban Centers in Thailand and Southeast Asia*, 175–80. Bangkok: White Lotus.

Yin Tianyu (尹天钰) (2001) 红河流域的青铜文化 ('The bronze culture along the Red River'), in 中国古代铜鼓研究会 (Association of Bronze Drums of Ancient China) (ed.), 铜鼓和青铜文化研究－中国南方及东南亚地区古代铜鼓和青铜文化第四次国际学术讨论会论文集 (*Bronze Drums and Bronze Culture: 4th International Conference on the Bronze Drums and Bronze Culture of Southern China and Southeast Asia*), 161–5. Guiyang: Guizhou Renmin Press.

Yunnan Provincial Institute of Cultural Relics and Archaeology (云南省文物考古研究所) (2008) 我国最大的史前"干栏式"建筑遗址： 剑川海门口 ('The most expansive prehistoric pile dwelling sites ever uncovered in our country: Haimenkou in Jianchuan county'), 云南日报 (*Yunnan Daily*), 13 June: 4.

Yunnan Provincial Musuem (云南省博物馆) (1995) 云南劍川海門口青铜时代早期遗址 ('The Bronze Age site at Haimenkou in Jianchuan, Yunnan'), 考古 (*Archaeology*) 9: 775–87.

Zhang, Herong and Luo, Erhu (张合荣 罗二虎) (2006) 论鸡公山文化 ('On the Jigonshan culture'), 考古 (*Archaeology*) 8: 57–66.

Zhang, Zengqi (张增祺) (1981) 滇西青铜文化初探 ('A study of the bronze culture of western Yunnan'), in Editorial Board (ed.), 云南青铜器论丛 (*Essays on the Bronzes from Yunnan*), 92–107. Beijing: Wenwu Press.

A study of the surface craft of weapons from the Ba-Shu region of ancient China

Zhihui Yao and Shuyun Sun

ABSTRACT Three samples from bronze weapons from the Ba-Shu region of ancient China, dating to the 5th–2nd century BC, displaying black mottled areas (*hu-ban-wen*, 虎斑纹) on the surface, were analysed using scanning electron microscopy (SEM), X-ray diffraction (XRD) and an optical microscope. The matrix of these samples shows a cast structure which has been heated. The main compositional elements of the mottled layer are SnO_2 and $Cu_{41}Sn_{11}$. The tin content of the matrix (about 12%) is less than that of the mottled layer (approximately 45%). The thickness of the mottled layer is about 20–40 μm. Our analysis indicates that the mottled layer is artificial and has an antioxidant function. The black colour of the mottled layer was probably acquired after burial. Three methods were chosen to simulate the formation of the *hu-ban-wen*, namely hot-tinning, mercury-amalgam tinning and coating with a Cu-Sn alloy solution. Hot-tinning followed by annealing is a suitable method to create *hu-ban-wen*. In addition, the fact that *hu-ban-wen* existed on the surfaces of a large number of weapons also indicates that hot-tinning was a mature craft at that time.

Introduction

Chinese archaeologists and historians use the term 'Ba-Shu' (巴蜀) to designate the eastern region of Sichuan in the pre-Qin period. A particular mottled pattern, also referred to as 'pattern-etched' or *hu-ban-wen* (虎斑纹), appears on the surface of some weapons dating from the 5th to the 2nd century BC (Figs 1 and 2). With regard to its formation, previously published studies have assumed that: (1) it was produced by casting (Ma 1998: 523); (2) it was formed naturally in the course of burial (Chase and Franklin 1979: 217); (3) it was obtained through mercury-amalgam tinning (Kossolapov and Twilley 1994: 257); (4) it was obtained by hot-tinning (He 1987: 46).

There are two suggestions as to how the black colour was achieved: created by a sulphuration treatment (Railway Archaeology Team of Jingsha, Hubei Province 1991: 417) or alternatively formed during burial similar to the *hei-qi-gu* (黑漆古) on the surface of ancient Chinese mirrors (Sun *et al.* 1992: 79; 1996: 179). Only one sword with *hu-ban-wen* from a chance find has been analysed and reported (Kossolapov and Twilley 1994: 257) and to date few studies of the craft of *hu-ban-wen* have been carried out.

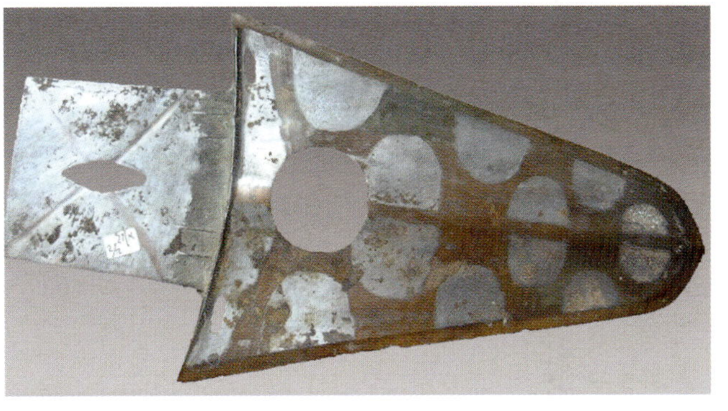

Figure 1 Silver-colour mottle on an ancient bronze dagger-axe.

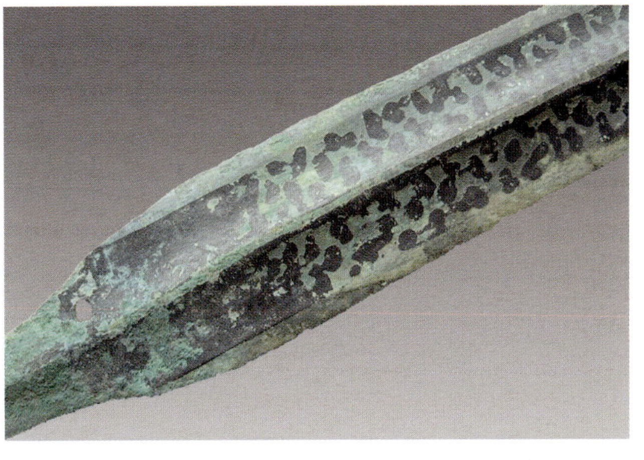

Figure 2 Black-colour mottle on an ancient bronze sword.

Analytical work

Three samples of weapons with *hu-ban-wen* from the Ba-Shu region, dating to the Warring States period (475–211 BC), have been analysed: two from Mianzhu and one from Chengdu. The irregular layer of mottling is not thick and is easily rubbed off. The samples were observed with an Olympus BH2-UMA polarisation microscope. Samples SMZ50 and SC51 were analysed using an X-ray diffraction (XRD) method. The XRD measurements were made with the use of a Bruker D8 Discover with GADDS employ-

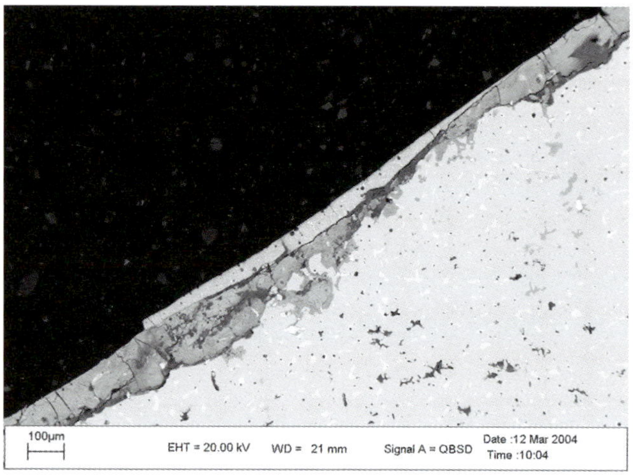

Figure 3 Backscattered electron image of sample SC51.

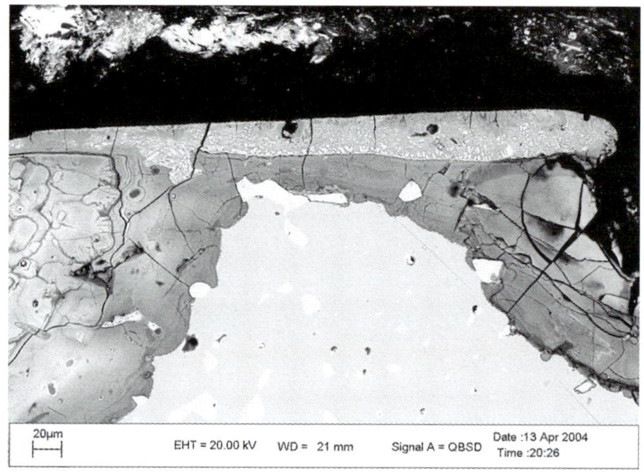

Figure 4 Backscattered electron image of sample SMZ50.

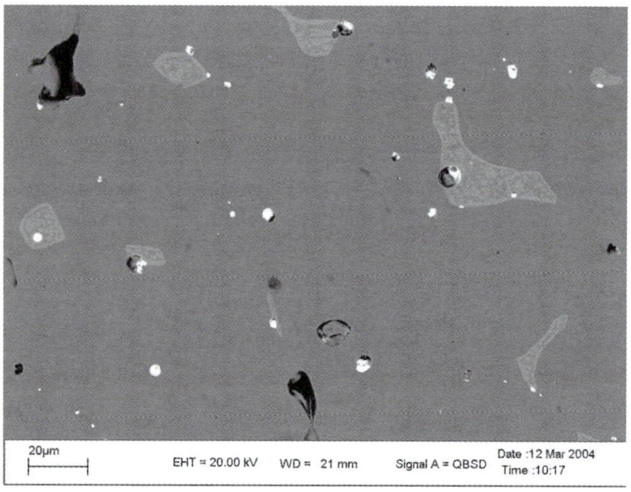

Figure 5 Backscattered electron image of the matrix of sample SC51.

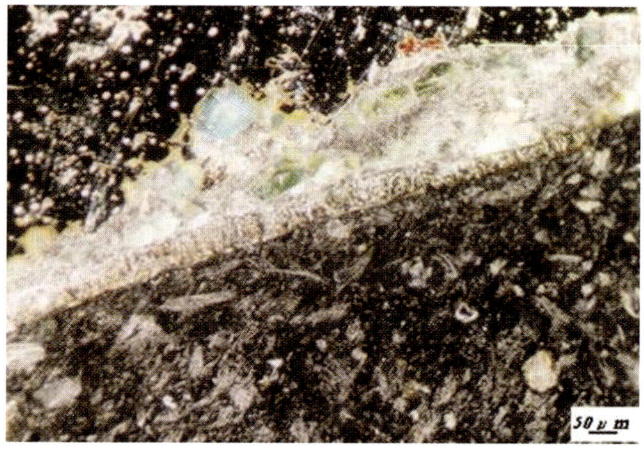

Figure 6 Microstructure of sample SC51 in the dark field.

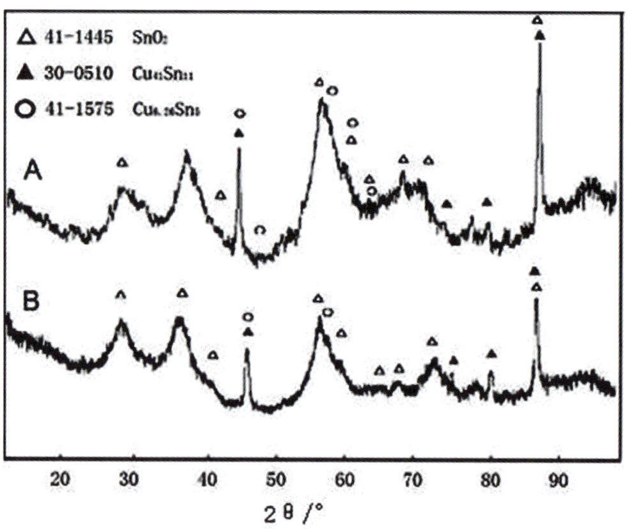

Figure 7 XRD pattern of the black mottles of samples SMZ50(A) and SC51(B).

ing copper K-alpha radiation at 40 kV and 40 mA for 120 seconds. Cross-sections of samples were prepared according to standard metallographic procedure and analysed in a scanning electron microscope (SEM) and energy-dispersive spectrometer (EDS), using a LEO-1450 SEM with KEVEX Sigma energy spectrometer at 20 kV for 50 seconds acquisition. The microstructure was observed through a metallographic microscope with NEOPHOT 21.

Results and discussion

The thickness of the mottled layer is about 20–40 μm. A clear and smooth interface (Figs 3 and 4) exists between the mottle and the matrix; the latter shows a cast structure. The segregation of the dendritic crystals formed during solidification of the molten alloy is not obvious. This suggests that the matrix might have been heated after casting (Fig. 5). Corrosion was always found on non-mottled areas, extending beneath the mottled layer (Fig. 3). Copper oxide, azurite, malachite and silicon are the main corrosion products between the mottled layer and the matrix (Fig. 6). Residual, uncorroded metal (copper-tin alloy phase) is present in the mottled layer. The non-mottled area is composed of cassiterite and copper oxides. About 40–46% tin is found in the mottled layer and 12% in the matrix, as determined by SEM–EDS analysis.

The XRD patterns of the mottled layers of the samples originating from different places are very similar (Fig. 7). The main components of the mottled layer are SnO_2, $Cu_{41}Sn_{11}$

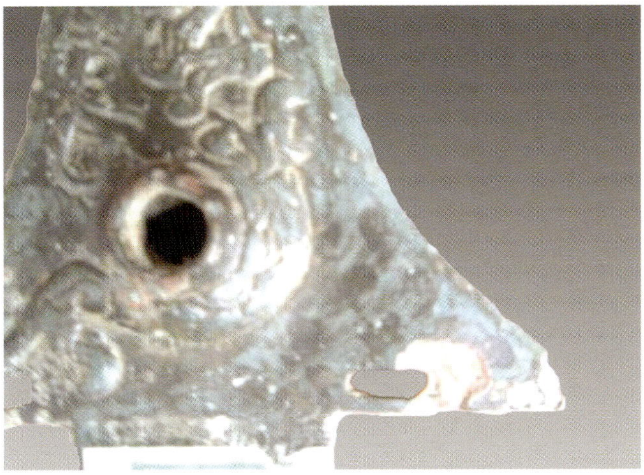

Figure 8 Dagger-axe (*ge*) 00286-0028.

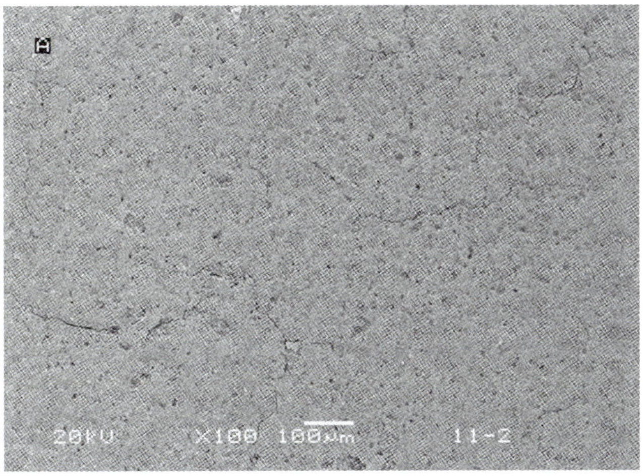

Figure 9 Secondary electron imaging of the *ge* (Zhou 11-2).

(δ phase) and $Cu_{6.26}Sn_5$ (near η phase), and analyses of the sample cross-sections also confirm that δ phase ($Cu_{41}Sn_{11}$) is present in the mottled layer. The extended diffraction peaks of SnO_2 probably obscured some diffraction peaks of other phases and had an effect on the analysis. The SnO_2 is most probably either the result of a long process of corrosion on the surface of the object or may be a sedimentary deposit on the surface of the matrix alloy (Wang and Lu 1995: 489).

The analytical results indicate that the mottled layer is a high tin bronze. There are three possible ways to produce tin enrichment on the surface of bronze: the first is negative segregation during casting; the second is selective corrosion from the environment; and the third is deliberate tinning on the surface of bronze.

In the first type of treatment, surface tin enrichment results from the segregation of the (α+δ) eutectoid phase during the cooling of a high tin bronze. One would therefore expect the (α+δ) eutectoid in the surface to be associated with the bronze matrix. However, in the samples analysed, a boundary between the mottled layer and the matrix was evident. In the second type of process resulting in tin enrichment, some burial environments would result in the corrosion of the α phase from the surface layer, leaving a tin-rich (α+δ) phase, with a silvery appearance. However such selective corrosion would be expected to occur evenly on the surface of the bronze rather than located only on the mottled areas. Thus, the first and the second type of enrichment do not explain our observations. The third type, deliberate tinning, was the most likely method used to make the mottled layer in our examples. The results of analyses of intact weapons using the non-destructive methods listed above (Figs 8 and 10) also verified the fact that tinning was probably the method used to form the mottled layer (Yao *et al.* 2005: 19).

The object in Figure 8 is a dagger-axe (*ge*) from Chengdu Museum. The mottled pattern and fine lines, clearly seen on its surface, have the same composition. We assume that the fine lines are the result of handiwork. Figure 9 shows a secondary electron image of another dagger-axe (Zhou 11-2) with silver mottled areas from Chengdu Museum. The grain boundaries can be clearly seen between possible equiaxed crystals of tin. The surface profile of the silver mottled areas is uneven and there are a large number of pores caused either by escaping gases or by shrinking during the condensation process. In general, these features were induced by the melting and condensation of the alloy. The analyses of cross-sections of samples with black mottled surfaces also confirmed this hypothesis. Figure 4 shows some large columnar crystals growing from the matrix into the mottled layer. Such columnar crystals could only grow that large if, on the surface, liquid metal came into contact with the heated matrix and the cooling velocity was slow, allowing the crystals to grow. At the same time, only liquid metal could run into the sand holes of the bronze matrix and fill them.

Was the *hu-ban-wen* etched? Some surface decorations on weapons dating from the Eastern Zhou period might have been made by etching with a ferric chloride solution (Chase and Franklin 1979: 217). We analysed a *pi* (铍, a kind of weapon) dating from the Western Han period from Shizishan, Xuzhou, Jiangsu province. An irregular black mottled surface, similar to *hu-ban-wen*, was observed on the surface of the *pi* (Fig. 10). This layer contains not only high tin, but also 6% ferric material. The structures of the mottled areas and of the non-mottled areas are clearly connected (Fig. 11). However, both the components and texture of the mottled layer are different from that of the *hu-ban-wen* from the Ba-Shu region. It is apparent that, unlike the *hu-ban-wen*, the mottled layer of the *pi* was probably etched using the ferric chloride solution mentioned above.

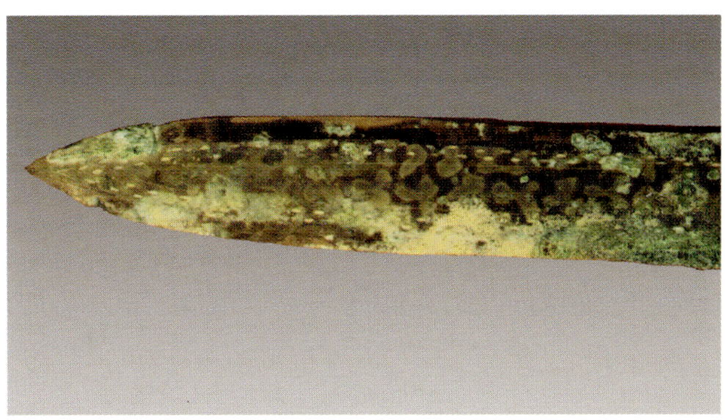

Figure 10 The *pi* from Shizishan, Xuzhou.

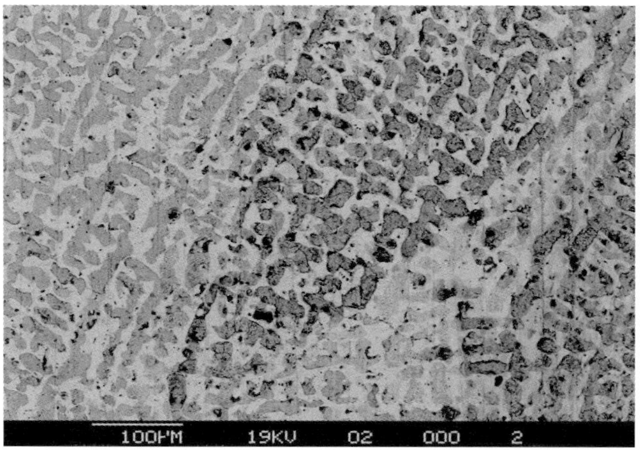

Figure 11 Backscattered electron images of mottled area and non-mottled area.

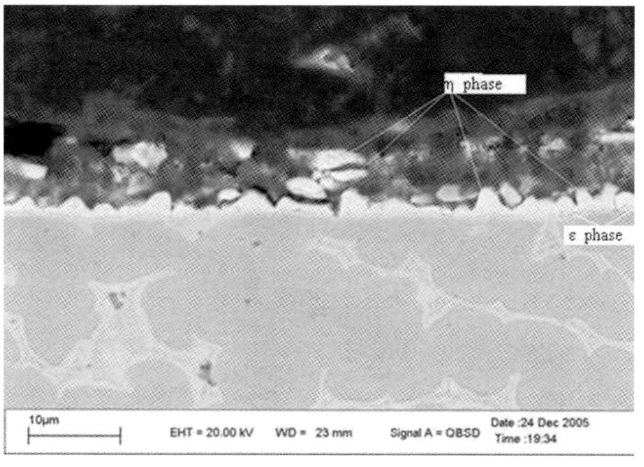

Figure 12 Backscattered electron image (SEM) of simulated sample by hot-dip tinning.

Simulation

On the basis of the analyses above, some simulation experiments were performed. Using hot-tinning followed by annealing it was possible to create the *hu-ban-wen*. Methods of operation included coating a copper rod with melted tin, coating with tin leaf or sheet, and hot-dip tinning. Clear boundary and columnar crystals could be seen growing when each of these methods was tested. Different heating conditions resulted in variations in the thickness of the coating (Fig. 12).

The appearance and microstructure of the mottle produced by the simulation were similar to those of the ancient samples from the Ba-Shu region, the only difference being that δ phase was generated in ancient samples and η, ε phases were generated in our facsimiles (see Meeks 1986: 133 for an explanation of this feature). If tinning coats were treated by heating-annealing at or above 450 °C, δ phase should be generated. Thus the process of heating-annealing could explain the homogeneous structure of the bronze substrate.

Simulation experiments of mercury-amalgam tinning and coating high tin copper alloy solution did not produce the *hu-ban-wen*.

Conclusions

We can draw some conclusions from the work described in this paper. The *hu-ban-wen* mottling is corrosion resistant. The structure of the non-mottled areas is friable but the structure of the mottled areas is relatively compact. The main component of the mottled layer, SnO_2 is a stable compound and protects the matrix from corrosion. The black colour is acquired through the effect of humic acid in the burial environment. The *hu-ban-wen* was not formed by etching, but possibly by hot-tinning and annealing. In the process of hot-tinning, new phases will be generated. The columnar crystals growing from the matrix into the coating are probably caused by the dissolution and diffusion of copper into liquid tin.

The large number of weapons with *hu-ban-wen* indicates that tinning was a mature craft at that time. Did the craft appear locally or was it 'imported' from other areas? Was it related to the Chu culture and North Steppe culture? Where was the source of tin ore located? While these and other questions are not yet addressed, we hope that our research may provide a start. Further work should be done on this subject, including analysing and comparing more examples of tinned bronzes from different areas.

Acknowledgements

We would like to express our gratitude to Professor Rubin Han, University of Science and Technology, Beijing, for many helpful suggestions; to Professors Zhongmao Zeng, Institute of Cultural Relics, Sichuan Province, and Tangkun He, Institute of Archaeology, Chinese Academy of Social Sciences, for providing samples; to Mr Yawei Dong, Ezhou Museum, Hubei Province, for providing a test setting and help; to Lin Xiao and Yulong Bai, Chengdu Culture Relics and Archaeology Institute, Sichuan Province, for assistance in our search for objects.

References

Chase, W.T. and Franklin, U.M. (1979) 'Early Chinese black mirrors and pattern-etched weapons', *Ars Orientalis* 11: 217–57.

He, Tangkun (何堂坤) (1987) 部分四川青铜器的科学分析 ('Analysis of some bronzes from Sichuan'), 四川文物 (*Sichuan Cultural Relics*) 4: 46–8.

Kossolapov, A. and Twilley, J. (1994) 'A decorated Chinese dagger: evidence for amalgam tinning', *Studies in Conservation* 39: 257–64.

Ma, Chengyuan (马承源) (1998) 中国青铜器 (*Chinese Bronzes*). Shanghai: Shanghai Ancient Books Press.

Meeks, N.D. (1986) 'Tin-rich surfaces on bronze: some experimental and archaeological considerations', *Archaeometry* 2: 133–62.

Railway Archaeological Team of Jingsha, Hubei Province (湖北省荆沙铁路考古) (1991) 包山楚墓 (*Baoshan Chu Tomb*). Beijing: Wenwu Press.

Scott, D.A. (1991) *Metallography and Microstructure of Ancient and Historic Metals*. Los Angeles: Getty Conservation Institute.

Sun, Shuyun, Ma, Zhaozeng and Han, Rubin (孙淑云 马肇增 韩汝玢) (1992) 土壤中腐殖酸对铜镜表面"黑漆古"形成的影响

('Influence of soil humic acid on the surface of *hei-qi-gu* mirrors'), 文物 (*Cultural Relics*) 12: 79–87.

Sun, Shuyun, Zhou, Zhongfu and Li, Qainmao (孙淑云 周忠福 李前懋) (1996) 铜镜表面"黑漆古"的痕像研究 ('Study of mineralised ghost pseudomorphic micro-structures in the surface of a *hei-qi-gu* mirror'), 自然科学史研究 (*Studies in the History of Natural Sciences*) 2: 179–88.

Wang, C.S. and Lu, B. (1995) 'Structural and element analysis on the nanocrystalline SnO_2 on the surface of ancient Chinese black mirrors', *Nanostructured Materials* 5: 489–96.

Yao, Zhihui, Sun, Shuyun and Xiao, Lin (姚智辉 孙淑云 肖磷) (2005) 成都博物院几件院藏兵器的分析研究 ('Study of bronze weapons in Chengdu Museum'), 文物保护与考古科学 (*Science in Conservation and Archaeology*) 2: 19–23.

Production of signature artifacts for the nomad market in the state of Qin during the late Warring States period in China (4th–3rd century BCE)

Katheryn M. Linduff

ABSTRACT In 1991 in the northern suburbs of Xi'an in the territory of the ancient state of Qin, objects of high historical value, including ceramic models for making bronze-casting moulds, pottery, bronze objects, iron artifacts, lacquer ware, stone objects, and other items, were unearthed. Among these finds were 25 ceramic models for casting 'steppe-style' bronze plaques of the type and iconography typical of the Ordos region located under the Great Bend of the Yellow River, as well as other categories of bronze object parts. Not only do the contents of this single tomb confirm what several scholars have previously proposed – namely that the Chinese were producing objects for their pastoral neighbors – it also suggests that the metal industry formed a significant piece of the Qin state-level as well as the local economy and its 'foreign' policy. The socio-economic and political implications of the industry and its significance to the Qin state are explored below.

Introduction

In December 1991, the Institute of Archaeology of Shaanxi Province carried out an excavation of a late Warring States tomb (*c*.350–221 BCE) in Beikang village, in the northern suburbs of Xi'an in the ancient state of Qin (Fig. 1). Objects of high historical value were unearthed including ceramic models for making bronze-casting moulds, pottery, bronze objects, iron artifacts, lacquer ware, stone objects, and other items. Among these finds were 25 ceramic models for casting 'steppe-style' bronze plaques of the type and iconography typical of the Ordos region located under the Great Bend of the Yellow River, as well as other categories of bronze object parts. A preliminary report of the excavation was published in 2003 (Yue 2003) and a full report was made available in 2006 (AISP 2006). Not only do the contents of this single tomb confirm what several scholars have previously proposed – namely that the Chinese were producing objects for their pastoral neighbors (Bunker 1983, 1991) – it also suggests that the metal industry formed a significant piece of the Qin state-level as well as the local economy and its foreign policy. It is the socio-economic and political implications of the industry and its significance to the Qin that is explored here. Readers should be aware that in this paper the term 'Ordos-style' (used by the excavators both to describe a visual style found in the Northern Zone and *beifang*) has been replaced with 'steppe-style'; 'Ordos' is used to designate the specific location under the Great Bend of the Yellow River.

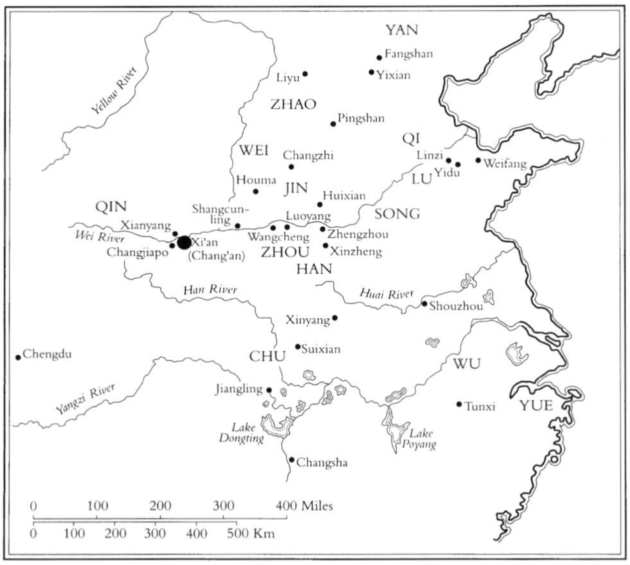

Figure 1 Map of the late Warring States (based on Hermann 1966: 8), with insert of Xi'an西安, Beikang北康 (after Yue 2003: 4).

The tomb and its contents

The tomb (Fig. 2) was constructed with a straight, but angled-in, burial shaft and a wall niche, containing the actual interment, and is oriented east–west with the burial chamber on the east wall; it is unlike tombs in other states at this time that followed the Han Chinese north–south orientation and pit tomb shape (Falkenhausen 2004: 156). The

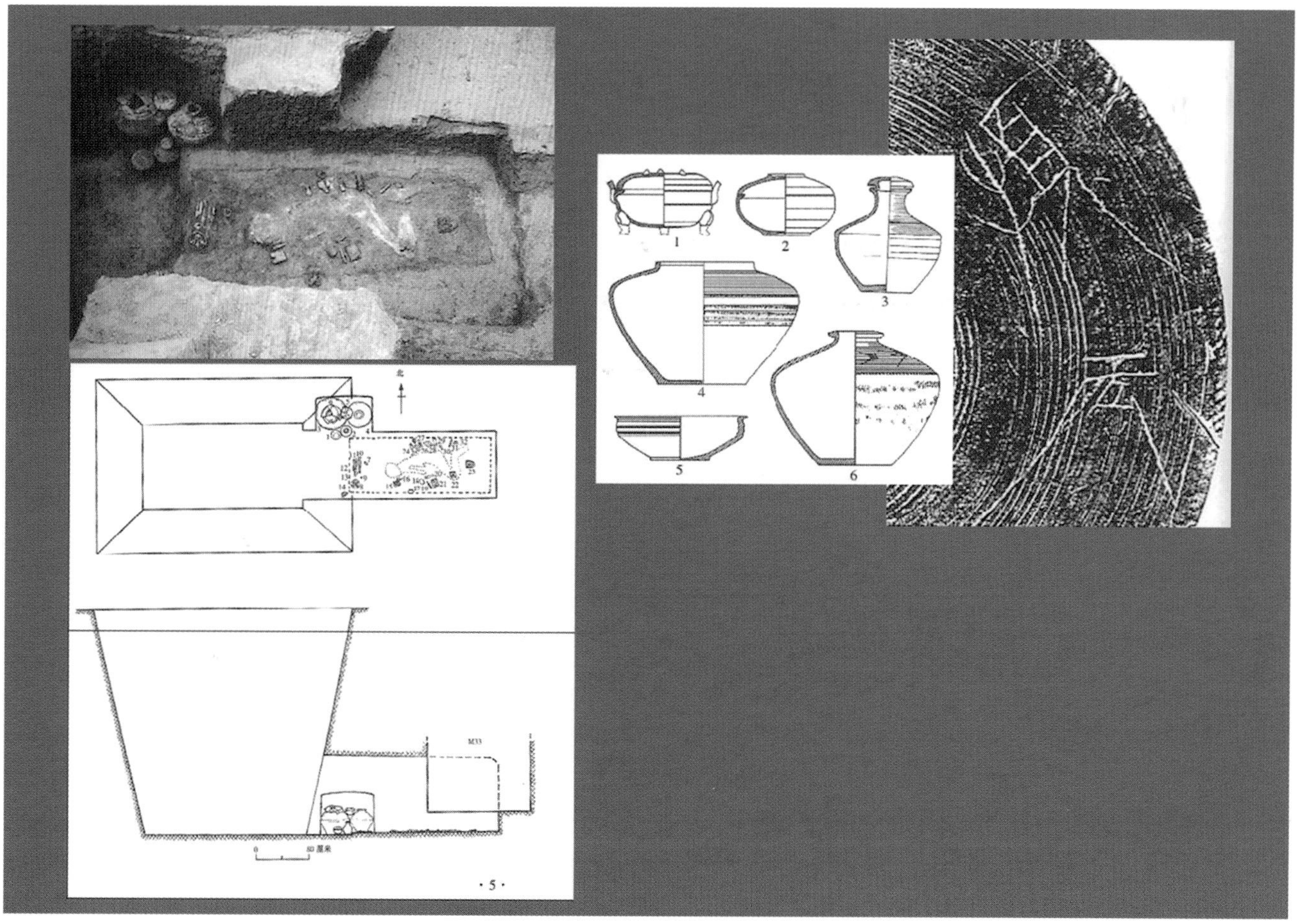

Figure 2 Line drawings of the Beikang tomb; pottery vessels (after Yue 2003: 5, 11, 13).

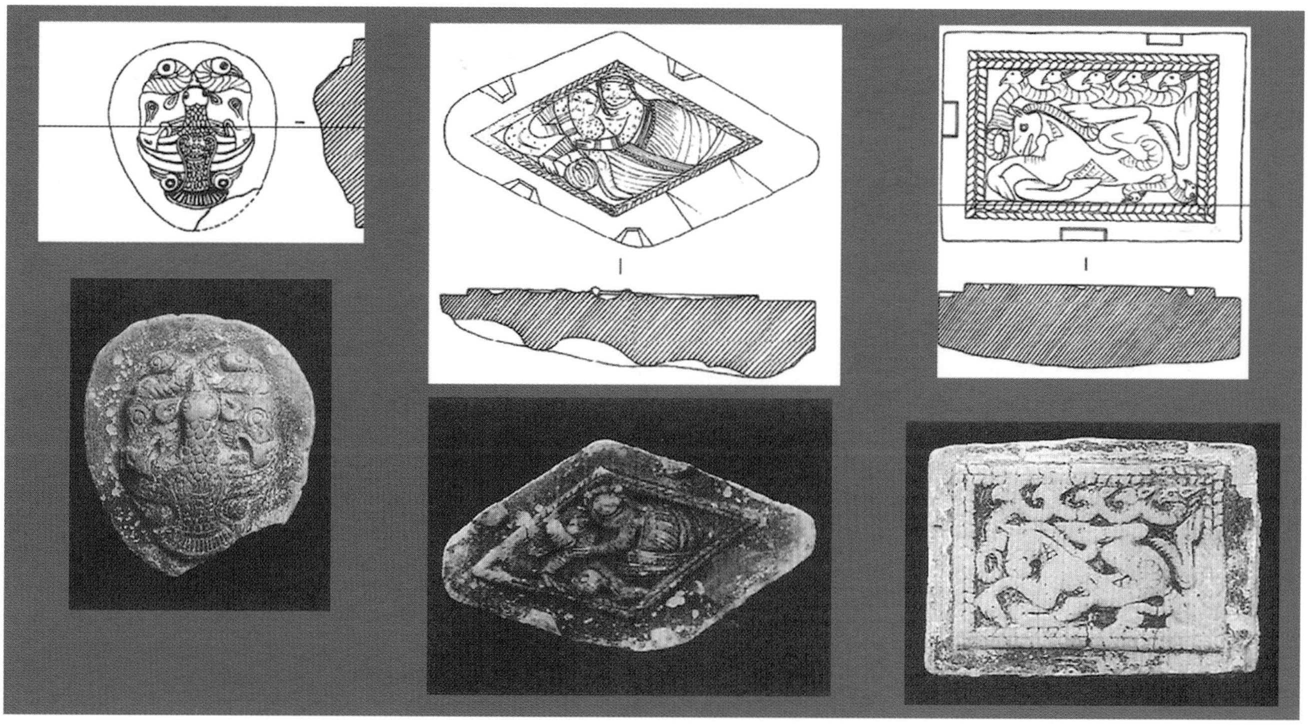

Figure 3 Images of clay casting models excavated from Beikang: scenes with human activity, a hawk and feline, and a horse (after Yue 2003: 6, 7, 8).

body, a deceased male of undetermined age, was placed in a flexed position on his right side in a wooden coffin with his head to the west. Burial goods were quite plentiful and included ceramic models for casting bronze objects, pottery, bronze, iron and stone objects, as well as lacquer ware – a total of 41 objects. Of these, the pottery was placed within

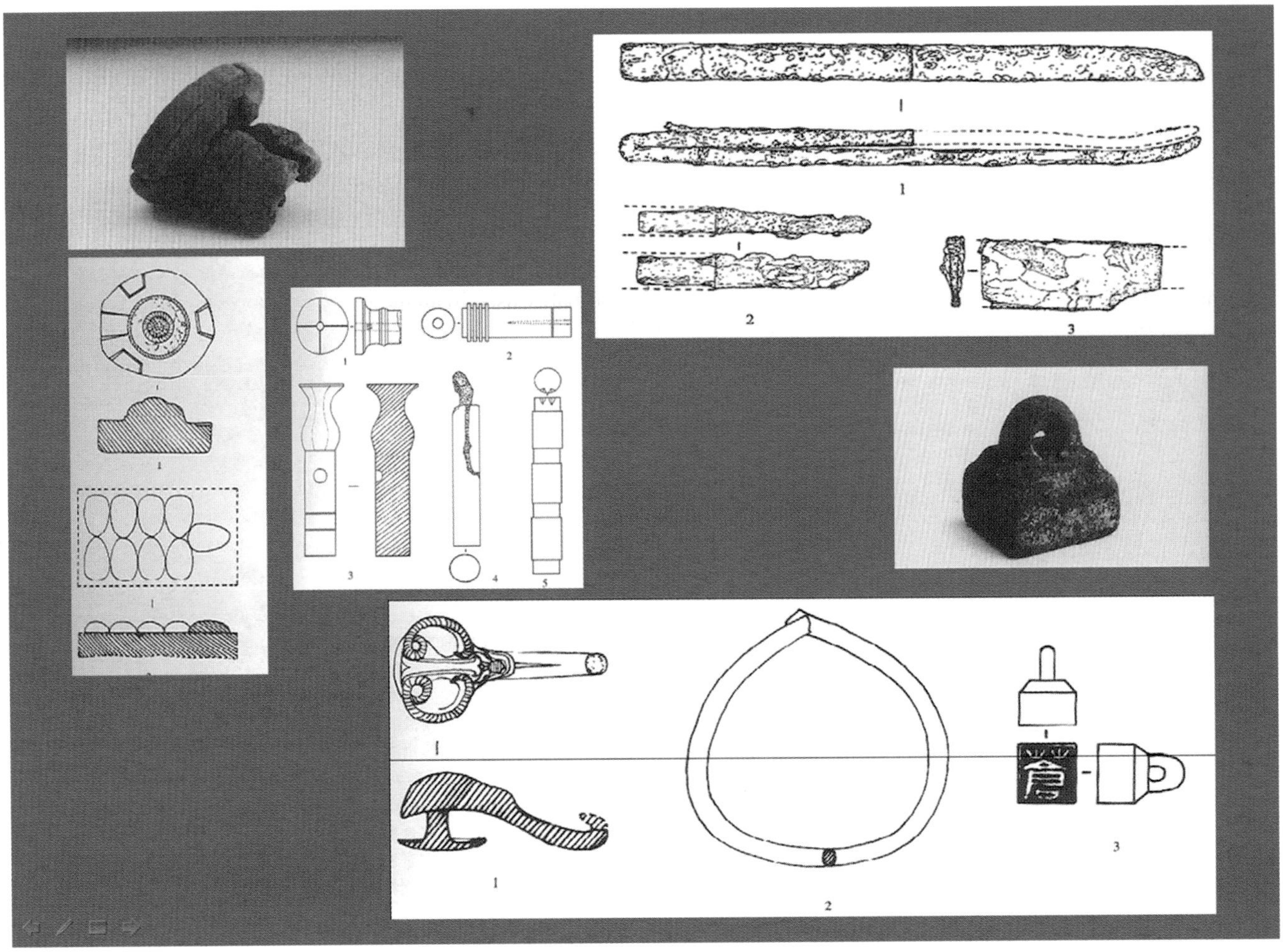

Figure 4 Images of chariot fittings, bosses, the name seal, iron tools from Beikang (after Yue 2003: 11, 13).

the wall niche, whereas the bronze seal and ring, the bronze-fitted lacquer ware and the iron-engraving knife were placed to the west of the head of the deceased inside the coffin. The ceramic casting models, whetstone, and other tools of his trade were placed on either side of the body or at his feet. Several animal bones (badly decayed and unidentified by the excavators) were found at the entrance of the tomb chamber outside of the coffin.

The most extraordinary find in this tomb for our purposes here is the group of the 25 ceramic models made for casting bronze or gold decorative plaques and other types of objects (Fig. 3). They include: a model with human figure designs (1 example); models for plaques with animal decoration (4 examples), including single and paired horses, paired rams, and one of a hawk fighting a feline; models of bubble-like bosses (10 examples); and models of other items including carriage equipment (10 examples), such as yoke bars and finials, a parasol finial, axles; as well as models for miscellaneous items such as a crossbow trigger and bronze vessel components. The six ceramic vessels in the tomb include a *ding*, *he*, *hu*, *guan*, *fou* and a *pen*. The three bronze items include a belt hook with a ram's head terminal, a bronze name seal with the intaglio graph *cang*, and a ring. Iron objects – tongs, a chisel and a knife – were severely corroded, but were made by hammering (wrought) and were likely part of the toolkit of the deceased. A stone whetstone and a lacquer beaker were also included (Fig. 4).

Tombs of this vertical shaft type, in which the pit opening is wider than the base, and a niche is present in the chamber wall, are frequently seen in tombs of the Qin state dating from the late Warring States period (*c*.4th–3rd century BCE) (Teng 1993; Shelach and Pines 2006: 206–16). The flexed position of the body of the deceased was also common at this time in this region as are the types and décor of the pottery vessels. Because the burial goods included more than the usual daily-use items, together with a large number of ceramic models for bronze casting, tools, and a name seal, the excavators proposed that the deceased was probably a bronze caster and one with a particular social status (Yue 2003), a point to which I will return later.

What is unexpected about this find is that this is the first time ceramic models for casting 'steppe-style' bronze plaques have ever been discovered intact in Shaanxi, or within the ancient territory of the state of Qin (Fig. 3). In fact, they are rarely found elsewhere inside ancient Zhou Warring States territory. For instance, at the unequalled excavation of a complete foundry at Houma, near the capital of the neighboring state of Jin in present-day Shanxi and dating from between 585 and 403 BCE, or somewhat earlier than our tomb, a very few models of 'steppe-style' objects were found in the debris (Li and Liang 1996: fig. 764). The overwhelming numbers of models and moulds excavated were made there to mass-produce high quality decorated objects in the 'Chinese' style. And although many have noted that the Houma decorative repertoire, such as the objects showing animal combat or composite animals, were inspired by knowledge of artifacts and/or imagery from as far afield as western Asia (Bagley 1996a; Bunker 1991, 1992; Li and Liang 1996), the resulting designs 'are stripped of any association with real predators and converted into a fantastic frolic' or eccentric composite (Bagley 1996b: 74). Most bear little resemblance to their steppe counterparts, and the new imagery is incorporated into the dominant Chinese aesthetic.

The tomb and its steppe-like contents from Beikang village near Xi'an, on the other hand, provide invaluable concrete data for studying the Qin state bronze culture, distinctive casting techniques, and cultural interaction between the Qin and peoples who must have commissioned and used such items and who probably lived in the Ordos region to the north of the capital of the emerging pre-imperial Qin polity. The designs and their execution are finely drawn and several are unusual, such as the hawk and tiger combat motif, suggesting that the foundry for which the deceased presumably worked was sophisticated, prosperous and produced a line of objects for discerning foreign consumers. These objects and their iconography do not compromise the 'steppe-aesthetic', are not 'sinicized' and remain distinct and, one must assume deliberately, 'foreign' or 'other'. Whether or not the Chinese were willing to sell 'Chinese-style' objects to the 'barbarians', perhaps restricted by the strict sumptuary laws in place from the mid to late Zhou, or if these peoples were unwilling or uninterested in using other artifacts than the 'steppe-style' ones, is not clear. This is an intriguing question to be investigated at another time.

Casting technology

Interestingly, the newly excavated ceramic models were probably used in an indirect lost-wax casting process, a relative latecomer in the bronze industry in China, and one not used as far as we know at Houma during its heyday in the 5th and 4th centuries BCE, where a consistently complex system of piece-mould casting was used (Bagley 1995). Models for the indirect casting process are known to have been made of clay, wood and bone, and all three materials were carved at Houma, but the excavators report only the use of piece-mould casting technology at the Houma foundry (Li and Liang 1996). Clay models such as those from Beikang are known from examples in collections such as the Metropolitan Museum of Art (New York) and the Freer/Sackler (Washington, DC) in the USA, and Emma Bunker described the casting process in 1995 based on the study of those examples (Bunker 1995: figs 62, 62.1). The process is somewhat complex, but made mass production relatively easy since after the model was formed a 'mother-mould' was made by pressing the model into a piece of soft, damp clay. When the clay dried, molten wax was poured into

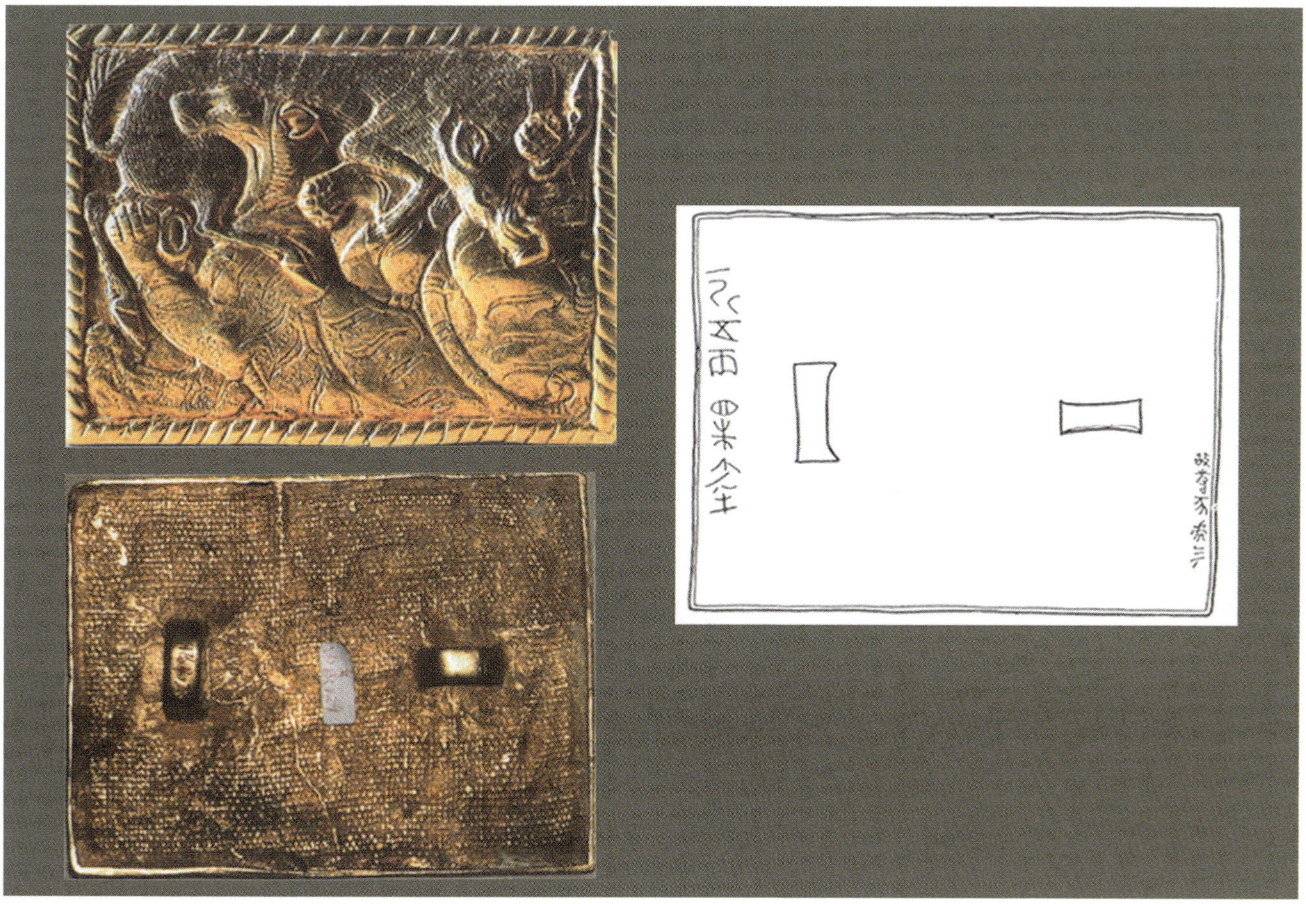

Figure 5 Xigoupan西沟畔 gold plaque front and reverse; Xigoupan西沟畔 reverse (line drawing) with Chinese characters denoting weight and subject matter (after Tian and Guo 1986: 353).

the open section of the mother-mould, and then removed when it hardened. The resultant wax model was packed in clay and baked at a high temperature leaving a clay mould. When the molten metal was poured into the mould, the wax burned out, but of course the clay model still remained and could be reused again and again (So and Bunker 1995: 60–61) (Fig. 5).

Sometimes these wax models were reinforced with a piece of cloth so that plaques could be made very thin, and this method, known as lost wax and lost textile, is a distinctive variation of the indirect lost-wax process and was responsible for numerous belt ornaments (Bunker 1988; 1994: figs 24, 25). Plaques cast in this manner can easily be recognized due to the appearance of woven pattern remains on their reverse. Such examples, found in Xigoupan in the Ordos and dating from the 3rd century BCE, as well as from sites as far away as in Tomb 30 at Xinzhuangtou, Yi Xian, Hebei province, the site of the southern capital of the Yan state in the late 4th or 3rd century BCE, are evidence of the extent of the market and clientele. The presence of Chinese characters on the reverse of another plaque from Xigoupan, also in the Ordos region, that give its weight and subject matter, attest to Chinese manufacture (in this case from the state of Zhao) of luxury items that met the demands of their pastoral owners (Tian and Guo 1986: 353; Li 1985: 1–22; Bunker 1992).

The Chinese favored the process that began with ceramic models, presumably because they had used clay models and moulds in bronze casting since the 2nd millennium BCE. Many 'steppe-style' ornaments cast in foundries beyond China were made by using models differently, such as those discovered in Pokrova in the southern Urals in Russia for instance, dated between the 6th and the 4th century BCE. A bone model of a crouching feline and two very similar gold, feline-shaped plaques found there demonstrate the practice of hammering a thin sheet of metal over a model made from a plentiful material such as bone or horn (Yablonsky 1993: figs 37.16, 76.12-13; bone model from kurgan 2, burial 5; two hammered plaques from kurgan 2, burial 1). It seems, therefore, that this indirect lost-wax and lost-textile process was a Chinese invention that was aimed particularly at the efficient production of objects for the foreign market.

Our bronze caster's tomb, however, contains the first direct evidence of the production of metal products for populations who favored objects that reflected the distinctive styles of the pastoral peoples living beyond the Chinese heartland. The historical circumstances that allowed and perhaps even encouraged such foreign exchange are, in the case of this tomb master, probably explicitly tied to the political and economic aspirations and ambitions of the pre-Empire state of Qin during the late 4th and 3rd century BCE.

Historical circumstances

The Qin state underwent comprehensive changes in structure and in all walks of life during the pre-imperial period, and especially with the creation of a new type of state in the mid-4th century BCE, or at just the same time as our tomb. For instance, before the mid-4th century, tombs of the elite and commoners from the state of Qin show burials of the dead performed in a manner parallel to that of the royal Zhou center (Falkenhausen 2004: 154–7). This was both prescribed by the Zhou ritual system and also a sign of allegiance to, or at least membership of, the larger Zhou political entity. In the mid-Spring and Autumn period, both bronze and ceramic ritual vessels were made according to the Zhou typology, but by the middle of the Warring States period there was a rapid abandonment of the use of those bronze and ceramic forms (Shelach and Pines 2006: 212). At that time, as Shelach and Pines convincingly argue, members of the lower strata of society entered the socio-political arena of the state and changed the 'high culture' of the elite. They argue that under the reign of Lord Xiao (r.361–338 BCE), and the sway of reformer Shang Yang, Qin developed into a highly independent and self-sufficient state, one with aspirations far grander than simply to be a member of the Zhou confederacy. Changes in the social, military and economic structure of the Qin state were emphasized with the purpose of consolidating and strengthening Qin from within. This had consequences for Qin cultural identity. Along with the demise of aristocratic culture and the decline of the Zhou ritual system, as evidenced by the changes in burial patterns, Qin came into its own as an independent unit. Although this situation was not unique to the Qin, it was rapid and sweeping there.

The changes created an 'otherness' for the Qin, which was the outcome of deep cultural and political prowess triggered by Shang Yang's reforms. As one of the Warring States, their power was consolidated as a result of (1) the impact of the demise of the hereditary aristocracy; (2) the emergence of a tightly organized state, which mobilized the populace for economic production and warfare; and (3) the far-reaching impact of mass conscription of men for the military (Shelach and Pines 2006: 220).

Perhaps the most conspicuous evidence of change in the mid-4th century BCE was a reliance on indigenous or local traditions in burial patterns, including material culture and aesthetic norms that guided the construction of tall stamped earth platforms and decorations for public buildings on a monumental scale at the new capital at Xianyang, just south of the river in present-day Xi'an (Shelach and Pines 2006: 214–16). Certainly this new identity was a by-product of political change, but ideology must also have played more than a passive role. The process of building a new and coherent identity was complex, and demanded internal economic integration, as well as encouraging the commercialization of the economy thereby increasing regional independence (Shelach and Pines 2006: 223).

Burial patterns as a barometer of change

Burial patterns are of particular interest here and are worth examining as an example of the Qin integrative process as well as a context for our tomb of the bronze designer. With the rapid abandonment of Zhou-style north–south tomb orientation, funerary artifacts and ritual by the mid-4th century BCE, burials of all social strata in the Qin territory began to display an east–west orientation and a flexed position for the body of the deceased. In addition, the placement of the corpse in a separate chamber, or catacomb, of the tomb, and the inclusion of small pointed stone tablets (*gui*) (Teng 1992) distinguished their burial customs from those of the other Warring States. Both the east–west orientation and the construction of catacomb tombs were not new in the region, and were in fact known from at least as early as the 5th century BCE at such sites as Yanglang and Daodunzi in the 2nd century BCE in present-day Ningxia, and in the 4th and 3rd centuries BCE in Gansu (Pengbao, Shilacun in Guyuan County), or in territory eventually incorporated into the Qin state. In other words, the east–west orientation was a local style of burial. The contents of these tombs mix bronze objects that are linked in style to the Central Plains (*ge* daggers) with those modeled on steppe taste, such as antennae daggers, horse and chariot fittings, as well as sacrifices of horses, bovines, and sheep (Linduff 1997: 33–54; Di Cosmo 1999: 938–40).

The 4th and 3rd centuries BCE were a period of northern expansion for the Zhao, Yan and Qin, whose peoples were mostly settled and who exported their products to a new type of 'foreigner' (*hu*) who according to early Chinese written records introduced the cavalry into China. The well-known and much repeated story of the king of Zhao adopting cavalry in 307 BCE was as much to give him a military advantage over his peer states as to provide effective means of contact with the so-called barbarians, or *hu*. From the large amounts of gold found in the debris of the so-called Xiongnu in the Ordos – for example at Xigoupan where the lost-wax lost-textile method was used – it can be seen how the *hu* leadership engaged in trade as well as profited greatly from commerce with several of the Chinese states, including Qin. Relations between the Ordos region and the states to the south were stable from the mid to the late 4th century BCE, as evidenced by records of material and diplomatic exchange (Di Cosmo 1999: 952); this stability was advantageous for both sides. The closing-in between the northern cultures and the Chinese zone on the one hand accelerated late Warring States' blending of symbols as witnessed in the Houma designs for Chinese consumption, while on the other also included direct importation of 'steppe-style' artifacts by the nomads. The latter were clearly a staple of the foundry business, at least in the northern states of Zhao and Qin. Finds in the Ordos such as those at Xigoupan and Aluchaideng are dated from this later period (late 4th to the 3rd century BCE), when a tendency toward standardization of artifacts such as ring-handled knives and ceramic pots with wavy décor below the neck can be documented. At the same time, there was a decisive increase in burials in the Ordos that included 'steppe-style' belt buckles and plaques, including ones depicting animal and human activities (Bunker 1973) – these subject matters are seen on the models found in the tomb at Beikang.

Luxury articles such as those found at Aluchaideng and Xigoupan presumably express the taste of a new class of aristocracy in nomadic society – no longer did weapons represent them in death, but rather chariots and horses and objects made from precious materials such as gold and silver, implying that pageantry including lavish adornment of the body was preferred. Shows of wealth and power were certainly at the center of this display, but the enhancement of personal possessions also points to a new class of individuals whose position was proportional to their success in managing relations with the Chinese and other neighbors. These relations were not only political and diplomatic, but also highly commercial. Some sections of the pastoral peoples were presumably transformed from a warrior aristocracy into diplomatic and commercial agents who monopolized and/or controlled border exchanges with China to their own profit. Coins were in regular use by this time and were uniform in type throughout the state of Qin, for instance. This manner of exchange was at its highest during the late Warring States period.

Discussion

When in 221 BCE the first emperor of the Qin ordered a campaign against the *hu*, or Xiongnu, in the Ordos, the policy of territorial expansion and resettlement of defeated military and refugees from other regions encroached on the Ordos, ending the period of stability. In this context of political and military emergency in the Ordos, the situation was compounded by an economic crisis among the pastoralists deprived of their pasturelands, and in that atmosphere the unification of the Xiongnu into an empire took place (Di Cosmo 1994). At that point the Chinese were no longer desirable suppliers of goods and artifacts and that chapter in the history of metallurgical technology closed.

But from the late 4th century BCE until the occupation of the Ordos by Qin troops in 221 BCE, as is shown at the tomb at Beikang, even artisans were entitled to burial in the contemporary Qin fashion, and to be buried with items that identified their profession and position in the local state economy. Clearly 'foreign' trade was part of the new push for economic and political independence, but it diminished with the aggressive expansion of Qin troops to the north and the taking over of precious pasturelands on which the economy of that exchange depended. Following that, the foundries produced for local Qin consumption, but bronze was no longer the primary expression for funerary ritual. Our bronze-caster lived during a high point of metal production and exchange with the outside world that would not be equaled again for centuries. And at least in the region of the Ordos, metallurgy was a local affair, but one that engaged the Qin consumers as well as suppliers in exchange with peoples whom they thought defined their cultural as well as political borders.

Acknowledgements

I am grateful to Anthony Barbieri-Low who brought the excavation report of the burial of the bronze-caster to my attention when he was preparing materials for his course on Qin-Han History at the University of Pittsburgh. I am also indebted to the excavator of the tomb, Yue Lianjian of the Shaanxi Institute of Archaeology, who accompanied me to the field station to see the materials first-hand in June 2006. I am especially appreciative to Emma Bunker who has once again saved me from misreading the objects by commenting on a draft of this paper.

References

AISP (Archaeology Institute of Shaanxi Province) 陕西省考古研究所 (2006)西安北郊秦墓(*The Qin Tomb in the Northern Suburb of Xi'an*). Xi'an (西安): Sanqin Press (三秦出版社).

Bagley, R.W. (1995) 'What the bronzes from Hunyuan tell us about the foundry at Houma', *Orientations* 26: 46–54.

Bagley, R.W. (1996a) 'Debris from the Houma foundry', *Orientations* 27: 50–69.

Bagley, R.W. (1996b) 'Introduction', in Li Xiating (李夏廷) and Liang Ziming (梁子明) (eds), *Art of the Houma Foundry*. Princeton, NJ: Princeton University Press.

Bunker, E.C. (1973) 'Anecdotal plaques of the eastern Eurasian region', in P. Denwood (ed.), *Art of the Eurasian Steppeland*, 121–42. London: Percival David Foundation.

Bunker, E.C. (1983) 'Sources of foreign elements in the culture of Eastern Zhou', in G. Kuwayama (ed.), *The Great Bronze Age of China: A Symposium*, 84–93. Los Angeles: Los Angeles County Museum of Art.

Bunker, E.C. (1988) 'Lost wax lost textile', in R. Maddin (ed.) *The Beginning of the Use of Metals and Alloys*, 222–7. Cambridge, MA: MIT Press.

Bunker, E.C. (1991) 'Sino-nomadic art: Eastern Zhou, Qin and Han artifacts made for nomadic taste', in *International Colloquium on Chinese Art History: Proceedings, Part 2*, 269–90. Taipei: National Palace Museum.

Bunker, E.C. (1992) 'Significant changes in iconography and technology among ancient China's northwestern pastoral neighbors form the fourth to the first century BC', *Bulletin of the Asia Institute* 6: 99–115.

Di Cosmo, N. (1994) 'Ancient inner Asian nomads: their economic basis and its significance in Chinese history', *Journal of Asian Studies* 53(4): 1092–126.

Di Cosmo, N. (1999) 'The northern frontier in pre-imperial China', in M. Loewe and E. Shaughnessy (eds), *The Cambridge History of Ancient China: From Origins of Civilization to 221 BC*, 885–966. Cambridge: Cambridge University Press.

Falkenhausen, L. von (2004) 'Mortuary behavior in pre-imperial Qin: a religious interpretation', in J. Lagerway (ed.), *Religion and Chinese Society*. Volume 1: *Ancient and Medieval China*, 109–72. Hong Kong: Chinese University of Hong Kong Press.

Hermann, A. (1966) *An Historical Atlas of China*. Chicago: Aldine Publishing Co.

Li, Xiating and Liang, Ziming (1996) *Art of the Houma Foundry*. Princeton, NJ: Princeton University Press.

Li, Xueqin (1985) *Eastern Chou and Qin Civilizations*, K.C. Chang (trans.). New Haven, CT: Yale University Press.

Linduff, K.M. (1997) 'An archaeological overview', in E.C. Bunker, T. Kawami, K.M. Linduff and Wu En (eds), *Ancient Bronzes of the Eastern Eurasian Steppe from the Arthur M. Sackler Collections*, 18–98. New York: Arthur M. Sackler Foundation, distributed by Harry M. Abrams.

Shelach, G. and Pines, Y. (2006) 'Secondary state formation and the development of local identity: change and continuity in the state of Qin (770–221 BC)', in M.T. Stark (ed.), *Archaeology of Asia*, 202–30. Malden, MA/Oxford/Carlton, Victoria: Blackwell Publishing Ltd.

So, J.F. and Bunker, E.C. (1995) *Traders and Raiders on China's Northern Frontier*. Seattle and Washington, DC: Arthur M. Sackler Gallery, Smithsonian Institution, University of Washington Press.

Teng, Mingyu (滕铭予) (1992) 关中秦墓研究 ('A study of the Qin tombs in the Guanzhong region'), 考古学报 (*Acta Archaeologica Sinica*) 3: 281–300.

Teng, Mingyu (1993) 论关中秦墓中洞室墓的年代 ('On the dating of catacomb burials among the Qin tombs in the Guanzhong region'), 华夏考古 (*Huaxia Archaeology*) 2: 90–97.

Tian, Guangjin (田广金) and Guo, Suxin (郭素新) (1986) 鄂尔多斯式青铜器 (*Ordos-type Bronzes*). Beijing: Wenwu Press.

Yablonsky, L. (1993) *Kurgany Levoberezhnogo Ileka*. Moscow: Institute of Archaeology, Russian Academy of Science.

Yue, Lianjian (岳连建) (2003) 西安北郊战国铸铜工匠墓发掘简报 ('Preliminary excavation report of the Warring States period tomb of a bronze caster in the northern suburbs of Xi'an'), 文物 (*Cultural Relics*) 9: 4–14.

Ancient iron and steel technologies in Asia

An early iron-using centre in the ancient Jin state region (8th–3rd century BC)

Rubin Han and Hongmei Duan

ABSTRACT The earliest finds of man-made iron artefacts in China and cast iron in the world are discussed in this paper. So far, 42 iron artefacts dating to the period before the 5th century BC and more than 730 pieces dating to the period before the 3rd century BC have been excavated. It is worth noting that 90% of the 60 early iron artefacts were made of cast iron. Some smelting sites of the Han dynasty and other remains have also been found in the Jin cultural region. These finds lead us to believe that the southern region of Shanxi province – the ancient Jin state area – is one of the centres for the early use of iron artefacts.

Archaeological materials recovered from the cultural area of the ancient Jin state have now provided an opportunity for the systematic study of the use of iron artefacts, as well as the technology of iron- and steel-making, during the period between the 5th and the 3rd century BC. An investigation of the history of iron-steel technology in China has shed light on the importance of iron- and steel-making in the development of economic (such as cultivation in irrigated fields) and socio-political aspects in ancient China. The technology was one of the significant factors that eventually led to the emergence of a unified China. It was also of assistance in sustaining the continuity of Chinese civilisation.

Introduction

The Jin state of China was one of the prominent polities of the Eastern Zhou era, which consisted of two phases: the Spring and Autumn period (770-476 BC) and the Warring States period (475–221 BC). It existed during a unique period of time in which apparent social and economic changes progressed rapidly. During the Spring and Autumn period, the Jin state was headed by Duke Wen (晋文公), who was regarded as the second figure among the five most powerful overlords. As documented in Zuozhuan (左传), the capital of the Jin state had been established at different locales over time: it was once in an area near present-day Quwo (曲沃) and subsequently in the Xintian (新田) area of Houma (侯马), both situated in present-day Shanxi province (Fig. 1).

The state of Jin was established in the 11th century BC when the younger brother of King Cheng of Western Zhou (周成王), known as Yu (虞), was granted the title of duke, and appointed to govern the Tang region – present-day Taiyuan (太原) in Shanxi (山西) province. In 715 BC, Duke Wu (晋武公) of the Jin state conquered 17 small states, including the Guo state (虢国). In 632 BC, Duke Wen (晋文公) became the head of the Jin state. After expanding the state's territory, he was recognised as the second among the most powerful overlords of the allied groups (Fig. 2). This was concurrent with a time when the existing and newly established states were setting up new administrative rules and tax laws to bring about political and economic

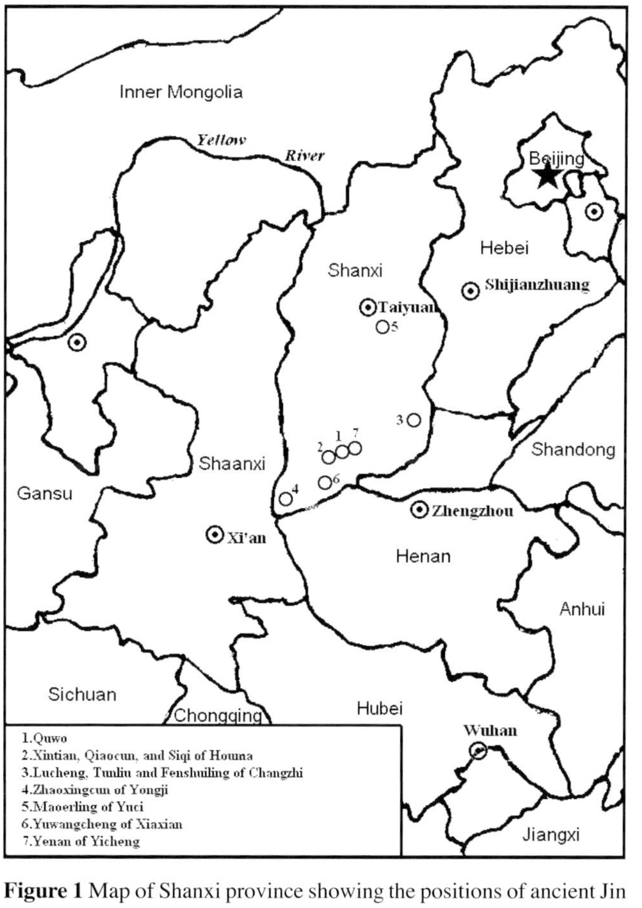

Figure 1 Map of Shanxi province showing the positions of ancient Jin capitals in different times.

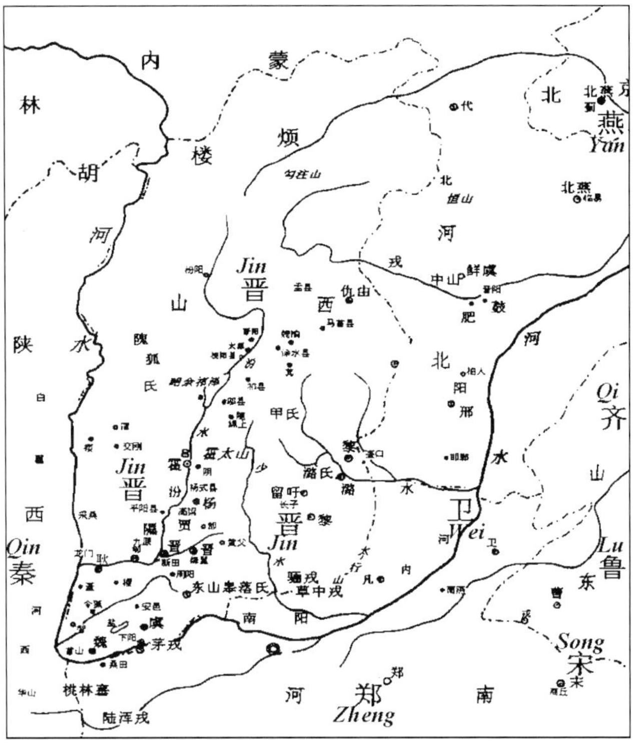

Figure 2 Map showing the territory of the Jin state in the Spring and Autumn period.

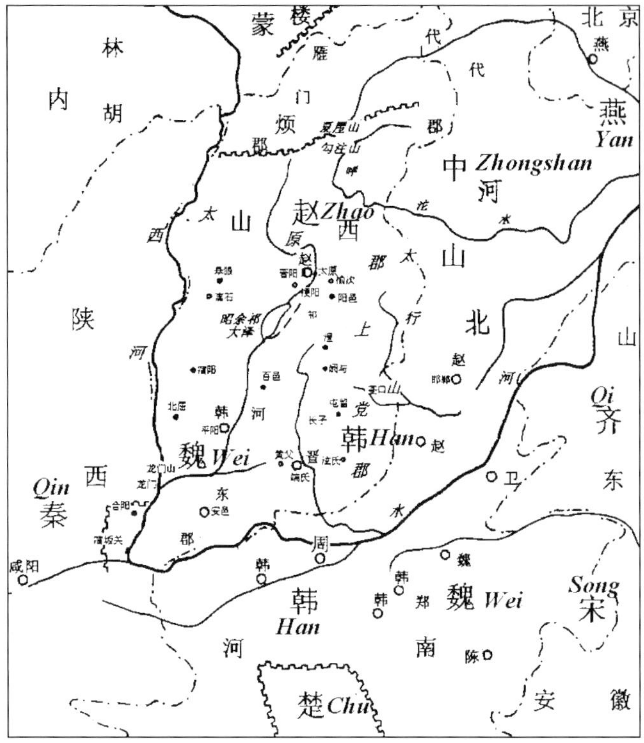

Figure 3 Map showing the territories of the Han, Zhao and Wei states in the Warring States period.

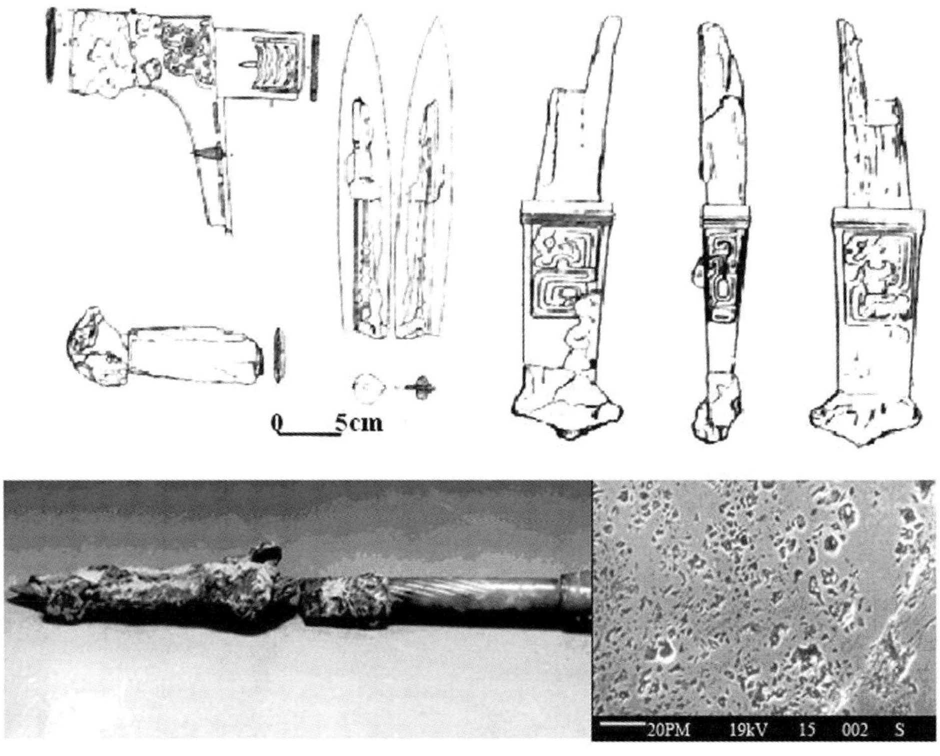

Figure 4 Top: illustrations of six iron objects with bronze handles excavated from tombs M2001 and M2009 in Guo state. Below: iron sword with a jade handle excavated from the M2001 tomb in Guo state and the microstructure of the body sample of the iron sword.

reforms. Such practices were recorded in a text on criminal law, which was said to have been cast onto the surface of an iron tripod (铸刑鼎).

The political power of the duke of Jin, however, waned during the 5th century BC and eventually led to the downfall of the Jin states in 405 BC. Soon the Jin territory was divided among the Han (韩), Zhao (赵), and Wei (魏) ministries, all of whose rulers eventually succeeded in acquiring the title of duke as well (Gu and Zhu 2001) (Fig. 3). According to Zou Heng's study, the Jin state lasted for a total of about 370 years and its early capital was possibly built in the Qucun (曲村) region of Tianma (天马) (Zou 1994).

The Jin state (in present-day southern and south-eastern Shanxi province) was situated in a strategic area that covered a large territory, having both a favourable climate for cultivation and rich mineral resources. Archaeological discoveries in the region revealed different types of ancient stone tools (adzes, axes, and millstones), some of which date back to 16,000 BC, suggesting that this area had been a popular habitat in ancient antiquity. Excavated stone agricultural implements from the Taosi site (陶寺遗址) of Xiangfen (襄汾) and the Dongxiafeng site (东下冯遗址) of Xia county (夏县) also attest to the fact that Neolithic inhabitants continued to populate the southern and south-eastern parts of Shanxi province. It appears that the ecosystem and natural resources in the area contributed to the political pre-eminence of the Jin state.

Archaeological investigation in southern Shanxi, part of the ancient Jin state, has now produced evidence of the earliest iron artefacts known in the Central Plains of China. The items include three iron artefacts made with meteoritic iron and three iron weapons fitted with bronze handles. All of them, dating to the end of the 9th century BC, had been buried in tombs at the Sanmenxia (三门峡) site (Fig. 4). These graves are thought to have belonged to the Guo state, which was annexed by the Jin state during the 8th century BC.

Among these artefacts are two *ge* daggers, uncovered from tombs M2001 and M2009, respectively. Both of them, showing similar bronze handles and turquoise inlays, are believed to have been personal belongings of the nobles of the Guo state. Test results reveal that two kinds of iron were used to make their blades: meteorite and man-made iron. These data set the earliest date for the use of man-made iron in China to the 9th–8th century BC (Han *et al.* 1999). Further studies reveal that the incipient stage of iron smelting in China was the use of a bloomery process, which was similar to the method used in other parts of the world at the beginning of the 1st millennium BC. The same production method continued to be used for steel-making until the 1st century BC. Unfortunately, the physical remains of the bloomery furnaces in China have yet to be discovered.

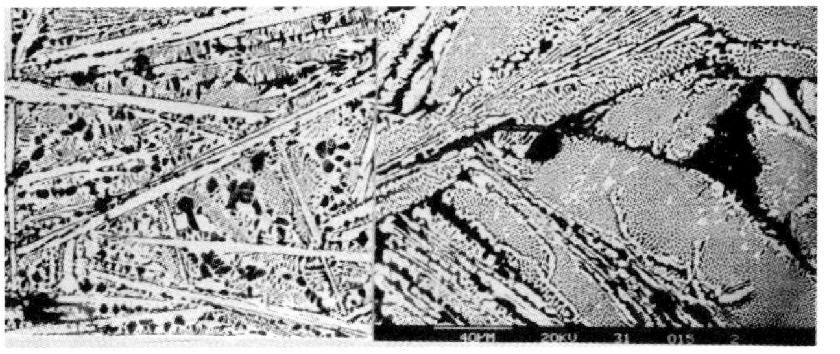

Figure 5 Micrograph of two fragments of cast iron from the site at Tianma-Qucun in Shanxi province.

Shortly after bloomery iron came into existence, the casting of iron was invented. Two fragments of cast iron dating to the 8th century BC have been reported at the Tianma site in the Qucun area of Shanxi province (Han 1998), although the exact circumstances associated with their discovery is not clear. They may very well be the earliest examples of cast iron pieces in the world. Figure 5 shows the microstructure of cast iron fragments from the Tianma site, Qucun.

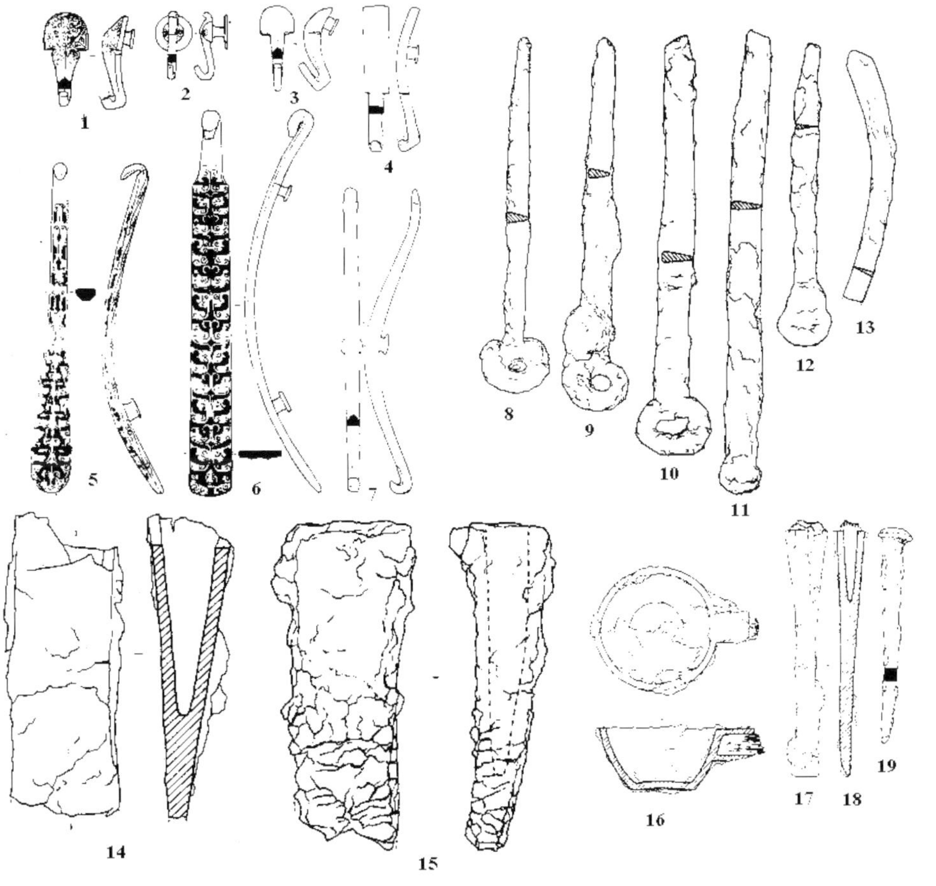

Figure 6 Bronze belt hooks and ironwares excavated from the Qiaocun cemetery.

Additional archaeological finds of two kinds of man-made iron artefacts have also been reported at an Eastern Zhou tomb in Liuhe (六合) city of Jiangsu (江苏) province. These examples date to the 8th–7th century BC (Beijing University of Iron and Steel Technology, Archaeometallurgy Group 1978; Qiu 1985). There are also more than 150 pieces of smelted iron artefacts obtained from sites dating no later than the 5th century BC, the majority excavated from sites in southern and south-eastern Shanxi province, which for some time was part of the Jin state. The iron objects excavated from 14 provinces and produced earlier than the 5th century BC had tallied, and so had the iron objects excavated from 20 provinces dating between the 5th and the 3rd century BC. The iron production of the later period appears to be considerably more significant than that of the earlier period (Fig. 6) (Duan 2001).

These data imply that by the 5th century BC, the availability of cast iron had altered several aspects of Chinese societies. The study indicates that iron was circulating widely due to the invention of a malleabilising process that turns white iron or pig iron into malleable iron, wrought iron or steel. New technical features had evidently enhanced the productivity and expanded the supply of metal, while minimising its production cost. These included casting iron artefacts with liquid metal and incorporating an annealing process, along with the making of wrought iron or steel from either pig or cast iron. The accessibility of iron farming implements in the Warring States period can be seen in Table 1, which lists iron farming implements excavated from graves in six provinces. The data also find corroboration in the textual materials of Mencius (孟子, 390–305 BC), who asked his pupil Chenxiang (陈相) 'Is Xuzi (许子) farming with iron?'. This passage also serves as the earliest reliable literary source regarding the use of iron in agriculture (Mencius).

The study of the Jin cultural areas and the findings of new archaeological materials provide the possibility of systematically studying the use of iron artefacts and the advancement of iron and steel technology from the 5th to the 3rd century BC. Archaeologists from the Archaeological Institute of Shanxi cooperated in the investigation of 10 graves and sites dating to the period before the 3rd century BC. The position of sites such as Qiaocun (乔村) and Siqi (虒祁), the graves of Houma (侯马), Lucheng (潞城), Tunliu (屯留), Fenshuiling (分水岭), Changzhi (长治), the Zhaoxingcun tombs (赵杏村) of Yongji (永济), Maoerling (猫儿岭) of Yuci (榆次), two iron-smelting sites of Yuwangcheng (禹王城) and Xiaxian (夏县) and other iron-smelting sites of Yenan (冶南) and Yicheng (翼城), are shown in Figure 1. A total of 751 iron artefacts were excavated from more than 1,000 graves of common people in the Jin cultural region. The ironware included ploughshares, shovels, adzes, pick-axes, sickles, weapons, casting belt hooks with inlaid silver and gold, and so on. About 60% of the 751 artefacts unearthed was iron belt hooks (448); four kinds of bronze belt hooks and iron parallels have been found, examples of which are shown in Figure 6. In the reinvestigation of these graves and sites of the Warring States period in Shanxi province, 130 samples of different kinds of iron artefacts were examined by metallographic analysis. The samples chosen for analysis are shown in Table 2. Due to the fact that many iron samples had been badly corroded and could not be analysed, the metallographic results of 60 iron samples are shown in Table 3. All of the samples were etched with 4% nital.

The results of metallographic analysis show that 90% of 60 iron samples were cast from the liquid state and 22 (about 41%) were white iron softened by annealing. The studies of farming implements are shown in Table 4. Among 19 samples of cast iron, 9 implements (about 47%) were made by iron casting and annealing. Some of the microstructures are shown in Figures 7–9.

Two methods of steel weapon production were used in the Jin cultural region: decarburisation of cast iron in the solid state, used either in the cast shape or forged, and the bloomery iron method (shown in Table 3). The products

Table 1 Iron farming tools excavated from graves in six provinces.

Location	No. excavated	No. of tools	Percentage of tools	No. of farming tools	Percentage of farming tools	Reference
Lianhuabao, Fushun, Liaoning province	80+	77	96.2	68	88.3	Wang Peixin (王培新) 1964
Fenshuiling, Changzhi, Shanxi province	36	31	86.1	21	67.7	Shanxi Provincial Cultural Relics Office (山西省文物管理委员会) 1957; Shanxi Provincial Cultural Relics Office (山西省文物管理委员会) 1964; Bian Xiucheng (边修成) 1972
Gudonggou, Xinglong, Hebei province	87	85	97.7	52	61.2	Zheng Chaozong (郑绍宗) 1956
Guweicun, Huixian, Henan province	93	*c.* 69	74	58	84.1	Huixian 1956
61 Chu graves, Changsha, Hengyang, Hunan province	70	21	30	17	80.9	Hunan Provincial Cultural Relics Work Line (湖南省文物工作队) 1956
Yinshanling, Pingle, Guangxi province	184	170	93	91	53.5	Guangxi Zhuang Autonomous Region Cultural Relics Work Line (广西壮族自治区文物工作队) 1978

Table 2 Specimens (5th–3rd century BC).

	Houma		Changzhi (Han)			Maoerling of Yuci (Zhao)	Zhaoxingcun of Yongji (Wei)
	Qiaocun graves	Siqi graves	Tunliu	Zhangzi	Lucheng		
Tool	adze (3), shovel (4), sickle (3)	ploughshare (1), shovel (1), chisel (1), adze (1)		shovel (2)		axe (1), adze (3)	adze (1)
Weapon	knife (4), sword (2)	knife (7), arrowhead with iron handle (2)		knife (5), sword (2), dagger (1), arrowhead (1)		sword (7), knife (2)	
Ornament	belt hook (6)	belt hook (33)		belt hook (17)		belt hook (7)	
Vessel		container (1), jar (1)					container (1)
Others	collar (1), iron piece (1)	wares (4)		iron ring (1), nail (1), ironware (1)		small knife (?) (1)	
Total	76			31		21	2

Table 3 Summary of objects and their material types.

Microstructure	Number and variety of iron samples excavated in several graves and sites				
	Houma	Changzhi	Yuci	Yongji	Total
White iron	adze (3), shovel (2), belt hook (9), jar (1), ironware (1), container (1)	belt hook (11)	pickaxe (1), adze (2)	container (1)	32
Decarburised casting iron	shovel (1), ploughshare (1), knife with round hoop (1), belt hook (1)	shovel (2), belt hook (3)		adze (1)	10
Solid state decarburised steel	adze (1), belt hook (1)	knife with round hoop (3), sword (2), dagger (1), iron ring (1)	sword (2), knife with round hoop (1)		12
Bloomery iron or its steel		knife (1), nail (1)			2
Undetermined	belt hook (1)	belt hook (1), ironware (quenched) (1)	belt hook (1)		4
Total	24	27	7	2	60

Table 4 Manufacturing techniques of objects excavated from the Jin cultural area.

Location	No. excavated	No. sampled	Manufacturing technique			
			White iron	Decarburised casting iron	Solid state decarburised steel	Undetermined
Houma	31	14	adze (3), shovel (4)	shovel (1), adze (1), ploughshare (1)	sickle (1)	3
Changzhi	35	9	pickaxe (1), adze (2)	pickaxe (1), scraper (1), shovel (2)		2
Yuci	5	1				1
Yongji	2	1		adze (1)		–
Total	73	25	10	8	1	6

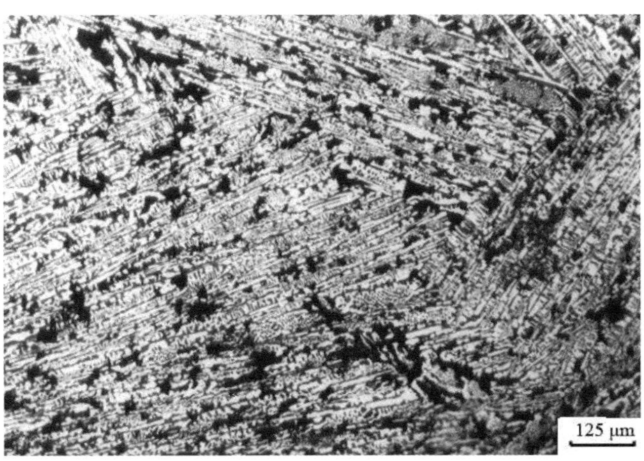

Figure 7 Microstructure of the adze from M5150 of the Qiaocun cemetery, Houma: ledeburite and pearlite.

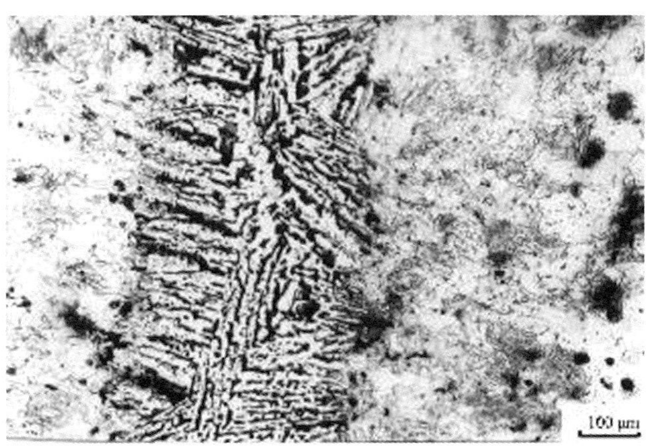

Figure 8 Microstructure of the shovel from M636 of the Qiaocun cemetery, Houma. Inner layer: ledeburite; outer layer: ferrite and pearlite.

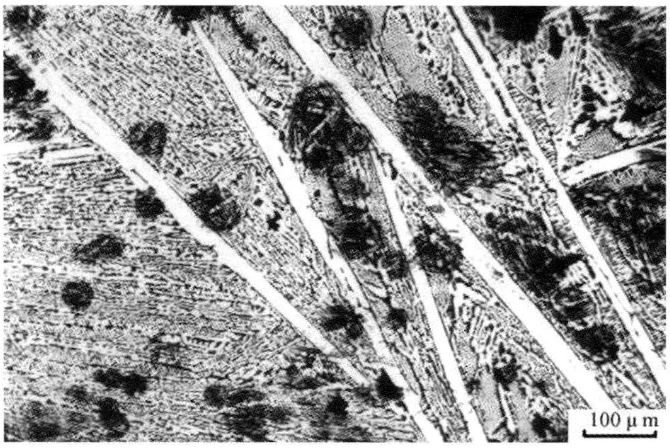

Figure 9 Microstructure of the ploughshare from M624 of the Qiaocun cemetery, Houma: proeutectic cementite and nodular graphite in ledeburite.

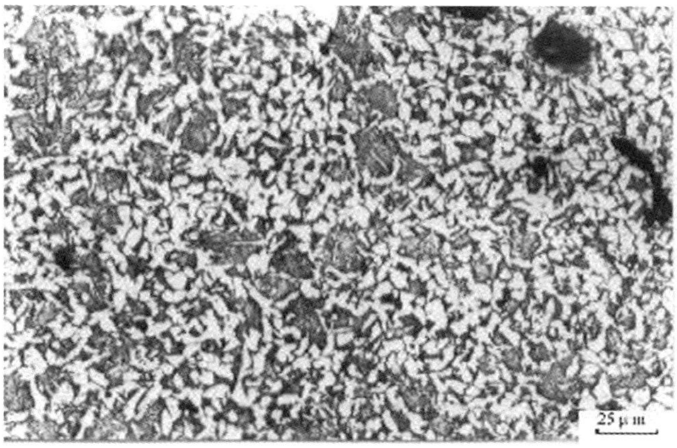

Figure 10 Microstructure of the knife with round hoop from M2163 of the Siqi site, Houma: pearlite and ferrite (decarburised steel).

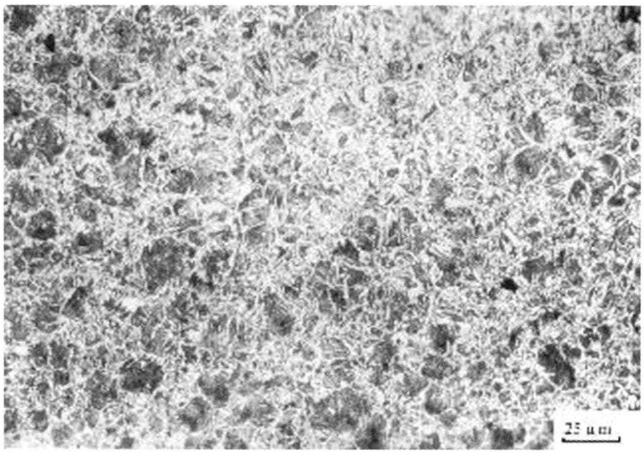

Figure 11 Microstructure of the knife with round hoop from M6 of the Tunliu site, Changzhi: pearlite and ferrite (decarburised steel).

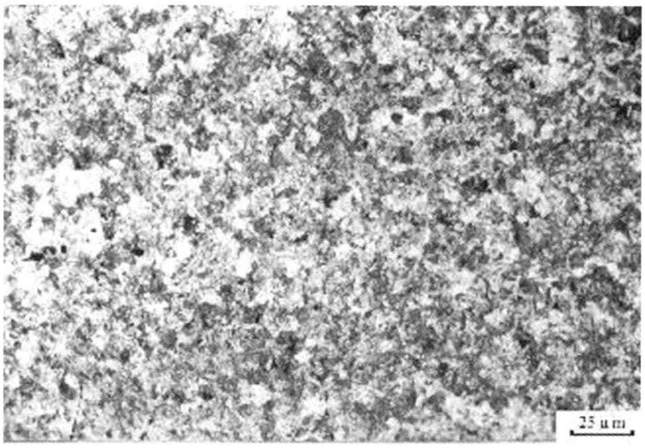

Figure 12 Microstructure of the knife from M43 of the Tunliu site, Changzhi: quenched martensite.

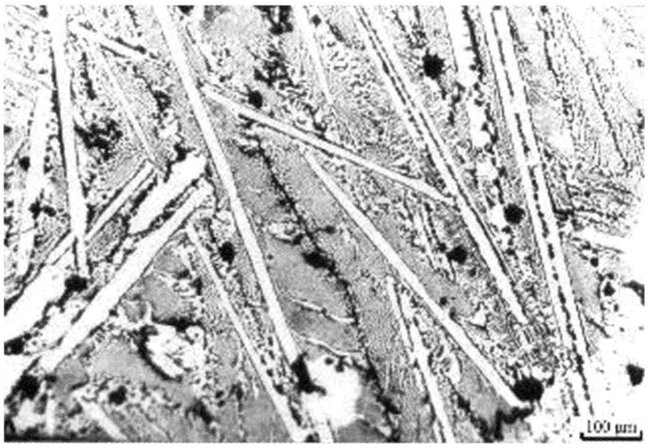

Figure 13 Microstructure of the belt hook from M778 of the Qiaocun cemetery, Houma: proeutectic cementite and pearlite in ledeburite.

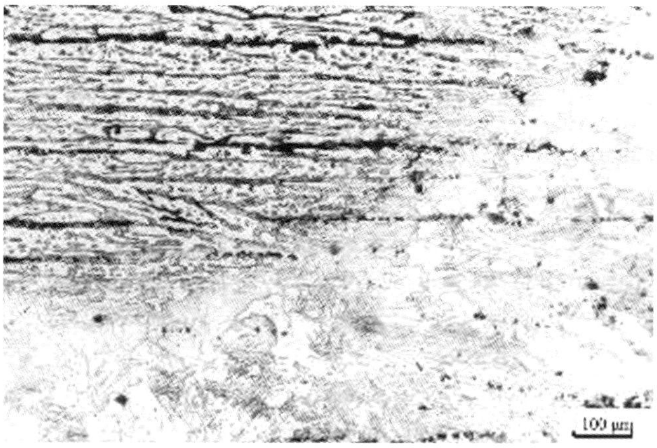

Figure 14 Microstructure of the belt hook from M2298 of the Siqi site, Houma. Upper left: ledeburite; right: ferrite and pearlite.

of both these processes can be distinguished by the composition, constituents and distribution of the inclusions. The micrographs of the samples are shown in Figures 10 and 11. The microstructure of a knife (M43) from Tunliu (Changzhi site) is shown to be quench-hardened, as seen in Figure 12.

In Table 3 the analysis of 28 irons belt hooks shows that 25 are made of cast iron (89%); among them, five objects have additionally been annealed (Figs 13 and 14). A total of 70 samples were found to be heavily corroded upon close examination, but some traces of the original microstructure still remained.

Discussion

Many types of iron farming implements excavated from the Jin cultural region were made by casting from a liquid state and showed the use of an annealing technique for improving the brittleness of pig iron. Cast iron implements with certain toughness were found in large quantity. This technology enabled the production of large ploughshares for deep furrowing, and sharp and strong but inexpensive sickles for harvesting the crops, the yields of which were at that time increasing. The results of the study of recent archaeological artefacts by metallography reinforced the above hypothesis. This represented an important factor for advancing the productive forces of an agricultural economy and developing the society of the Jin state.

The results are seen in weapons excavated from these regions, examined by metallographic analysis and shown to have been made by decarburisation in the solid state. Their high quality steel exceeds that of M44 weapons excavated from Yixian (易县) of Yan (燕) state, Hebei (河北) province. At the same time, most samples of weapons from Yixian show that they were made from bloomery iron and carburised steel (for the results of their analysis see Beijing University of Iron and Steel Technology, Metal Working Laboratory 1974). The high quality iron weapons gradually replaced bronze ones. The strengthening of the national economy and military power was a significant factor in the formation of the Jin state and of the independent Han, Zhao and Wei states.

Another feature is the absolute number of belt hooks found in these regions, which account for 65% of iron findings excavated from ordinary graves of Qiaocun, Houma. Examination of the inlaying technology of iron belt hooks with gold, silver or gold-silver alloys revealed that four different styles were used. Additionally, 200 bronze objects were recovered, mostly belt hooks, an aid in determining the position and status of their buried masters.

The iron mines of the southern Shanxi province are rich in natural resources. There are at least 12 iron-mining areas and the magnetite ore has a high grade of iron. Many iron-smelting remains, including iron slags, furnace walls, ironware, salamander, iron ore, clay moulds, etc. have been collected and studied. Smelting sites of the Han dynasty in Yuwangcheng (禹王城) covered an area of 7,000,000 m^2 and were divided up into areas: some produced iron castings, some produced tools or coins, and others produced vessels. Numerous baked clay stack-moulds have been found, some containing the handwritten word *Dongshan* (东三), i.e. the third iron and steel factory of the Eastern River District, and the inscription Iron Office (铁官).

Many huge iron castings after the 7th century AD, such as Buddhist figures, warriors, lions, oxen and so on, are still standing tall and upright in many places in China, especially in Shanxi province. For example, the earliest iron Buddha head (铁佛头) was made of cast iron in AD 632 at Linfen (临汾), Shanxi province. It is 6 m high, 5 m wide and weighs about 10 tons, as seen in Figure 15. The hollow head of the iron Buddha was used for storing the sutras (Duan 2001). This era also witnessed the first use of cast iron for building a bridge across the Yellow River – the excavation site of large artefacts from the Tang (唐代) dynasty at Yongji (永济), Shanxi province. The iron sculptures on each side of the river bank were used to support the bridges. Six iron pillars under each iron ox, 40 cm in diameter and 3.6 m long, were used as anchors for the floating bridge, as shown in Figure 16 (Liu 2005).

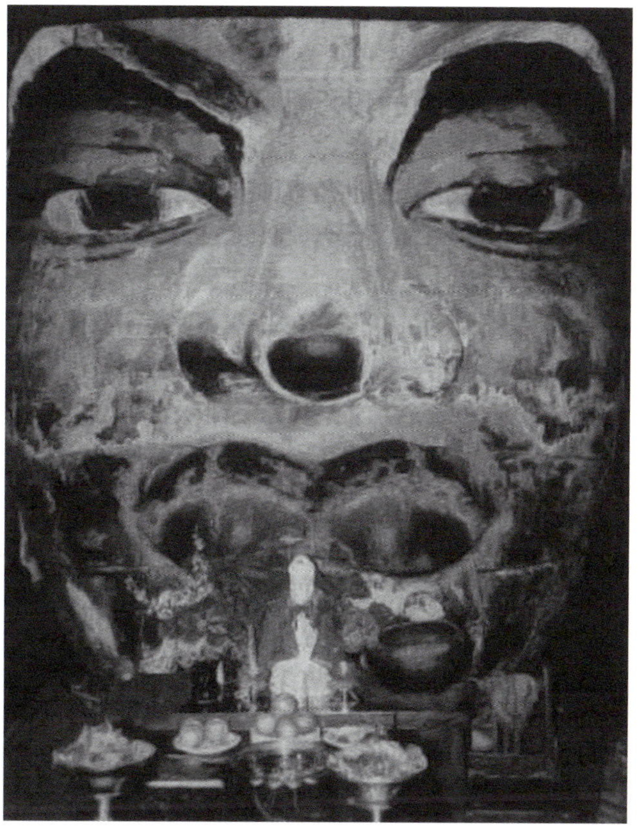

Figure 15 The earliest Buddha head of cast iron (*c.*AD 632) in Linfen, Shanxi province.

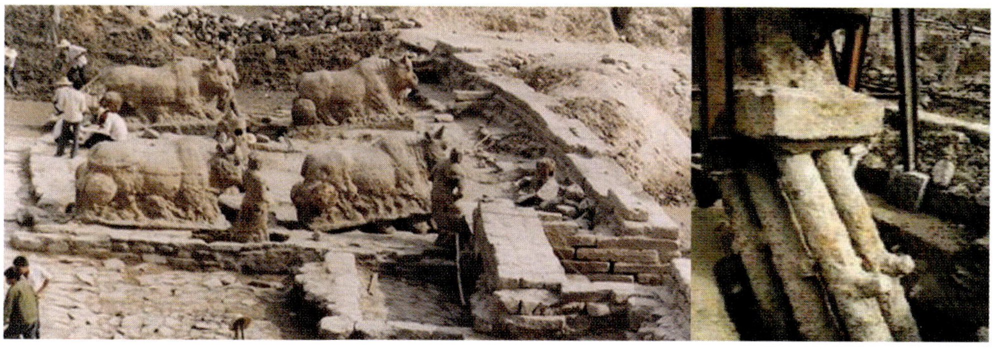

Figure 16 The group of iron castings (oxen and pillars) for the bridge across the Yellow River at the Pujindu ferry site, Yongji city, Shanxi province.

It is reported that Shanxi has the greatest number of large sculptures in China. This is further evidence that the ancient Jin cultural region of the Shanxi province played an important role in the use and development of cast iron- and steel-making in ancient China.

Acknowledgements

We wish to thank Professors Zhang Qingjie, Wu Zhenlu, Xie Yaoting, Tao Zhengang and Fan Wenqian of the Shanxi Institute of Archaeology, for cooperating in the study of iron objects excavated from the Jin cultural region, and for providing us with useful and important background information. Grateful thanks are also extended to Professors Chiou-Peng TzeHuey, Mei Jianjun and Chen Jianli, as well as our students, Lin Yixian and Jiang Xi, for their support and help.

References

Beijing University of Iron and Steel Technology, Archaeometallurgy Group (北京钢铁学院中国冶金简史编写组) (ed.) (1978) 中国冶金简史 (*A Concise History of Metallurgy in China*). Beijing: Science Press, p. 55.

Beijing University of Iron and Steel Technology, Metal Working Laboratory (北京钢铁学院压力加工实验室) (1974) 易县燕下都44号墓铁器金相考察初步报告 ('A metallographic study of iron artefacts unearthed at no. 44 tomb at Yixian county, in Hebei'), 考古 (*Archaeology*) 4: 241–3.

Bian Xiucheng (边修成) (1972) 山西长治分水岭126号墓发掘简报 ('Excavation of Warring States tomb no. 126 at Fenshuiling in Changzhi city, Shanxi'), ,文物 (*Cultural Relics*) 4: 38.

Duan, Hongmei (段红梅) (2001) 山西侯马地区出土战国铁器的金相学研究 (*Metallographic Study of the Iron Artefacts of the Warring States Period Unearthed in the Houma Region, Shanxi*) (unpublished doctoral dissertation, University of Science and Technology Beijing).

Gu, Derong (顾德荣) and Zhu, Shunlu (朱顺禄) (2001) 春秋史 (*A History of the Spring and Autumn Period*). Shanghai: People Publishing House, pp. 27, 86, 356–8.

Guangxi Zhuang Autonomous Region Cultural Relics Work Line (广西壮族自治区文物工作队) (1978) 平乐银山岭战国墓 ('The Warring States tombs at Yinshanling county in Pingle city'), 考古学报 (*Acta archaeologia sinica*) 2: 211–50.

Han, Rubin (韩汝玢) (1998) 中国早期铁器的金相学研究 ('A metallographic study on early ironware of China'), 文物 (*Cultural Relics*) 2: 87–96.

Han, Rubin (韩汝玢), Jiang, Tao (姜涛) and Wang, Baolin (王保林) (1999) 虢国墓出土铁刃铜器的鉴定与研究 ('Examination of copper artefacts with iron edges excavated from the cemetery of the Guo state'), in Henan Institute of Cultural Relics and Archaeological Research *et al*. (河南省文物考古研究所等) (eds), 三门峡虢国墓地 (*The Cemetery of the Guo State in Sanmenxia*). Beijing: Cultural Relics Press, pp. 559–73.

Huixian (1956) 辉县发掘报告 (*Excavations in Huixian County, Henan*). 北京：科学出版社 (*Science Publishing House*), p. 108.

Hunan Provincial Cultural Relics Work Line (湖南省文物工作队) (1956) 长沙、衡阳出土战国时代的铁器 ('Excavation of iron artifacts of Warring States period at Changsha city and Hengyang city, Hunan'), 考古通讯 (*Archaeological* Bulletin) 1: 79.

Li, Yuanqing (李元庆) (1997) 三晋古文化源流 (*Origins of Ancient Cultures in the Three Jin Regions*). Taiyuan: Shanxi Guji Chubanshe, pp. 1–2.

Liu, Yongsheng (刘永生) (2005) 山西考古研究所汇报资料 (*Research Reports presented to the Shanxi Institute for Archaeological Research*).

Qiu, Lianghui (丘亮辉) (1985) 河南汉代铁器的金相普查 ('A metallographic survey of Han-period iron artefacts in Henan province'), in 科技史文集 (*Studies in the History of Science and Technology*). Shanghai: Shanghai Science and Technology Press, p. 107.

Shanxi Provincial Cultural Relics Office (山西省文物管理委员会) (1957) 山西长治市分水岭古墓的清理 ('Cleaning up of Warring States tombs at Fenshuiling in Changzhi city, Shanxi'), 考古学报 (*Acta Archaeologica Sinica*) 1: 118.

Shanxi Provincial Cultural Relics Office（山西省文物管理委员会）（1964）山西长治分水岭战国墓第二次发掘 ('The second excavation of Warring States tombs at Fenshuiling in Changzhi city, Shanxi'), 考古 (*Archaeology*) 2: 132, 133, 137.

Wang, Peixin (王培新) (1964) 辽宁抚顺市莲花堡遗址发掘简报 ('Excavation of sites at Lianhuaao in Fushun city, Liaoning'), 考古 (*Archaeology*) 6: 287.

Zheng, Chaozong (郑绍宗) (1956) 河北兴隆县发现战国生产工具铸范 ('Mould casting production implements discovered at a Warring States period site at Gudonggou in Xinglong county, Hebei'), 考古通讯 (*Archaeological* Bulletin) 1: 29.

Zou, Heng (邹衡) (1994) 论早期晋都 ('A view of the early capitals of Jin state'), 文物 (*Cultural Relics*) 1: 29–32.

From western Asia to the Tianshan Mountains: on the early iron artefacts found in Xinjiang

Wu Guo

ABSTRACT This paper discusses the dates and origins of early iron artefacts found in Xinjiang, and proposes some hypotheses regarding the origin of iron metallurgy in ancient China. Having examined early iron finds excavated to this date in Xinjiang, it is argued that iron came into use in Xinjiang during the 9th–8th century BC, a date later than the first appearance of iron in western Asia. Several pieces of archaeological evidence are presented to demonstrate that some cultural connections existed between Xinjiang and western Asia during the beginning of the 1st millennium BC. Considering that iron appeared in western Iran around the 10th–8th century BC, it is then suggested that the appearance of iron in Xinjiang around the 9th–8th century BC could have resulted from cultural interaction between Xinjiang and western Asia. As for the origin of iron in central China, it is suggested that iron might have been introduced into central China during the 9th century BC from Xinjiang and other frontier regions by the steppe people, although no decisive archaeological evidence supporting this hypothesis has been found.

Introduction

What is the origin of the early iron artefacts discovered in ancient cemeteries in the Xinjiang region, such as those at Yanbulaq, Yanghai, Charwighul, Chong Bagh, and so on? Was the technique of iron-making invented in ancient China itself or was it transmitted from the outside world through Xinjiang? Some scholars insist that the technology is an indigenous Chinese invention (Kong and Feng 1988; Zhu 2001: 53; Bai 2003: 308), while the majority suggest that it was introduced from the West (Chen 1989; Tang 1993: 563–4; Zhao 1996: 289; An 1996; Mei 2000: 74–5; Wagner 2003; Liu 2004). In this paper, this question is revisited, focusing on the date of the iron items found in Xinjiang and the background of cultural exchanges between Xinjiang and western Asia, or the region of the Pontic steppes, in the early part of the 1st millennium BC. Some hypotheses are put forward regarding the origin of iron metallurgy in ancient China.

Date of the early iron artefacts unearthed in Xinjiang

Originally, scholars discussed this question based on the finds at the Yanbulaq cemetery in Hami county (XIA 1989a). There are two tombs in the cemetery (M75 and M31) from which some iron artefacts were unearthed (Fig. 1). Based on ^{14}C calibrated dates, tomb M75 dates from around 1000 BC, or even earlier. The other tomb, M31, dates from around 1312–1127 BC (3240±135, calibrated). However, after analysis of the features of the structure of

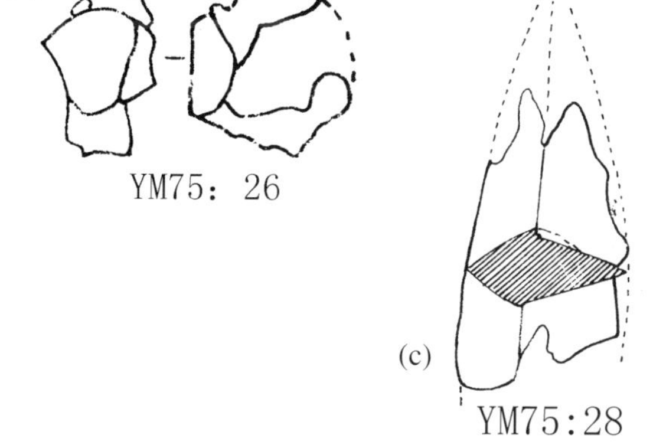

Figure 1 Iron artefacts excavated from Yanbulaq cemetery: (a) iron knife from tomb 31; (b) iron fragment from tomb 75; (c) damaged iron point of a sword from tomb 75 (after XIA 1989a: 347–9, fig. 26).

the tombs, Li Wenying proposed that the tombs belonged to a later period, around the 7th century BC (Li 1997: 27). This conclusion forced scholars to reconsider the calibrated ^{14}C dates and the original viewpoint.

Recently, there has been further important research on this question. Shao Huiqiu noticed that M75 can be divided into two stages, based on the relationship of the layers. The upper tomb, in which all the iron artefacts were found, belongs to a later stage, and the other burial objects, such as ceramic vessels, also differ from those in the lower layers (Shao 2006).

The M31 knife is typologically early, and, based on ceramic vessel types, according to Shao Huiqiu this tomb belongs to the early stage. However, the date of this tomb (*c.*1312–1127 BC) is too early to be believed, because it is earlier than the date of the iron excavated from the regions neighbouring Xinjiang, especially those found in Central Asia (Wagner 2003: 130). If only the characteristics of the painted pottery and the calibrated dates are considered, and ignoring the views of Li Wenying, the date of the iron knife excavated from this tomb might be around the 10th–9th century BC. This find is a very important piece of evidence, but it is the only one that has so far been discovered. If several examples like this are found, this would confirm the speculation that iron artefacts appeared in Xinjiang around the 10th–9th century BC. More detailed archaeological excavations focusing on this culture need to be done in the future.

In recent years, more materials have been found in the Yanghai cemetery in Shanshan county, Turfan (XIA and CROT 2004). A bimetallic (bronze and iron) artefact unearthed from tomb M5 at Yanghai cemetery may be a toggle used for connecting rope. This tomb has been dated to around the 10th–9th century BC based on the types of ceramic vessels found. Moreover, some items within the inventory, such as the bridle bits and the cheek-piece (in the shape of the primitive ones), can be dated to a time around the 9th–8th century BC, fixing the date of this tomb to this period (Fig. 2). Furthermore, there are obvious cultural relations between the Yanghai cemetery and the Yanbulaq culture in Hami county, in the Yanbulaq cemetery where perhaps the earliest iron artefact has been found in Xinjiang. Finally, there are similar finds in necropolis B of the Sialk site in Iran, which was dated to around the 10th–9th century BC (Ghirshman 1938–39: 220, pl. XXIX: 6, 7). Several similar artefacts were also unearthed from tomb 1 in Arzhan, in Tuva, which was dated to around the end of the 9th century BC. Therefore, this artefact is the most important evidence for seeking the origin of the early iron objects discovered in ancient cemeteries in the Xinjiang region.

From the 8th century BC, the quantity and types of burial goods made of iron increased rapidly in some regions in Xinjiang, especially in the Tianshan Mountains region. Many iron objects were found at Chong Bagh cemetery (Fig. 3), the date of which has typologically been placed around the 8th–7th century BC (Tang 1993: 561); this is also confirmed by the ^{14}C calibrated dates (IA and CROB 1987, 1991).

Some iron artefacts, including several knives and a sword shard, were excavated from the Qiongkeke cemetery in Nileke county in 2001 and 2002 (XIA 2002). The excavators suggested that these iron artefacts could be dated to around 1000 BC. The iron knife from 2002YNQM03 grave was even dated to a period around the 12th century BC (Liu 2004). However, Lü Enguo argued that this cemetery may not be as early as the ^{14}C calibrated date suggested (Lü 1999). According to Chen Jianli who did the ^{14}C dating measurement at Peking University, the reliable dates for these iron objects from Yili should also be placed around the 8th–7th century BC. In any case, the discovery of the Qiongkeke cemetery and the whole region of the Yili River valley require further research, considering that the region links Xinjiang with Central and western Asia.

In summary, on the basis of archaeological evidence available so far, especially the bimetallic artefact from M5 in the Yanghai cemetery, iron artefacts would certainly seem to appear in Xinjiang around the 9th–8th century BC. Only

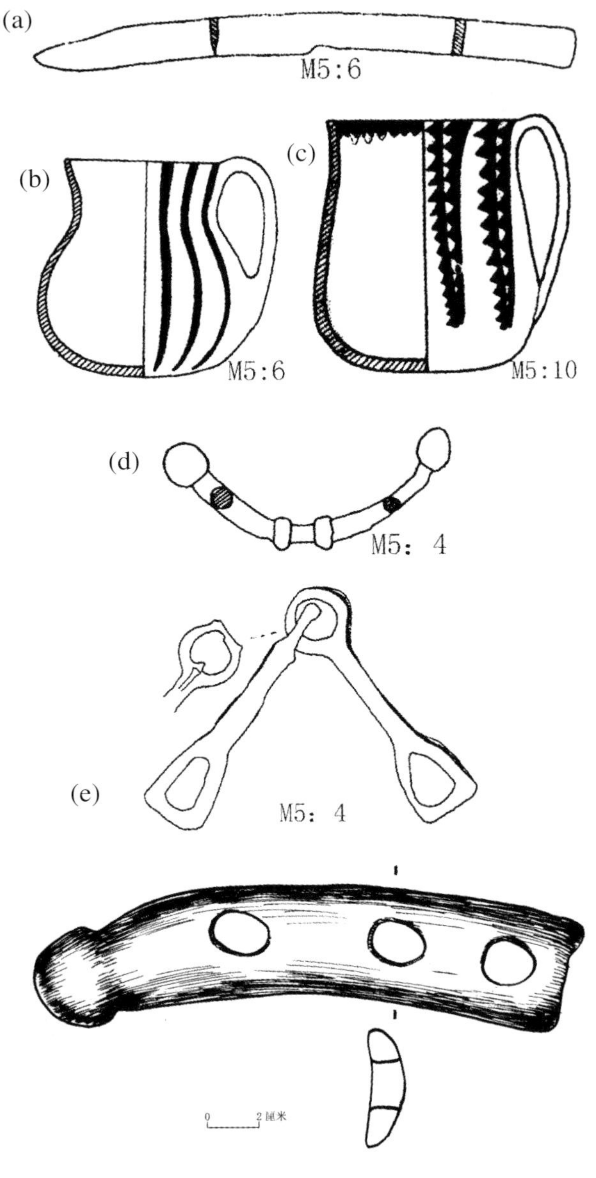

Figure 2 Burial objects from tomb M5, Yanghai cemetery: (a) bronze knife; (b) and (c) ceramic vessels; (d) bimetallic toggle; (e) bronze bridle bit and bone cheek-piece (after XIA and CROT 2004: fig. 42). (Drawings of the bridle bits and cheek-pieces: Lü Enguo.)

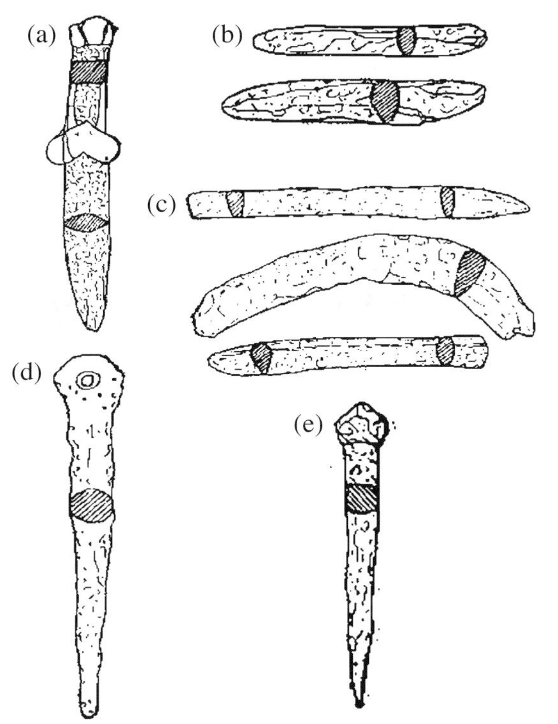

Figure 3 Iron artefacts found at Chong Bagh cemetery: (a) iron sword from tomb 17, cemetery I; (b) iron knives from tomb 27, cemetery I; (c) two iron knives and a sickle from tombs 4 and 10, cemetery II, and tomb 7, cemetery I; (d) iron awl from tomb 12, cemetery II; (e) iron awl from tomb 27, cemetery I (after IA and CROB 1991: 693, fig. 14; 701, fig. 24).

the iron knife unearthed from tomb M31 in the Yanbulaq cemetery might be dated to the 10th–9th century BC, a date that still needs to be confirmed by future discoveries.

Relationships between western Asia and Xinjiang from the late 2nd to early 1st millennium BC

Iron appeared in western Asia earlier than it did in Xinjiang. A large amount of data concerning iron metallurgy exists for the region of western Iran, where iron appeared towards the end of the 2nd millennium BC (Iron I), *c.*1450/1350–1100 BC. Many iron artefacts dating from the period between 1100 and 800 BC (Iron II) have been found at several sites, such as the Sialk B necropolis and many sites in the Luristan region (Cuyler Young 1965, 1967; Pigott 1980: 417–52; Fleming *et al.* 2005: 41–2). Some of these discoveries indicate that there were relations between western Asia and Xinjiang during the period between the end of the 2nd and the early stage of the 1st millennium BC.

The face-to-face horse motif

Comparing the face-to-face horse motif decorating a pottery jug from the Sialk B necropolis in ancient Iran with the same motif on a rock in Hutubi near Urumqi in Xinjiang, the possibility that the respective people in these locations had

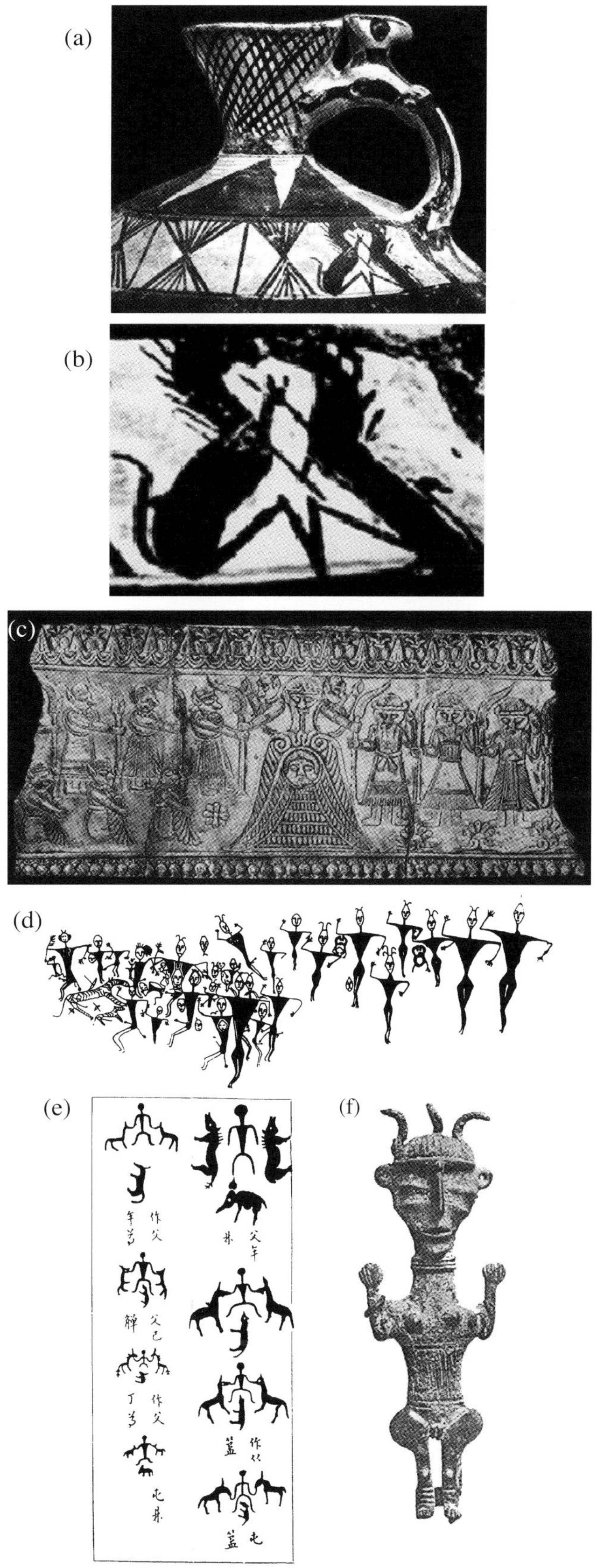

Figure 4 Motifs and emblems found in Iran and Xinjiang: (a) painted pottery jug unearthed from Sialk B necropolis in ancient Iran (10th–9th century BC) (after Ghirshman 1964: fig. 14); (b) detail of the face-to-face horse motif on the jug; (c) silver plaque from Luristan (10th–9th century BC) (after Ghirshman 1964: fig. 64); (d) the same motif on rock in Hutubi, near Urumqi, in Xinjiang (after Wang 1993: fig. 3); (e) emblems of the late Shang dynasty, 13th–11th century BC (after Lin 1988b: fig. 5); (f) Iranian bronze sculpture, 8th–7th century BC (after Pope 1981: VII, 73: D).

some interaction appears convincing. The date of this motif can be confirmed by some clan emblems of the late Shang dynasty, cast on the surface of ritual bronze vessels dated to the 13th–11th century BC (Su 1994: 559). The painted jug from the Sialk B necropolis has been dated to the 10th–9th century BC (Ghirshman 1964: 16, fig. 14), while the date of the bronze statue found in Luristan in western Iran is around the 8th–7th century BC. Thus, the date of the Hutubi petroglyph should be around the 13th–9th century BC, possibly closer to the 10th–9th century BC (Wang 1993). In particular, images with the same crown and posture, both on the Hutubi petroglyph and the bronze sculpture from Iran, and the same feature of a head drawn in a man's chest, both on the Hutubi petroglyph and the silver plaque from Luristan, would suggest that there were contacts between the two regions (Fig. 4).

The cult of the twin brothers in the form of two horses was widespread (Kuzmina 2003). Hutubi was a clan religious centre in this period. Furthermore, this religious dancing might have influenced the last emperor of Yin (the late Shang dynasty) (Lin 1998a,b). Hutubi was also very important geographically, located in the centre of the Tianshan Mountains at a time when cultures were shifting from the Bronze Age to the early Iron Age in Xinjiang. An indication of a relationship with western Iran in this period would be very useful for our understanding of the origin of iron metallurgy in Xinjiang. This location and chronology may offer an opportunity to explain the appearance of iron artefacts in the cemeteries of Yanbulaq and Yanghai.

Introduction of Neo-Assyrian culture elements into Central Asia

The New Assyrian Empire (935–612 BC) began to expand around the 9th century BC. There are some objects indicating that some Assyrian elements may have been introduced into Central Asia indirectly, such as *konghou* (horizontal angular harp), gold blossom ornaments, sticks and the half-stepped lozenge, all found in the Subeixi culture in Xinjiang. The harp from tomb 90 of cemetery I in Yanghai is very similar in shape to the seven-string Assyrian instruments carved on the wall relief at Nimrud (Lawergren and Gurney 1987; Lawergren 2003: 89–91) (Fig. 5). The gold blossom ornaments found in the cemeteries at Aidinghu, Yanghai and Goubei in Turfan are similar to those of the New Assyrian Empire (Fig. 6). There are many sticks made of wood in the cemeteries at Yanbulaq in Hami and Subeixi in Turfan. These discoveries suggest that the sticks found in Xinjiang may have something in common with those from the culture of the New Assyrian Empire (Fig. 7). The half-stepped lozenge appeared on Persepolis painted pottery, probably as a symbol of the lower heavens (Pope 1981: 833–4). The Assyrians used this motif to crown their buildings, and even painted it on their pottery (Namio 1963: 109, fig. 261). There is one type of ceramic vessel excavated from the cemetery at Yanghai in Shanshan county the erect handle of which is in the shape of just such a half-stepped lozenge. This is a unique phenomenon during the early stage

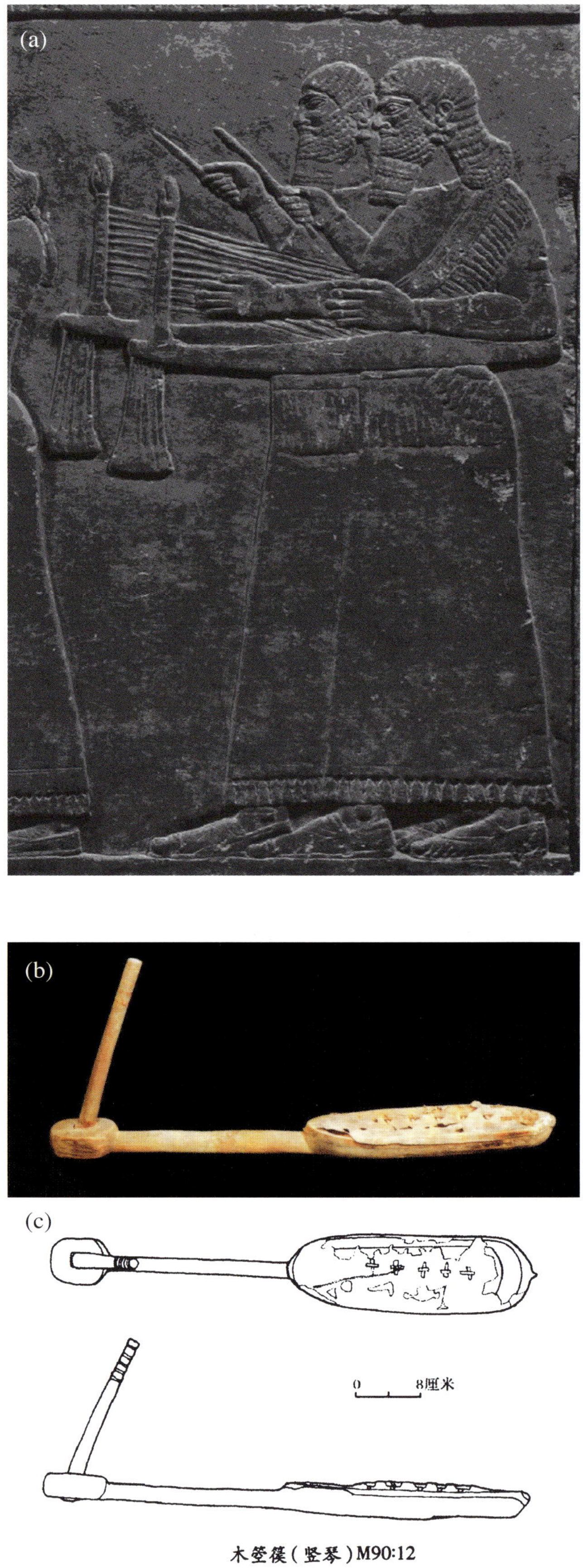

Figure 5 *Konghou/cank* (horizontal angular harp) of the New Assyrian Empire and Xinjiang: (a) wall relief of an Assyrian harp at Nimrud, 860 BC (Shanghai Museum 2006: plate 7); (b and c) harp found in tomb 90 of Yanghai cemetery I (after XIA and CROT 2004: fig. 39).

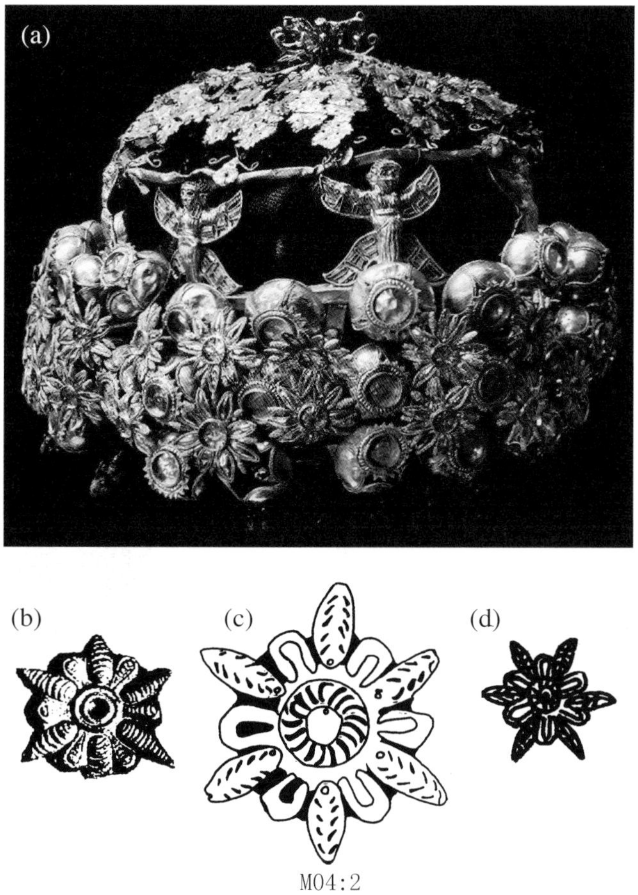

Figure 6 Gold blossom ornaments of the New Assyrian Empire and Xinjiang: (a) ornament of Melliso Mokhanas at Nineveh, Queen of Neo-Assyria, wife of Assurnazirpal II, 884–824 BC, from Nimrud; (b) gold blossom ornament found in Aidinghu cemetery in Turfan; (c) gold blossom ornament found in Goubei cemetery in Turfan; (d) gold blossom ornament found in Yanghai cemetery in Turfan (after MX and CROT 1982: 369, fig. 7, pl. 7; XIA 1989b: 39, fig. 6; OPRACJ 1998: 68, fig. 52.7).

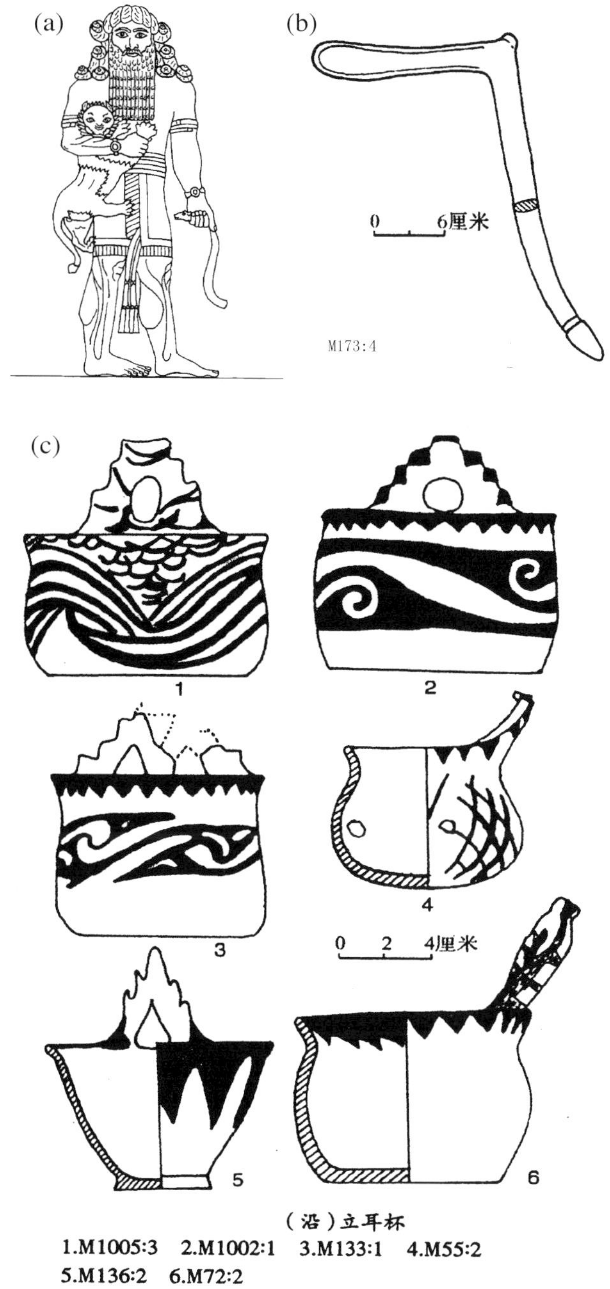

Figure 7 The stick and half-stepped lozenge of the New Assyrian Empire and Xinjiang: (a) sculpture from the gate of Dur Sharrukin of the New Assyrian Empire (end of the 8th century BC); (b and c) a wooden stick from tomb 173 and ceramic vessels from Yanghai cemetery I (after XIA and CROT 2004: fig. 37, 21).

of the Subeixi culture, based on all the finds of the whole Xinjiang region to date. It is conceivable that this design of handle could have imitated the battlements of western Asia.

Iron artefacts prevailed in the Neo-Assyrian Empire, Sialk B and Luristan around the 10th–8th century BC. For example, many iron artefacts were unearthed from the Luristan region, including a set of harness from the site of War Kabud Pusht-i Kuh, which was dated to the 9th–8th century BC (Haerinck and Overlaet 2004: 55; Stöllner *et al.* 2004, I: 310–37, II: 693–9). A few pieces of archaeological evidence suggest that people living in this area had some connection with the inhabitants of Xinjiang at that time, and that the iron knife from Yabulaq tomb 31 and the bimetallic artefact from Yanghai tomb 5 could possibly be dated to about the 10th–8th century BC. Therefore, it would seem that there were some connections between these places through the region of the north of Iran, the Oxus and Jaxartes rivers and the northern region of the Tianshan Mountains. The sites of Sialk and Luristan are considered to be the ruins of the civilisation of the early nomads who migrated to western Iran in the early 1st millennium BC. Not only did these peoples have opportunities for contact with those living in Central Asia and the steppes, but also with the Neo-Assyrian Empire. Thus, these nomads may have played a role in the spread of iron metallurgy between the West and the East.

Fragments of an iron sickle, excavated from the southern mound at Anau which belongs to the Yaz I culture, are perhaps pieces of the most ancient iron artefact known so far in western Central Asia (Masson and Sarianidi 1972: 160). If the decoration of the ceramics found in western Central Asia is compared with those from sites in Xinjiang, such as Banjiegou in Qitai county and Charwighul cemetery in Hejin county, many similarities can be noted. In fact, the cultures in which the handmade painted pottery prevailed in Central Asia may have been influenced by the cultures in Xinjiang (Francfort 2001; Mei 2000: 74–5). If the Anau sickle is earlier than those found in Xinjiang, it could be used to indicate a suitable point of the route for the spread of iron metallurgy.

Relationships between Xinjiang and the Pontic Caspian-Aral Sea steppes

When considering the route by which iron may have been transmitted to Xinjiang, the Eurasia steppe is another obvious focus of attention. Is there any evidence to show that iron metallurgy was taken to Xinjiang across the steppes? According to archaeological finds known so far, it seems that some relationship existed between the Pontic steppes and Xinjiang. An Olenniye *kamni*, dated to the 8th–7th century BC, was found in a Cimmerian tomb in northern Bulgaria. In another example, the weapon was carved on a Crimean stele from the village of Tselinniy (Belozor and Korpusova 1980). Several such objects were also found in the Pontic steppes and are considered to be Cimmerian (Chlenova 1975; Terenozhkin 1976). This type of Olenniye *kamni* was brought from the East, especially the region of the Altai Mountains in Xinjiang (Wang and Wang 1996: fig. 49). Other finds, such as harnesses, arrowheads with a barb at one side of the end of the socket, and cauldrons with a short knob on the top of each erect handle, may also have been the results of transfer to the northern Caucasus from the East (Fig. 8). The bronze cauldrons in particular had their origins in the northern region of China (Iessen 1954). Several such cauldrons have been found in the northern foothills of the Tianshan Mountains, suggesting that the people of this area had contact with those of the northern Caucasus (Guo 1999, 2002). Even though a sudden increase in iron can be observed around 750 BC, iron was used infrequently before 1100 BC in eastern Europe. The first iron artefacts appeared in Transcaucasia in the 12th–11th century BC, and by the 10th century BC a fairly appreciable range of iron objects came into being, including some in the northern Caucasus (Gzelishvili 1964: 9–19; Vosnesenskaya and Tekhov 1973). However, the initial appearance of iron had practically no effect on the volume of non-ferrous metallurgical production in the Caucasus, and a noticeable decline in the production of bronze only took place in the 8th–7th century BC (Chernykh 1992: 295). The Cimmerians might have learned iron metallurgy against this background, and many elements indicate relationships between the Pontic steppes and Xinjiang before the 8th century BC, so the possibility cannot be ignored that iron metallurgy might have been transmitted to Xinjiang from the region of the Eurasian steppes located to the north of the Caspian and Aral seas. However, it should be pointed out that the first appearance of iron in the region of the eastern Urals and southern Siberia did not happen very early, and so further evidence is needed to clarify this issue (Koriakova and Epimakhov 2007).

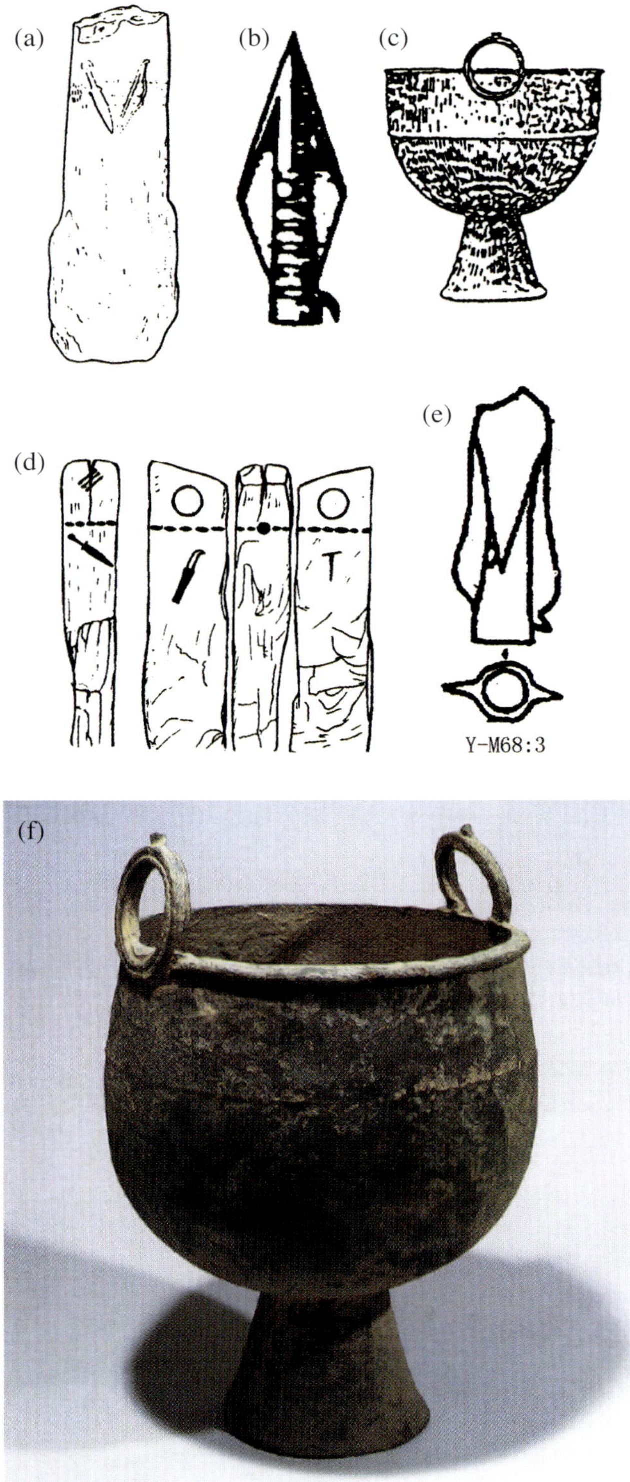

Figure 8 The relationship between the Pontic steppes and Xinjiang before the Scythian period: (a) stele from the Pontic steppes; (b) arrowhead from the Pontic steppes; (c) bronze cauldron from the northern Caucasus; (d) stele found in Xinjiang; (e) bronze arrowhead found in Xinjiang; (f) bronze cauldron found in Xinjiang (after Wang and Wang 1996: fig. 49; XIA 1989a: fig. 24:2; Chi 1993).

Conclusions

The earliest iron object found so far in Xinjiang is the bimetallic toggle excavated in tomb M5 at the Yanghai cemetery in Shanshan county, Turfan, which can be dated to the

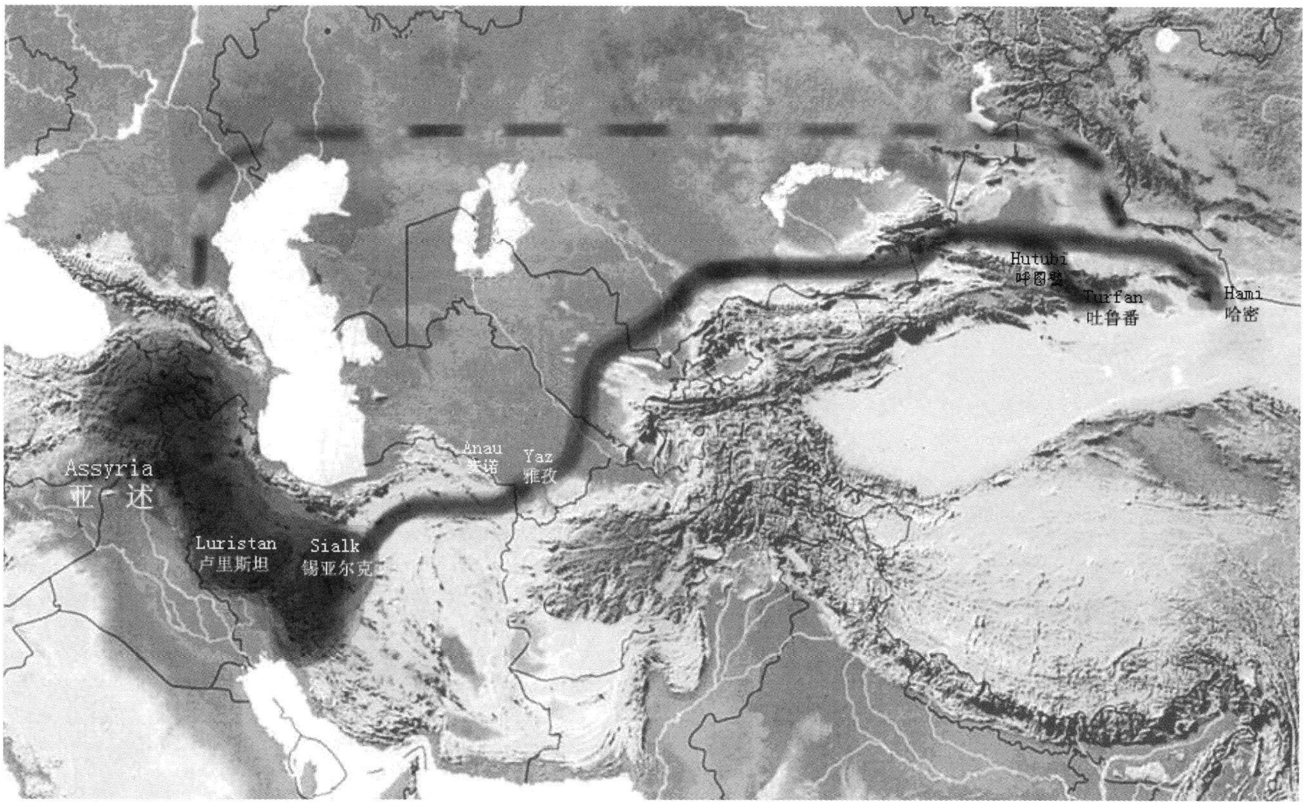

Figure 9 The possible route of the spread of iron metallurgy from western Iran to the Tianshan region in Xinjiang.

9th century BC (XIA and CROT 2004). It seems that iron was introduced to Xinjiang from north-west Iran during the 10th–9th century BC. Based on the iron artefacts found to date, a preliminary hypothesis may be put forward regarding the origin of iron technology in the Central Plains of China. The earliest iron artefact discovered there to date is a sword unearthed from tomb 2001 of the king of the Guo state in Henan province (Han 1998; Han *et al.* 1999), which has been dated to the early part of the 8th century BC (IA 1999: 126–8, 530; Peng 2006). Thus, iron metallurgy may have spread to the Central Plains of China around the end of the 9th century BC. The people living in oases along what was to become the Silk Road or in the steppe region may have played an important role in this rapid transmission of iron technology (Fig. 9). It has been noted that it was not common for iron articles to be buried in graves in the early period, and so the iron artefacts of the 10th–9th century BC may have not been found in Xinjiang for this reason. We can only hope for new discoveries in the future.

Acknowledgements

I am grateful to Professor Jianjun Mei of the Institute of Historical Metallurgy and Materials, University of Science and Technology Beijing, John Moffett of the East Asian History of Science Library, Needham Research Institute, Dr Zinovi Matskevich of the Department of Anthropology, Harvard University, and an anonymous reviewer for giving me invaluable comments on this manuscript.

References

An, Zhimin (安志敏) (1996) 塔里木盆地及周围地区青铜文化遗存 ('The remains of Bronze Age cultures in the Tarim Basin and surrounding areas'), 考古 (*Archaeology*) 12: 70–77.

Bai, Yunxiang (白云翔) (2003) 中国的早期铁器与冶铁的起源 ('On early iron artefacts and the origins of iron metallurgy'), 桃李成蹊集——庆祝安志敏先生八十寿辰 (*Essays in Honour of An Zhimin*), 308. Hong Kong: Chinese University Press.

Belozor, V.P. and Korpusova, V.N. (1980) 'Mogila kimmerijskovo voina u Dzhankoya v Krymu' ('The burial of the Cimmerian warrior near Dzhankoy in the Crimea'), *Sovetskaja Arkheologija* 3: 240–44.

Chen, Ge (陈戈) (1989) 新疆出土的早期铁器-兼谈我国开始使用铁器的时间问题 ('Early iron artifacts unearthed in Xinjiang: with a note on the earliest use of iron in China'), 庆祝苏秉琦考古五十五年论文集 (*Memoir for the Celebration of Su Binqi's Fifty-five Years of Archaeological Research*) 425–32. Beijing: Wenwu Press.

Chernykh, E.N. (1992) *Ancient Metallurgy in the USSR: The Early Metal Age*. Cambridge, New York: Cambridge University Press.

Chi, Wenjie (迟文杰) (1993) 庭州文物集粹 (*The Essence of the Relics in the Ting Administrative Region*). Ürümqi: Arts and Photography Publishing House of Xinjiang.

Chlenova, N.I. 1975) 'O svyazykh severo-zapadnovo Prichernomor'ya I Nizhnevo Dunaya s vostokom v kimmeriiskuyu epokhu' ('On the relations of the north-western Black Sea coast and the lower Danube with the Orient in the Cimmerian age'), *Studia Thracica: Frako-skifskie kul'turnie svyazi* (Studies of Thrace: Frako-Scythian Cultural Connections), 69–90. Sofia: Aedibus Academia Litterarum Bulgarica, Institum Thracicum.

Cuyler Young, T. Jr. (1965) 'A comparative ceramic chronology for western Iran, 1500–500 BC', *Iran* 3: 53–85.

Cuyler Young, T. Jr. (1967) 'The Iranian migration into the Zagros', *Iran* 5: 11–34.

Fleming, S.J., Pigott, V.C., Swann, C.P. and Nash, S.K. (2005) 'Bronze in Luristan: preliminary analytical evidence from copper/bronze artifacts excavated by the Belgian Mission in Iran', *Iranica Antiqua* 40: 35–64.

Francfort, H.-P. (2001) 'The cultures with painted ceramics of south Central Asia and their relations with the northeastern steppe zone (late 2nd–early 1st millennium BC)', in R. Eichmann and H. Parzinger (eds), *Migration und Kulturtransfer. Der Wandel vorder- und zentralasiatischer Kulturverhältnisse im Umbruch vom 2. zum 1. vorchristlichen Jahrtausend*, 231–6. Bonn: Habelt, Kolloquien zur Vor- and Frühgeschichte 6.

Ghirshman, R. (1938–39) *Fouilles de Sialk près de Kashan 1933, 1934, 1937, Vol. II*. Paris: Librairie Orientaliste Paul Geuthner.

Ghirshman, R. (1964) *Persia: From the Origins to Alexander the Great*. London: Thames and Hudson.

Guo, Wu (郭物) (1999) 青铜鍑在欧亚大陆的初传 ('The early spread of bronze cauldrons in Eurasia'), 欧亚学刊 (*Eurasian Studies*) 1: 122–50.

Guo, Wu (郭物) (2002) 论青铜鍑的起源 ('On the origin of bronze cauldrons'), 21世纪中国考古学与世界考古学 (*Chinese Archaeology and World Archaeology in the 21st century*) 395–6. Beijing: China Social Sciences Press.

Gzelishvili, I.A. (1964) *Zhelezoplavilnoe proizvodstvo v drevnei Gruzii* (The Iron-melting Industry in Ancient Georgia). Tbilisi: Metsniereba.

Haerinck, E. and Overlaet, B. (2004) *The Iron Age III Graveyard at War Kabud, Pusht-i Kuh, Luristan*. Leuven: Peeters.

Han, Rubin (韩汝玢) (1998) 中国早期铁器(公元前5世纪)的金相学研究 ('The study of the metallography of the early iron artifacts – 5th century BC – found in China'), 文物 (*Cultural Relics*) 2: 87–96.

Han, Rubin, Jiang, Tao and He, Songlin (韩汝玢 姜涛 何松林) (1999) 三门峡虢国墓地出土铁刃铜器的鉴定与研究 ('Appraisal and research on the bronze artefacts with iron blades uncovered from the cemetery of the Guo Kingdom in Sanmenxia'), in 中国社会科学院考古研究所 (Institute of Archaeology, Chinese Academy of Social Sciences) (ed.), 三门峡虢国墓地~(The Cemetery of Guo Kingdom in Sanmenxia), 559–73. Beijing: Wenwu Press.

IA (Institute of Archaeology, Chinese Academy of Social Sciences) (中国社会科学院考古研究所) (1999) 三门峡虢国墓地 (*The Cemetery of the Guo Kingdom in Sanmenxia*). Beijing: Wenwu Press.

IA and CROB (Institute of Archaeology, Chinese Academy of Social Sciences and Cultural Relics Office of Bayan Gol Mongol Autonomous District, Xinjiang) (中国社会科学院考古研究所，新疆巴音郭楞蒙古自治州文管所) 1987. 新疆轮台群巴克墓葬第一次发掘简报 ('The first excavation season of graves at Qunbake in Luntai county, Xinjiang'), 考古 (*Archaeology*) 11: 987–96.

IA and CROB (Institute of Archaeology, Chinese Academy of Social Sciences and Cultural Relics Office of Bayan Gol Mongol Autonomous District, Xinjiang) (中国社会科学院考古研究所，新疆巴音郭楞蒙古自治州文管所) (1991) 新疆轮台群巴克墓葬第二、三次发掘简报 ('The second and third excavation season of graves at Qunbake in Luntai county, Xinjiang'), 考古 (*Archaeology*) 8: 684–703.

Iessen, A.A. (1954) 'Nekotorye pamyatniki VIII–VII VV.do n.e' ('Some sites of VIII–VII century BC'), in S. Kavkaze (ed.), *Voprosy skfo-sarmatskoj arkheologii*, 112–31. Moscow: Izdatelbstvo Akademii nauk SSSR.

Kong, Lingping and Feng, Guozheng (孔令平 冯国正) (1988) 铁器起源问题 ('On the origin of iron'), 考古 (*Archaeology*) 6: 542–6.

Koriakova, L.N. and Epimakhov, A.V. (2007) *The Urals and Western Siberia in the Bronze and Iron Ages*. Cambridge/New York: Cambridge University Press.

Kuzmina, E.E. (2003) 'Origins of pastoralism in the Eurasian steppes', in M. Levine, C. Renfrew and K. Boyle (eds), *Prehistoric Steppe Adaptation and the Horse*, 203. Cambridge: McDonald Institute for Archaeological Research.

Lawergren, B. (2003) 'Western influences on the early Chinese Qin-Zither', *Bulletin of the Museum of Far Eastern Antiquities* 75: 79–109.

Lawergren, B. and Gurney, O.R. (1987) 'Sound holes and geometrical figures, clues to the terminology of ancient Mesopotamian harps', *Iraq* 49: 37–52.

Li, Wenying (李文瑛) (1997) 哈密焉不拉克墓地单人葬、合葬关系及相关问题探讨 ('Study on the relation between single occupancy graves and graves with two or more corpses'), 新疆文物 (*Xinjiang Cultural Relics*) 2: 26–7.

Lin, Meicun (林梅村) (1998a) 帝辛甲骨所见殷宫秘史 ('The inside story of the royal palace during the reign of Dixin, last emperor of the Yin dynasty, revealed by oracle-bone inscriptions'), 学术集林 (*Academic Quarterly*) 14: 184–222.

Lin, Meicun (林梅村) (1998b) 吐火罗神祇考 ('On Ñakte, the Tokharian god'), 国学研究 (Chinese Culture Heritage Research) 5:1–26.

Liu, Xuetang (刘学堂) (2004) 中国冶铁术的起源 ('On the origin of the metallurgy of iron in China'), 中国文物报 (*China Cultural Heritage*) 2 April.

Lü, Enguo (吕恩国) (1999) 察吾呼文化研究 ('Study on the culture of Chawuhu'), 新疆文物 (*Xinjiang Cultural Relics*) 3/4: 79–80.

Masson, V.M. and Sarianidi, V.I. (1972) *Central Asia: Turkmenia Before the Achaemenids*. London: Thames and Hudson.

Mei, Jianjun (2000) *Copper and Bronze Metallurgy in Late Prehistoric Xinjiang: Its Cultural Context and Relationship with Neighbouring Regions*. Oxford: Archaeopress.

MX and CROT (Museum of Xinjiang Uygur Autonomous Region, Cultural Relics Office of Turfan region) (新疆维吾尔族自治区博物馆，吐鲁番地区文管所) (1982) 新疆吐鲁番艾丁湖古墓葬 ('The ancient tombs in the region of Aiding Lake in Turfan in Xinjiang'), 考古 (*Archaeology*) 4: 365–72.

Namio, Egami (1963) *Outline of World Archaeology 11, West Asia II*. Tokyo: Heibonsha Limited, Publishers.

OPRACJ (Office of the Protection and Repair for the Ancient City of Jiaohe)(交河古城保护修缮办公室) (1998) 交河故城——1993年、1994年度考古发掘报告 (*Ancient City of Jiaohe: Report of the Examination in 1993 and 1994*). Beijing: Orient Press.

Peng, Yushang (彭裕商) (2006) 虢国东迁考 ('On the eastern migration of the Guo kingdom'), 历史研究 (*Historical Research*) 5: 12–22.

Pigott, V.C. (1980) 'The Iron Age in west Iran', in T.A. Wertime and J.D. Muhly (eds), *The Coming of the Age of Iron*, 417–61. New Haven, CT/London: Yale University Press.

Pope, A.U. (1981) *A Survey of Persian Art, II*. Ashiya: SOPA.

Shanghai Museum (上海博物馆) (ed.) (2006) 艺术与帝国——大英博物馆藏亚述珍品 (*Art and Empire: Treasures from Assyria in the British Museum*). Shanghai: Shanghai Painting Press.

Shao, Huiqiu (邵会秋) (2006) 从M75看焉不拉克墓地的分期 ('On the periodisation of the Yanbulaq cemetery based on research of tomb 75'), 新疆文物 (*Xinjiang Cultural Relics*) 2: 84–5.

Stöllner, T., Slotta, R. and Vatandoust, A. (2004) *Persiens antike Pracht: Bergbau, Handwerk, Archäologie: Katalog der Ausstellung des Deutschen Bergbau-Museums Bochum vom 28. November 2004 bis 29. Mai 2005*. Bochum: Deutsches Bergbau-Museum.

Su, Beihai (苏北海) (1994) 新疆岩画 (*The Petroglyphs of Xinjiang*). Ürümqi: Art and Photography Press of Xinjiang.

Tang, Jigen (唐际根) (1993) 中国冶铁术的起源问题 ('On the origin of iron metallurgy in China'), 考古 (*Archaeology*) 6: 556–65.

Terenozhkin, A.I. (1976) *Kimmijtsy* (Cimmerians). Kiev: Naukova Dumka.

Vosnesenskaya, G. A. and Tekhov, B.V. (1973) 'Tekhnologiya proizvodstva kuznechnykh izdelii iz Tliiskogo mogilnika (X–VI vv.do n.e.)' ('Technology of metal production from the Tliiskogo cemetery (X–VI century BC)'), *Sovetskaya Arkheologiya* 3: 153–62.

Wagner, D.B. (2003) 'The earliest use of iron in China', *Bulletin of the Museum of Far Eastern Antiquities* 75: 127–69.

Wang, Binghua (王炳华) (1993) 新疆天山生殖崇拜岩画初探（'A preliminary study on the motif of reproductive worship and belief on the petroglyphs in Xinjiang's Tianshan'），丝绸之路考古研究 (*Archaeological Research on the Silk Road*) 360–400. Ürümqi: Xinjiang People's Publishing House.

Wang, Linshan and Wang, Bo (王林山、王博) (1996) 中国阿尔泰山草原文物 (*The Relics of the Steppes in the Altay Mountains in China*). Ürümqi: Arts and Photography Publishing House of Xinjiang.

XIA (Xinjiang Institute of Archaeology) (新疆文物考古研究所) (1989a) 新疆哈密焉不拉克墓地 ('The cemetery of Yanbulaq in Hami'), 考古学报 (Acta Archaeologica Sinica) 3: 325–62.

XIA (Xinjiang Institute of Archaeology) (新疆文物考古研究所) (1989b) 鄯善县洋海、达浪坎儿古墓群清理简报 ('Brief report of the examination of cemeteries in Yanghai and Dalangkanger'), 新疆文物 (*Xinjiang Cultural Relics*) 4: 39.

XIA (Xinjiang Institute of Archaeology) (新疆文物考古研究所) (2002) 尼勒克县穷科克一号墓地考古发掘报告 ('Report of the examination of cemetery I in Qiongkeke in Nileke county'), 新疆文物 (*Xinjiang Cultural Relics*) 3/4: 13–53.

XIA and CROT (Xinjiang Institute of Archaeology, Culture Relics Office of Turfan Region) (新疆文物考古研究所，吐鲁番地区文物局) (2004) 鄯善县洋海一号墓地发掘简报 ('Brief report of the examination of cemetery I in Yanghai in Shanshan county'), 新疆文物 (*Xinjiang Cultural Relics*) 1: 1–27.

Zhao, Huacheng (赵化成) (1996) 公元前5世纪中叶以前中国人工铁器的发现及其相关问题（'On the discoveries of iron artefacts before the middle of the fifth century BC'），考古文物研究 (*Research on Archaeology and Relics*) 289–300. Xi'an: Sanqin Press.

Zhu, Zhongxi (祝中熹) (2001) 中国古代始炼铁及秦人用铁考述（'Discussion on the beginning of iron metallurgy of ancient China and the use of iron by the people of the Qin state'），陇右文博 (*Antiquity and Museum of Longyou*) 1: 50–53.

South Indian Iron Age iron and high carbon steel: with reference to Kadebakele and comparative insights from Mel-siruvalur

Sharada Srinivasan, Carla M. Sinopoli, Kathleen D. Morrison, Rangaiah Gopal and Srinivasa Ranganathan

ABSTRACT This paper is based on studies of the use and modes of production of high carbon iron alloys in relation to surface finds from Iron Age and early historic sites in southern India, in particular the site of Kadebakele where recent excavations have revealed finds of iron and steel, some of which according to preliminary studies, seem to be of a higher carbon content. Preliminary comparative studies are also made on surface finds of crucibles related to high carbon steel production at Mel-siruvalur.[1]

Introduction

The iron and steel artefacts uncovered from the site of Kadebakele, Karnataka in southern India span the early phases of occupation radiocarbon dated from 800 to 400 BC. This site has been excavated by a collaborative team from the Karnataka Department of Archaeology, the University of Michigan, and the University of Chicago. The site was occupied from at least the early 1st millennium BC until the early centuries AD. Radiocarbon dates from the Iron Age period span from *c.*800 to 400 BC and it is thought that iron/steel artefacts from these levels may rank among the very early well-dated examples of higher carbon steels.[2]

Steel and south India

High carbon iron alloys are known to have been produced in Asia (e.g. *wootz*, the traditional Indian crucible steel (Smith 1960: 14–24, Bronson 1986) and Chinese cast iron (Tylecote 1976: 85)) before they were widely used in Europe. Classical Mediterranean accounts make several references to the reputation of Indian iron and steel (Bronson 1986: 18). From the 17th century onwards European travellers have described the production of steel ingots by carburising iron, or melting constituents in crucibles in parts of south India e.g. Golconda/Nizamabad (Voysey 1832), Mysore (Buchanan-Hamilton 1807) and Salem (Buchanan-Hamilton 1807; Wood 1893) in the present-day states of Andhra Pradesh, Karnataka and Tamil Nadu respectively. Coomaraswamy (1956) has also described crucible steel processes in the Kandyan region of Sri Lanka. *Wootz* is thought to be a European corruption of the south Indian Telegu/Kannada word for steel, *ukku*. *Ukku* may derive from the related Tamil word *uruku,* meaning to make molten. It could also be related to *ekku*, the old Tamil word for sharpness/sharp spear.[3]

According to Smith (1960: 14–16; 1981: 67, 72) Indian *wootz* ingots were used to forge oriental Damascus swords found to be of a high (1.5–2%) carbon content with a typical 'watered silk' pattern. Damascus blades, made in Khorasan and Isfahan in Persia from south Indian *wootz* from Golconda were thought to have been the finest weapons then made in Eurasia, reputed to cut even gauze handkerchiefs (Bronson 1986: 22–3, 1). The properties of south Indian steel, which became synonymous with Damascus steel, were investigated and debated during the 19th and 20th centuries by several European scientists, including Michael Faraday, with the aim of industrial production (Smith 1960: 25–9). A typical *wootz* ingot analysed in 1804 by Mushet (cited in Smith 1960: 22) had 1.3% carbon and a dendritic structure.

Recent investigations of crucibles from surface collections from the known sites of Konasamudram, Nizamabad district, near former Golconda (Voysey 1832; Lowe 1990) and Gatihosahalli in Karnataka (Freestone and Tite 1986; Rao 1989; Anatharamu *et al.* 1999) indicate the existence of specialised semi-industrial *wootz* production processes, many of which, though not precisely dated, may range at least over the later medieval period, i.e. 16th–20th century. While these investigations have thrown some light on the ceramic material of the crucibles, studies remain to be done on the nature of the final product and metallurgical process. During the course of field investigations on copper mining and smelting in south India the lead author of this paper came across a previously unrecorded archaeometallurgical site in Mel-siruvalur, South Arcot district, Tamil Nadu. Investigations showed it to be a site for producing *wootz* crucible steel by what appears to be a process of carburisation of low carbon wrought iron (Srinivasan 1994, 1998, 2007; Srinivasan and Griffiths 1997). The discovery of this production centre supports the idea that *wootz* steel production was relatively widespread in south India, and further extends the known horizons of this technology.

High carbon steels from India and crucible steel from southern India

Pliny's *Natural History* mentions the import of iron from the Seres, which could refer to the ancient southern Indian kingdom of the Cheras, while the *Periplus of the Erythraean Sea* unequivocally mentions the import of iron and steel from India (Schoff 1915; Bronson 1986: 18). Certainly trade between the ancient Cheras and the Roman world is supported by finds of late Roman coins (mid-4th century AD) along with Chera coins from the river bed in Karur, the ancient capital of the Sangam era, while trade with 'Yavanas', i.e. Ionians or Mediterranean people, is mentioned in Sangam literature (Srinivasan 1994). Investigations on the iron-rich Iron Age sites of Tamil Nadu and Malabar (mid-1st millennium BC to early centuries AD) could be revealing. These fall within the domain and period of the Sangam Chera dynasty which might tie in with Roman accounts of iron from the Seres, possibly referring to the Chera kingdom. Excavations at an Iron Age site at Kodumanal, Tamil Nadu (3rd century BC), close to Karur, the capital of the Chera kingdom of the Sangam era (3rd century BC to the 3rd century AD) has revealed furnaces stacked with vitrified crucibles that were found separate from abundant iron slag (Rajan 1990: 98). Investigations by Sasisekaran (2004: 45, pl. 23.2) indicated the presence of high carbon steel, with a structure of about 1% carbon, in a projectile edge from Kodumanal, while investigations on a crucible fragment from Kodumanal show that it was iron-rich, which does not rule out its possible use in crucible steel production (reported in Srinivasan and Griffiths 1997; Srinivasan 2007).

Past observers of the manufacture of *wootz* steel in India have commented on the process of carburising iron to steel in crucibles where a batch of closed crucibles with the low carbon iron charge was stacked in a large furnace and fired in a long 14–24 hour cycle at temperatures of not less than 1200 °C in a strongly reducing atmosphere (Percy 1860–80: 773–6). The known sites of crucible steel production in south India, i.e. at Konasamudram and Gatihosahalli, date from at least the late medieval period (16th century).

Mel-siruvalur: archaeometallurgical evidence for high carbon crucible steel processing

In November 1991, field investigations were carried out[4] on old copper workings in Mamandur, South Arcot district, about 40 km from the nearest town of Tiruvannamalai. While searching for evidence of copper smelting activity, metallurgical debris was found near the village of Mel-siruvalur about 5 km from Mamandur, which is shown to be related to crucible steel production. Evidence of metallurgical activity came from a mound (25 × 8–9 m wide and up to 5 m high) just behind the village and from some trenches near the houses. The villagers had no memory of this metallurgical activity. Occupation of the area in antiquity is indicated by pottery sherds collected adjacent to an old canal, about ½ km away from the mound. Slag debris and crucible fragments were also found all around the canal site. Among the sherds were many large rim fragments, about 3 cm thick, belonging to huge storage jars approximately 60 cm in diameter. These had no slip, and were found to be tempered with rice hulls. Their resemblance to Iron Age storage jars of red ware without slip has been pointed out.[5] The Iron Age occupation in Tamil Nadu, which began in the 1st millennium BC, continued until as late as the 5th century AD. Iron Age dolmens have been found in Thiruvannamalai and Tirukoilur taluks in South Arcot district. Sasisekaran (2002) also observed signs of Iron Age occupation at the site.

Numerous crucible fragments were found at the Mel-siruvalur mound together with fragments of glassy slag, charge and debris (Fig. 1). The reconstructed fragments of crucibles resembled the aubergine-shaped closed crucibles used for *wootz* steel production known from other sites in south India such as Gatihosahalli and Konasamudram, indicating that the crucibles had been broken to retrieve the ingots. Also found were thick covering lids of a diameter of about 7 cm, which would have sealed the refractory vessel during firing with the iron charge. The exterior surface of the crucibles was covered with a thick black ash glaze indicating that the crucibles were fired under highly reducing conditions. Flat vertical ridges seen on some crucibles suggest that the crucibles had been stacked together. The base of the lid was uneven and bore imprints of a fibrous pattern suggesting that the crucible could have been stuffed with leaves and grass to act as carbonaceous material for carburising the charge.

Several curved crucible bases, about 0.8–1.5 cm thick, were also found. A lining of glassy slag with a honeycomb pattern covered the area below the fin to the crucible base, which would have been occupied by molten charge and solid ingot. Some of the slag linings on the crucible bases had rusty remnants of the charge attached to them. The dimensions of the various fragments indicated that the ingots were of a diameter of *c.*2.5 cm. Rusty patches and globules could

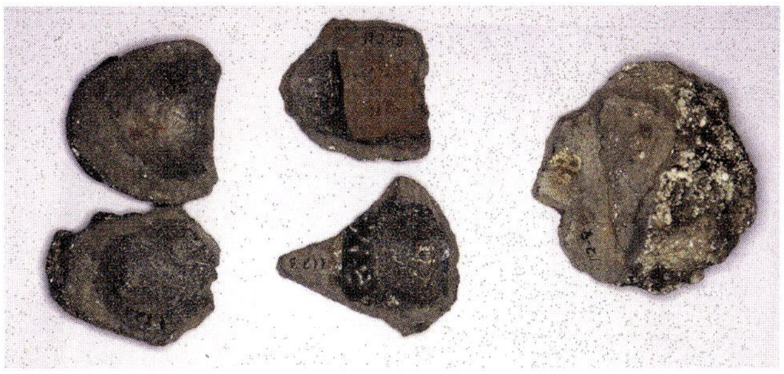

Figure 1 Assemblage of steel-making crucible fragments from surface finds at Mel-siruvalur, Tamil Nadu.

Table 1 Mel-siruvalur crucible: EPMA (wt%) analysis of prills in a section of a lid fragment.

Constituent	Fe (wt%)	Cu (wt%)	As (wt%)	S (wt%)	Total
Prill1	97.615	0.038	0.023	0.565	98.241
Prill 2	102.385	0.075	–	0.251	102.385

Analysed by EPMA on a polished and carbon-coated cross-section using a JOEL Superprobe JXA-8600 at 20 kV with ZAF correction, within instrumental accuracy of 1% over 100% using probe diameter of 50 µm. Trace Si and P were noted but were not analysed by EPMA; however these elements were found to be less than or around 1% using SEM with EDAX analysis (HITACHI S-570 with link AN-1000). Note: the results are as analysed and not normalised to 100%.

Table 2 EPMA-WDS (wt%) analysis of constituents in glassy slag fin on interior of crucible.

Constituent	Fe	Al	Si	Ni	Cu	Ti	P	S	Ca	Total
Prill	82.28	0.18	3.25	1.74	0.27	0.09	9.95	0.12	1.35	99.38
Glassy slag	0.32	8.62	24.46	0.0	0.0	0.89	0.02	0.0	7.13	43.79

Analysed by EPMA on polished, carbon-coated cross-section using a JOEL Superprobe JXA-8600 at 20 kV with ZAF correction, within instrumental accuracy of 1% over 100% with probe diameter of 5 µm. Lighter elements such as O, C, N etc. were not monitored

be seen scattered on the ring of slag fin, above this fin, along the crucible walls and even below the lid, almost certainly caused by splashing of molten charge possibly due to a carbon boil reaction. All of the above suggests that the ingot solidified from a molten charge.

Analytical results and discussion

The blackish glaze on the exterior suggests that the crucibles were fired under highly reducing conditions at high temperatures with heavy attack by fuel ash. Sections for microscopy were taken from different fragments of crucibles representing the lid, the glassy fin and the crucible base. The latter had a thin layer of glassy slag with a honeycomb pattern with some remnant rusty charge. Examination of the cross-section indicated that the rusty layer formed from the remains of the molten charge was extremely thin, no more than 10 µm thick. When this rusted layer was sectioned, tiny uncorroded metallic prills were detected by optical microscopy, but by the time fine polishing of the specimen was complete they had fallen out of the loose rust matrix. These uncorroded metallic prills were found, better preserved, in the glassy slag fins and slag coating on the interior of the crucibles, and the exterior black glaze in the lid section. While the prills in the exterior glaze may be attributed to factors such as the contamination by particles of charge or iron filings, which would also have been carburised, or the splashing of molten metal, the prills that were found trapped within the glassy slag fin and lining of the interior of the crucible can be linked directly to the final composition of the molten charge. The prills were of a diameter of less than 100 µm, with most prills in the glassy slag on the interior of the crucible being no more than 10–20 µm. Analysis of the prills using electron probe microanalysis with wavelength-dispersive X-ray spectrometry (EPMA–WDS) or scanning electron microscopy with energy-dispersive X-ray analysis (SEM–EDS) confirmed that they are steel prills (Tables 1 and 2). Instrumental accuracy is expected to be 1% at 100% of analysed component by EPMA–WDS and about 8% at 25% by SEM–EDS.

The etched microstructure of the largest detectable prill had a lamellar structure of fine pearlite eutectoid (of equilibrium composition of 0.8% carbon) inside original hexagonal grains of austenite (Srinivasan 1994). This suggests that it derives from a very good quality hypereutectoid high carbon (>0.8%) steel. The prill had a hardness of around 400 VPN which is within the range for normalised steel of approximately 1% carbon (Scott 1991: 82), suggesting that the crucible had not been quenched. The presence of much smaller amounts of a light unetched network between grains near the boundaries and as occasional needles in the pearlite was also noted. SEM–EDS analysis suggested that this was mainly cementite (iron carbide) with some phosphorus impurities, i.e. cementite-phosphide. Iron phosphide tends to form a ternary eutectic of steadite along with pearlite (Avner 1988: 439) and its presence could indicate a slightly higher carbon content of about 1–1.2%. Smaller prills in the lid also had a similar microstructure.

The prills trapped in the glassy slag fin of 3–4 mm thick (Fig. 2) and in the 1 mm thick slag lining of the interior of the crucible best reflect the composition of the molten liquid from which the ingot solidified. These also have the same structure as described above: lamellar pearlite eutectoid inside hexagonal grains surrounded by an intergran-

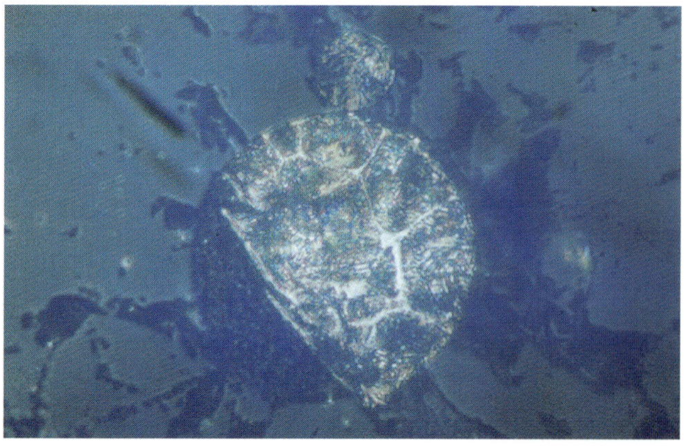

Figure 2 Mel-siruvalur crucible: high carbon steel prill (1.2% carbon) (50 µm at original magnification 400×) showing pearlite surrounded by a cementite network.

ular network of cementite with some cementite needles. EPMA and SEM–EDS analyses also indicated the presence of phosphorus in the cementite network, strongly suggesting that the final product or solidified ingot was a high carbon steel of around or over 1% carbon. The microstructure of the prills compares well with that of a 1.2% carbon steel illustrated in Avner (1988: fig. 7.14). The impression of the honeycomb pattern in the slag lining and fin may also be related to the two-phase microstructure of the molten metal with the carbide network surrounding the polygonal grains of high temperature austenite which forms pearlite at lower temperatures.

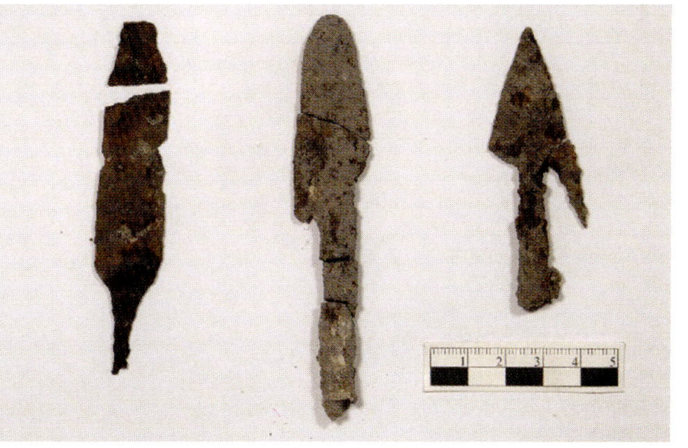

Figure 3 Projectile points from Kadebakele. The iron projectile (middle) (acc. No. 337, Level 6, -21.6E-204 N) has radiocarbon dates of *c*.400–350 and 310–210 BC from charcoal associated with the specimen.

Iron and steel artefacts from the Iron Age site of Kadebakele

The south Indian site of Kadebakele is located on the northern bank of the Tungabhadra River in Koppal district of northern Karnataka, India. More than 60 hectares in area, occupation at the site spanned from the 1st millennium BC Iron Age period until the 1st millennium AD.

The Early Historic Landscapes of the Tungabhadra Corridor (EHLTC) research project seeks to examine settlement and economy throughout this sequence, through a program of excavation at Kadebakele and contemporary village sites and a regional survey in a 35 km^2 area surrounding Kadebakele. Excavations in 2003 and 2005 focused on the Iron Age sequence, in excavations of habitation and ritual mortuary areas in an area of the site referred to as the 'upper terrace' – a well-defended zone on top of an outcropping hill that rises some 110 m above the river flood plain.

Excavations were carried out in six areas of the upper terrace, which encompass some of the diverse activities that took place at the site. These include: residential areas characterised by the remains of houses and associated features (Block B, and Unit 52E/-86N); areas of waste deposition (Unit 94.84E/-16.93N, upper levels of -21.6E/-204N); areas of megalith construction (Block A, and lower levels of -21.6E/-204N); and a water storage feature (8E/-142N). With the exception of the test trench placed within the bed of the reservoir feature, all of the areas excavated produced metal artefacts. In this paper, we report on studies on some of the objects. Some speculation is also made about the typological and other similarities of the objects to those from other Iron Age finds, and the environmental and other considerations related to the finds of ferrous artefacts and modes or places of production etc.

One of the artefacts studied is a projectile point excavated at Kadebakele. Charcoal found in association with the iron projectile (Fig. 3, middle image; acc. No. 337, found in Level 6, -21.6E-204 N) had radiocarbon dates of *c*. 400–350 and 310–210 BC. A cross-sectional sample was taken from the cylindrical shaft of the projectile, which had a 9 mm outer diameter and a 3 mm diameter inner hole. The microstructure (Fig. 4) at lower magnification shows a network of ferrite grains with some pearlite along grain boundaries – the structure of low carburised iron or steel. The slag inclusions, which are elongated in an arc-like fashion,

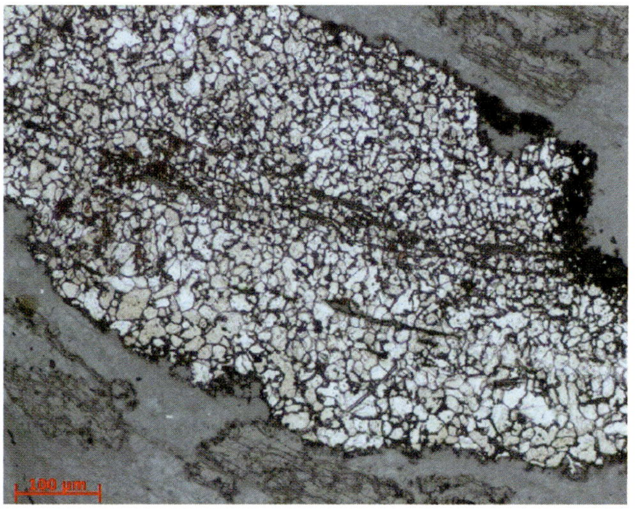

Figure 4 Microstructure of a broken specimen from the circular shaft iron projectile (acc. no. 337) from Kadebakele (*c*.400 BCE).

suggest that it was heavily worked in that direction along the circumference. The hardness measurements were in the range of Hv (Vickers hardness) 165 (whereas for iron it is in the range of 65–75 Hv). This follows the parameters of mild steel that can be forged, which is borne out by the microstructure.

A small iron ring shown in Figure 5 (acc. no. 900, 22E-28 N, Level 7) was found above a plaster surface while charcoal from this level was radiocarbon dated to 800–440 BC. What is particularly interesting is that the microstructure of the ring (Fig. 6) consisted almost entirely of a pearlitic structure with a fine lamellar network suggesting a composition close to the eutectoid composition of 0.8% carbon. Given the secure dates of the layer from which it is excavated of 800–440 BC this could be one of the earliest known examples of pearlitic or higher carbon steel found anywhere in the world. The structure is not inconsistent with that found in the crucible fragment from Mel-siruvalur, which is confirmed to be from high carbon crucible steel production. The possibility that it may have solidified from the viscous if not molten state is suggested by the absence

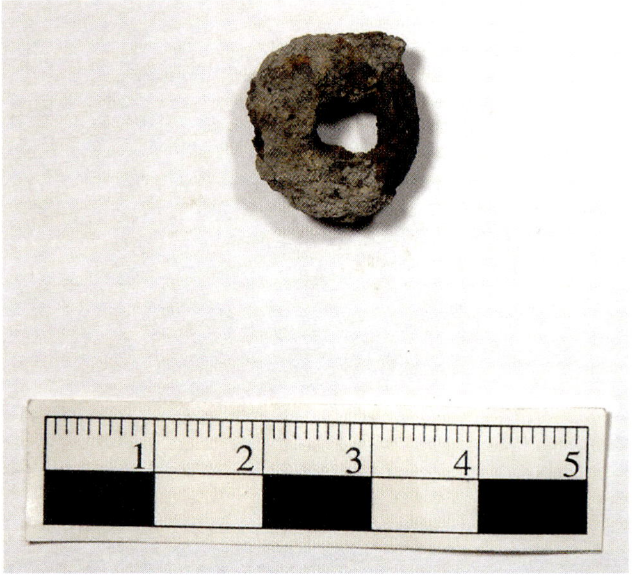

Figure 5 Iron ring (acc. no. 900) from Kadebakele from the level radiocarbon dated to 800–440 BCE.

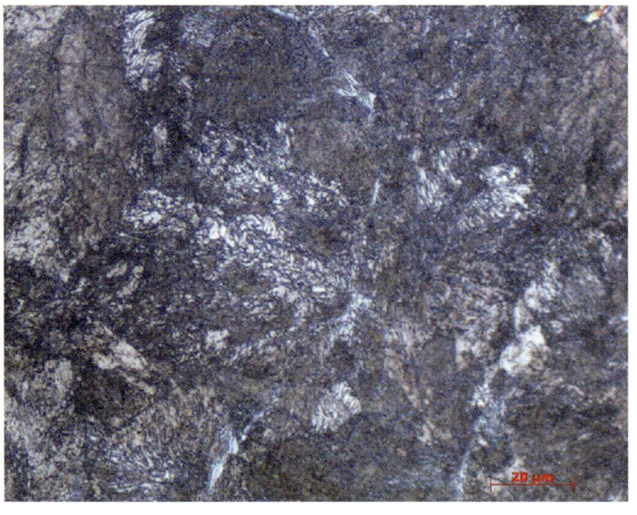

Figure 6 Photomicrograph of iron ring, acc. no. 900 (800–400 BCE) (original magnification 50×) showing pearlite with some cementite needles and some ferrite.

of any evidence of working, implying a possible link with the kind of crucible steel processes discussed in the context of southern India. However, further investigations are required to understand these aspects.

Conclusions

The investigations reported here indicate that at the Iron Age site of Kadebakele in Karnataka, there is preliminary evidence for the use of steel of a higher carbon content in a tiny ring and certainly forged mild steel in a projectile shaft, ranking among the earliest known higher carbon steels anywhere. Perhaps the finds at Kadebakele suggest some kind of precursors to the crucible steel process for high carbon steel, as seen clearly in evidence at the site of Mel-siruvalur. Although this latter site for crucible steel production, first identified by Srinivasan (1994), has not yet been properly investigated or excavated from an archaeological point of view, it shows potential for possible megalithic evidence as commented elsewhere by Sasisekaran (2002).

Acknowledgements

The authors would like to thank Mr Padaikathan, Department of Metallurgy, Indian Institute of Science, Bangalore for assistance in the metallurgical studies on Kadebakele artefacts, R. Krishnamurthy of the *Dinamalar Daily* and Dr Purushottam, formerly of the Department of Mines and Geology, Madras for helping Sharada Srinivasan with the site visit to the Mel-siruvalur/Mamandur area in 1991. We also express our appreciation to the late N. Seeley, D. Griffiths, J. Merkel and K. Reeves of the Institute of Archaeology, London for support in the studies on the Mel-siruvalur crucibles. We also thank Thelma Lowe, Dr Parthasarthy of the Geological Survey of India-Madras Circle, Dr Radhakrishan, the late Dr C. V Seshadri and Professor Balasubramaniam for their interest and assistance.

Notes

1. First identified by Sharada Srinivasan.
2. The metallurgical studies were made in collaboration with the National Institute of Advanced Studies, Bangalore with Sharada Srinivasan as principal investigator.
3. John Marr, pers. comm.
4. By Sharada Srinivasan with the help of the State Department of Mines and Geology, Madras, and Geological Survey of India-Madras Circle.
5. By the late C.S. Patil (pers. comm.) of the Mysore Archaeological Survey.

References

Anatharamu, T.R., Craddock, P.T., Nagesh Rao, K, Murthy, S.R.N. and Wayman M.L. (1999) 'Crucible steel of Ghattihosahalli, Chitradurga District, Karnataka, southern India', *Historical Metallurgy* 33(1): 13–25.

Avner, S.H. (1988) *Introduction to Physical Metallurgy*. Singapore: McGraw-Hill.

Bronson, B. (1986) 'The making and selling of wootz – a crucible steel of India', *Archeomaterials* 1(1): 13–51.

Buchanan-Hamilton, F. (1807) *A Journey from Madras through the Countries of Mysore, Canara and Malabar*. London: T. Cadell and W. Davies.

Coomaraswamy, A.K. (1956) *Mediaeval Sinhalese Art*. New York: Pantheon Books.

Freestone, I.C. and Tite, M.S. (1986) 'Refractories in the ancient and preindustrial world', in W.D. Kingery (ed.), *Ceramics and Civilisation*, 35–63. Pittsburgh, PA: American Ceramic Society.

Lowe, T.L. (1990) 'Refractories in high-carbon iron processing: a preliminary study of the Deccani wootz-making crucibles', in W.D. Kingery (ed.), *Ceramics and Civilization IV: Cross-craft and Cross-cultural Interactions in Ceramics*, 237–50. Pittsburgh, PA: American Ceramic Society.

Percy, J. (1860–80) *Metallurgy*, vol. 2, part 3. The Netherlands: De archaeologische pers Nederland.

Rao, K.N.P. (1989) 'Wootz: Indian crucible steel', *Metal News* 11: 1–6.

Rajan, K. (1990) 'New light on the megalithic cultures of the Kongu region, Tamil Nadu', *Man & Environment* 16(1): 93–102.

Sasisekaran, B. (2002) 'Metallurgy and metal technology in ancient Tamil Nadu', *Indian Journal of History of Science* 37(1): 17–29.

Sasisekaran, B. (2004) *Iron Industry and Metallurgy: A Study of Ancient Technology*. Chennai: New Era Publications.

Schoff, W.H. (1915) 'The eastern iron trade of the Roman empire', *Journal of the American Oriental Society* 35: 224–39.

Scott, D.A. (1991) *Metallography and Microstructure of Ancient and Historic Metals*. Marina del Rey, CA: Getty Conservation Institute/ J. Paul Getty Museum in association with Archetype Books.

Smith, C.S. 1960. *A History of Metallography. The Development of Ideas on the Structure of Metals before 1890*. Chicago: University of Chicago Press.

Smith, C. S. (1981) *A Search for Structure: Selected Essays on Science, Art and History*. Cambridge, MA: MIT Press.

Srinivasan, S. (1994) 'Wootz crucible steel: a newly discovered production site in south India', *Papers from the Institute of Archaeology* 5: 49–61.

Srinivasan, S. 1998. 'Highlights of ancient south Indian metallurgy: technical evidence for the early use of high-tin bronzes, high-carbon steel, metallic zinc, bronze smelting and solid cast image', in *Proceedings of the 4th International Conference on the Beginning of the Use of Metals and Alloy (BUMA-IV)*, 79–84. Sendai: Japan Institute of Metals.

Srinivasan, S. (2007) 'On higher carbon and crucible steels in southern India: further insights from Mel-siruvalur, Kodumanal and Pattinam', *Indian Journal of History of Science* 42(4): 673–95.

Srinivasan, S. and Griffiths, D. (1997) 'Crucible steel in south India: preliminary investigations on crucibles from some newly identified sites', in P. Vandivar, J. Druzik, J. Merkel and J. Stewart (eds), *Material Issues in Art and Archaeology V*, 111–27. Warrendale, PA: Materials Research Society.

Srinivasan, S. and Ranganathan, S. (2004) *India's Legendary Wootz Steel: An Advanced Material of the Ancient World*. Bangalore: National Institute of Advanced Studies and Indian Institute of Science.

Tylecote, R.F. (1976) *A History of Metallurgy*. London: Metals Society

Voysey, H.W. (1832) 'Description of the manufacture of steel in southern India', *Journal of Asiatic Society of Bengal* 1: 245–7.

Wood, C. (1893) 'Discussion of Turner's paper on production of iron in small furnaces in India', *Journal of the Iron and Steel Institute* 44(2): 177–80.

Survival of traditional Indian ironworking

Vibha Tripathi and Prabhakar Upadhyay

ABSTRACT India was well known for its high quality iron and steel production in the ancient world. Recent radiocarbon dates from Iron Age sites in India have pushed back the beginning of iron-making in India to the middle of the 2nd millennium BC. More importantly, the sites yielding such early dates are situated in the Vindhya-Ganga Plain in the heartland of India, dating back to *c.*1600 BC. The technology of iron manufacturing developed to great heights and became famous all over the ancient world. Massive wrought iron structures, such as the Delhi iron pillar, from the 4th–5th century AD onwards, testify to the mastery of the artisans. During the British period some iron was even exported to Britain, especially for highly significant works including important steel structures such as the Britannia tubular bridge and London Bridge. However, gradually, with the beginning of industrialisation and the import of pig iron from Britain, the traditional Indian iron industry declined. We are now faced with a big question regarding what happened to the traditional Indian iron technology. This paper seeks to answer the question: did traditional Indian ironworking vanish altogether without leaving any legacy behind?

Introduction

Iron technology in India has a long history – India was well known for its high quality iron and steel production in the ancient world. Recent radiocarbon dates from Iron Age deposits in India have pushed back the beginning of iron-making in India to the middle of the 2nd millennium BC. More importantly, the sites yielding such early dates are situated in the Vindhya-Ganga Plain in the heartland of India. These dates are much older than those reported from the border of India and beyond. This is almost clinching evidence suggesting an earlier and independent beginning of ironworking in India. Starting from *c.*1600 BC with wrought iron technology, the metallurgy attained a level of expertise enabling the production of steel at least as early as the 6th century BC. Much has been written about it in recent years (Banerjee 1965; Chakrabarti 1985, 1992; Sahi 1979, 1994; Prakash 1997; Tripathi and Tripathi 1994; Tripathi and Mishra 1997; Tripathi 2001, 2007; Balasubramaniam 2002; Srinivasan and Ranganathan 2004). The prominence of high quality iron and steel being produced in India was acknowledged by the civilisations of the ancient world. Ironworkers of ancient India displayed ingenuity in the innovation of the manufacture of sharp-edged objects such as daggers and swords. This aspect is testified by the records of foreign historians, for example Herodotus (5th century BC), who noticed iron-tipped arrows being used by Indian soldiers fighting in the battle of Thermopylae. Almost at the same time, Ktesias, the Greek ambassador, gratefully acknowledged the gift of Indian swords that he received at the Persian court, a gift made to him by the king and his mother. Also noteworthy is a gift of 100 talents of steel in the form of ingots as a tribute by the vanquished Indian ruler of the Taxila region of north-western India during

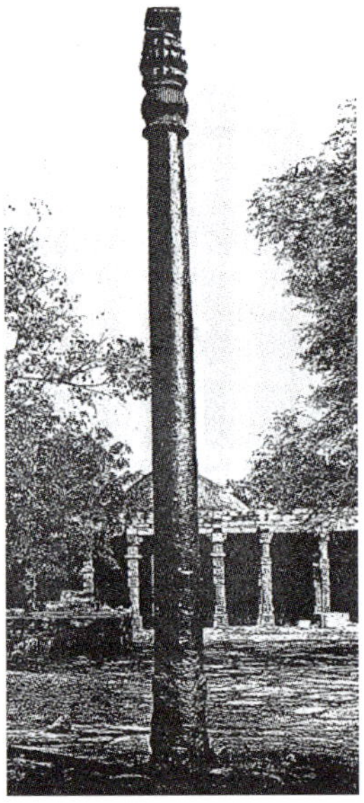

Figure 1 The Delhi iron pillar.

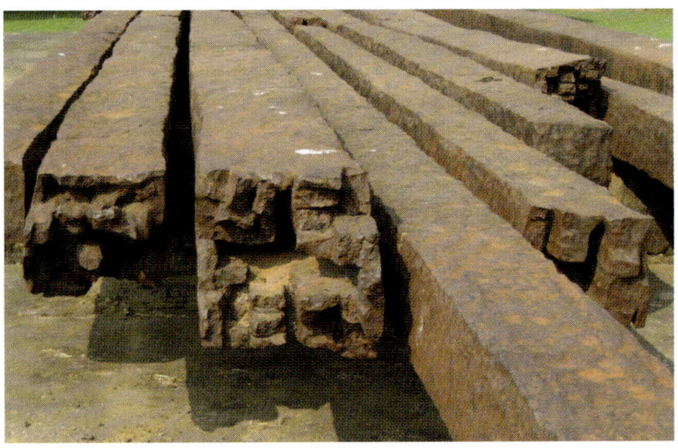

Figure 2 Konark beams.

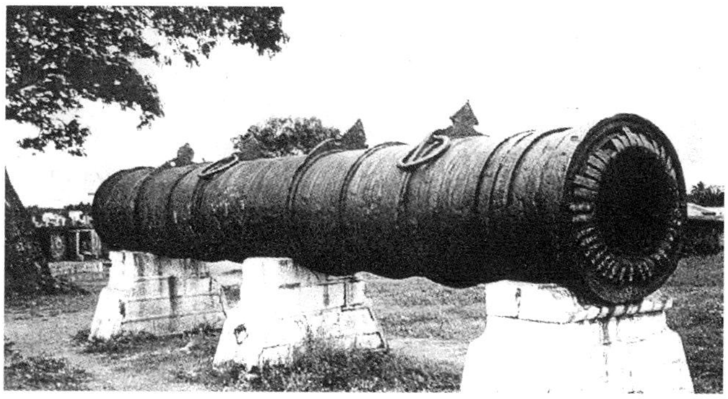

Figure 3 Iron cannon.

Alexander's expedition to the Indian subcontinent in 326 BC. Arrian (200 BC), another authority on the period, mentions the import of Indian steel to an Abyssinian port. Thus, Indian iron and steel was exported to different parts of the Graeco-Roman world from ancient times. The famous *wootz* or Damascus steel was a much-acclaimed commercial commodity all over the world until the pre-modern period.

By the early centuries of the Christian era, the engineering skill of ironworkers was sufficiently developed to produce a seven-ton victory pillar. The Delhi iron pillar (Fig. 1) has been referred to as a 'rustless wonder' by an eminent modern metallurgist (Anantharaman 1996). The process of manufacturing such iron continued until much later, as evidenced by massive beams once used in the Sun Temple at Konark (9th–10th century AD) (Fig. 2) or the colossal pillar at Dhar (11th century AD), the tallest iron pillar in the world. During medieval times, the emphasis shifted more towards weapons, for example the manufacture of guns and cannons. The Moghuls displayed ingenuity in cannon-making and many of their masterpieces have survived the vagaries of time (Fig. 3).

Traders from Middle Eastern countries came to south Indian ports to obtain *wootz* steel ingots used in making swords and daggers which were then marketed to the world as Damascus steel swords. Besides this, a large amount of other iron was shipped (generally through intermediaries) to many European countries. The French were regular traders of Indian iron and steel, and they later attempted to establish their own production in southern parts of the country (Krishnan 1952; Alam 2002). Efforts were also made by British and Swedish engineers to set up their own iron-producing units in several states of India – from Kumaon-Garhwal to the states in the far south.

The British, impressed by the excellent quality of Indian iron and steel, imported it for the construction of the Britannia tubular bridge across the Menai Straits at the beginning of the 19th century AD. Likewise, 50 tons of Indian steel are recorded to have been used in the manufacture of the famous London Bridge (La Touche 1918: 233). Iron production in India reached its zenith during the 19th century and continued until the first quarter of the 20th century. Iron manufacturing was widespread – almost a household industry with a large number of people involved.

However, the subsequent period saw a decline in the indigenous iron- and steel-making industries, various factors being responsible for the demise of the once flourishing iron industry. The last blow to the indigenous iron industry was dealt with the import of cheap pig iron from Britain. Gradually, with industrialisation and the advent of modern blast furnaces, traditional iron-making finally withered away. However, the legacy of traditional iron- and steel-working in India was carried on until recently by certain pre-industrial societies, for example the Agarias and the Asurs, the traditional Indian ironworkers, but iron smelting was almost completely discontinued towards the latter half of the 20th century. Efforts have been made during recent years to locate, identify and arrange demonstrations of the traditional ironworking that has in some way survived as a valuable legacy in certain remote parts of India.

There was a sizeable population of the Agaria, Asur, Lohra, Gadulia Lohars etc., residing in different parts of India, who were involved in iron smelting and forging. Most of these societies occupied forested zones near the ore deposits. They worshipped deities related to their craft and had their own socio-religious customs (Elwin 1942; Leuva 1963). They had limited relations with nearby villagers – generally they interacted with them through intermediaries who functioned as itinerant traders selling their product. The Agarias and the Asurs occupy a large tract stretching from the southern part of Uttar Pradesh, Madhya Pradesh, Bihar, Jharkhand, Chhattisgarh, Orissa, Andhra etc., up to the Chhotanagpur plateau that is known for its ore deposits and forests (Tripathi and Tripathi 1994; Tripathi 2001). This region is known as the Agaria belt (Fig. 4). Traditional ironworking communities also inhabited parts of other states such as Assam and Bengal, especially around Darjeeling, Himanchal Pradesh, Uttarakhand in the north, and parts of Andhra Pradesh, Karnataka, Kerala, Pondicherry, Tamil Nadu etc. in the south. Ironworking in the south not only had a great antiquity but also a rich tradition of use in megalithic burials dated from 1500/1400 BC to the 1st–2nd century AD or even later. It was the southern region which is believed to be the home of the famous *wootz* or Damascus steel. Thus there is a consistent history of the prevalence of iron- and steel-making in India, which was widespread both in time and space.

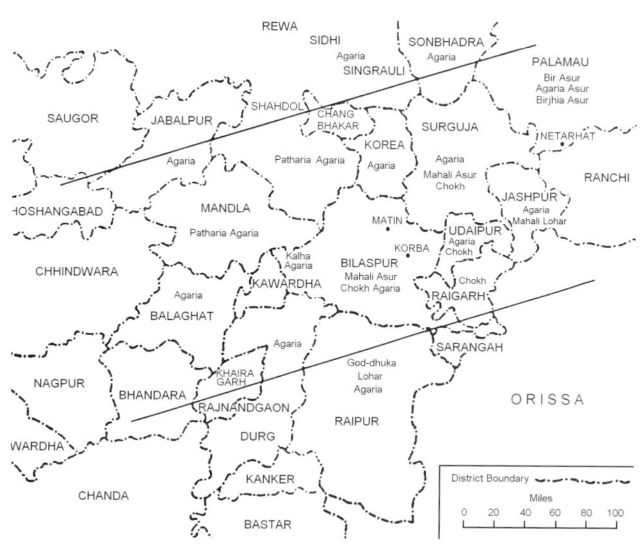

Figure 4 The Agaria belt.

Current research

Recent investigations carried out into the fate of traditional Indian iron technology during the early parts of the 20th century have brought to light certain ethnic groups known for their association with iron production for generations. We have been able to pick up the strands of a nearly lost technological skill, and encourage the elders of the society to recall and produce iron in small furnaces, similar to those used by their ancestors for generations. This paper focuses on the Agarias residing in southern Uttar Pradesh and adjacent areas of Madhya Pradesh in the Sonbhadra-Sidhi region (Fig. 5).

Agarias are traditionally known as ironworkers even today (the word Agaria means 'those working with fire'). But iron smelting has been virtually discontinued in these parts for the last several decades and it is difficult to find a young Agaria who has seen an iron-smelting furnace in operation. However, the relatively senior members of the Agaria society did have some memory of the process. In our efforts to revive the technology of iron smelting and forging, we persuaded a group of Agarias to run a furnace to smelt the iron ore that is present in abundance there. The operation was successful and a small iron ingot was produced at the conclusion of the smelting (Fig. 6). Observing the metallurgical process of the Agarias provided us with an insight into the age-old practice of ironworking.

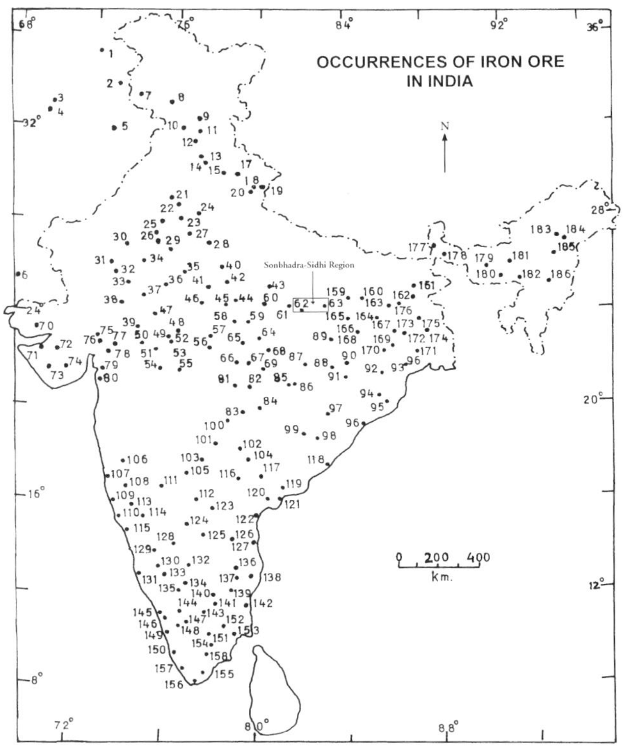

Figure 5 The Sonbhadra-Sidhi iron belt.

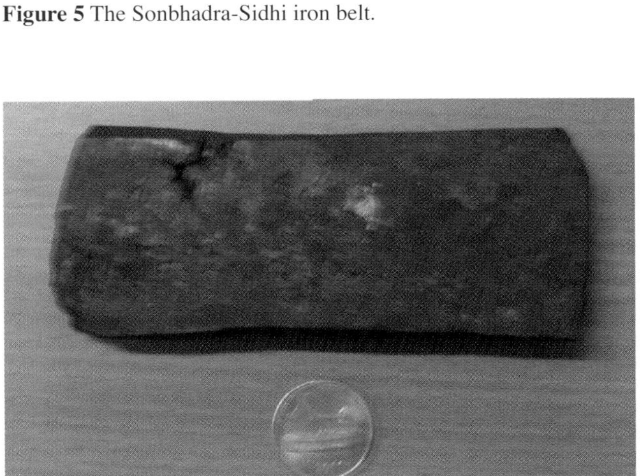

Figure 6 Iron ingot.

Traditional ironworking in the Sonbhadra-Sidhi region

A close examination focused on ironworking by the Agaria community (an ethnic community of the Sonbhadra-Sidhi region traditionally known to consist of iron smelters) at a village called Pipara in the Sidhi district in the state of Madhya Pradesh. We observed the community going through the entire process of iron production, from ore selection, mining, and blessing of the ore to iron smelting, forging and smithing. The ironworking process is described below.

Ore selection/mining

Iron-bearing deposits are distributed profusely on the ground surface therefore little mining is involved in the procuring of iron ore. This must also have been the case in the geographical zones occupied by the Agarias and the Asurs. Nodules are generally picked by hand from the surface by these traditional ironworkers; this would have been even easier in ancient times when such ore was available in greater abundance. The Agarias of these regions show a preference for low-grade ores. Thus the deposits considered to be of little significance by modern geologists were greatly favoured by the pre-industrial ironworkers. Because of the small size of deposits and inferior quality, most such deposits do not even appear in modern geological literature. But the early workers found them sufficiently adequate for a variety of reasons, including easy accessibility, close geographical proximity, and the fact that the silica in such ores acted as a fluxing agent and made smelting at a lower temperature easier.

The ore was selected and picked up in the form of nodules from exposed surfaces near *nullahs* or rain gullies. Shallow digging was resorted to if required. In areas where iron was present in the form of veins in the parent rock, they were retrieved by fire-setting followed by water thrown on the hot surface. The desired mineral was also separated using long chisels in certain cases.

The traditional workers seem to have an extremely good feel for minerals. Even when a variety of ores is available in the vicinity, they opt for precisely the kind and type of ore suitable for their specific purpose. For instance, the Agarias of Chhattisgarh have exploited the locally available hematite. However, magnetite was also worked for specific purposes. The Kamis (Nepali ironworkers of Darjeeling) opted for magnetite, the black coloured and hard to smelt ore, because it was more suitable for the production of sharp-edged weapons such as the *khukri* (dagger) or *ban* (arrow). Significantly, both magnetite and hematite are found together in several of the mining belts. Some of the smelters of the Pratappur block of Chhattisgarh prefer to work with magnetite sand from the local river beds, as this is free from gangue and is capable of producing good quality high carbon iron (Tripathi and Mishra 1997). They manage to produce white iron out of this ore, which is in demand for specific purposes requiring a superior variety of metal. The Agarias at Pipara preferred to work with low-grade hematite ore.

Ore preparation

The tribal smelters take meticulous care in preparing the ore for smelting; the success of the process depends on detailed preparation and planning, ore dressing being an integral part of the whole process. The ore is collected in fairly large amounts, sufficient to last for at least 4–5 days. The entire group consisting of several families is involved in ore collection, and they cover an area of about 15–20 km^2 a day. The ore is first washed, crushed into pea-sized pieces, winnowed, sifted and finally roasted for removal of further impurities such as carbonaceous material and moisture. At the ancient sites, we came across large stone querns and pestles bearing red-coloured marks left behind by the crushing of ores, and discarded pieces of waste. Excavations also yield such remains. Frequently, their significance is not properly understood or perhaps not even realised by the excavators. The thoroughly cleaned and crushed ore is roasted to remove further impurities and then set aside for smelting.

Iron smelting

A real insight into the ironworking process, however, may be gained only through an observation of pre-industrial smelting. The tradition of ironworking was a long one spanning centuries, transmitted from generation to generation without much change, alteration, experimentation or improvisation, possibly due to the sanctity and secrecy associated with the process. On superficial examination, the ironworking methods of the traditional societies appear generally to be very simple and elementary in nature. But attempts at laboratory simulation tend to indicate that it requires experience and expertise of a high order, perfected through trial and error over a long period. Even the slightest diversion from the usual practice leads to failure or inadequate production out of a smelt. Therefore, to ensure success, the master artisan keeps a close vigil on the ironworking operation.

Furnaces

Small furnaces are used among most traditional societies (Fig. 7), including the Agarias in question. Locally available clay is used for preparing the furnaces. Straw or rice-husk is mixed with the clay, soaked for a long time and kneaded thoroughly. The smelters do not use any scales for measurement but use their fingers (*anguls* or finger width) instead. To prepare the furnace, first a pit of about 20–22 cm diameter is dug in the ground. A channel of the same width is dug joining this pit at its right angles to the main opening of the furnace under preparation. The wall of the furnace is about 15–18 cm thick. The outer diameter of the furnace is about 60 cm at the base and tapers towards the top. The mouth of the furnace at the top is 12 cm diameter at the opening and the furnace becomes broader towards the base. To slide the charge into the mouth of the tapering furnace, a channel-like platform (about 35 cm wide and 65 cm long) is erected. Additional support is provided by two bamboo sticks at the end. Another opening (about 20 cm high) is made at the surface level. This is used both for inserting the bellows through a tuyère and for removing the bloom at the conclusion of the work. During the operation, this opening is sealed with a mixture of ash, charcoal powder and mud. After the furnace structure is ready, it is allowed to dry. At this stage, a paste of some smooth, yellowish clay of a special nature is applied on the inner side of the structure. The workers believe that this paste affects the smelting process as well as the quality and quantity of the end product. This paste appears to have the properties of a fluxing agent but the clay of this paste has yet to be analysed for its properties.

Bellows

Bellows are made of cow or buffalo hide, mounted on a hollow block of wood of about 25–27 cm in diameter. Twin bellows are linked, attached to a clay tuyère by bamboo tubes. The bellows are operated either with the feet, or with sticks, with a frequency of 120 strokes/minute, giving a flow rate of 100/per minute.

The smelting process

The process begins with some religious rites and offerings to their deity, and prayers for the successful conclusion of

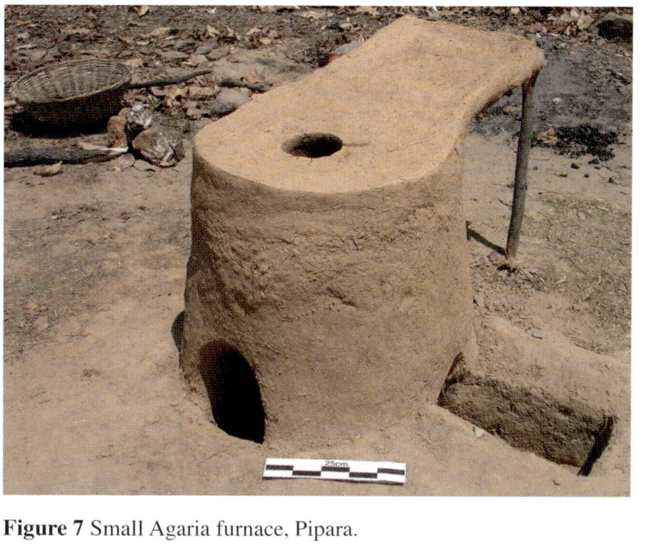

Figure 7 Small Agaria furnace, Pipara.

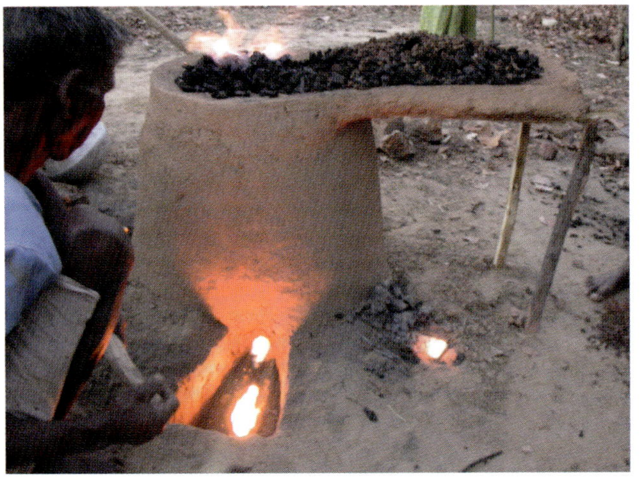

Figure 8 Agaria furnace in operation at Pipara.

the operation. The furnace is charged with alternate layers of charcoal and ore. After the furnace has been ignited and the bellows start blowing air at full blast, additional charges of ore, alternated with charcoal are made. Piles of crushed ore and charcoal are arranged on the platform (this probably allows some pre-heating and removes moisture) ready for charging as required during the course of smelting. Approximately 12 kg of ore and 15 kg of charcoal are used in one operation lasting for about 4–5 hours. At intervals, the slag is tapped using a thin stick to open a channel in the side of the furnace. The molten slag runs out of this channel in a red-hot condition (Fig. 8). This opening is sealed in the same way as the aperture for the bellows. The bellows are operated at a faster and quicker pace at the peak of the operation, gradually slowing the pace towards the end. When the master artisan has assessed that the iron is ready, the bellows, tuyères and the temporary seal of loose clay and ash are taken away and the bloom is then removed with a pair of tongs through this opening. Generally, approximately 3–4 kg of iron is produced, depending on the quality of the ore and the success of the operation.

Forging

An open hearth, which is simply a shallow pit full of red-hot burning coal, is worked with bellows; the bloom is heated to 700–800 °C and hammered with a heavy sledgehammer. This process is repeated several times. The impurities trapped in the form of sintered slag are pushed out of the bloom, giving it compactness and homogeneity as well as some hardness on the surface because of a prolonged direct contact with charcoal, i.e. carbon, which diffuses into the surface of the iron. Iron thus produced by these traditional smelters is preferred by local people to factory-produced iron for its superior qualities of durability, ductility and capacity to provide better cutting edges.

Observations

The observations above are made primarily about the Agarias of the Sonbhadra area, but they are also generally true of several other ironworking communities. Similar studies have been conducted on the Asur Birjia of the Gumla region in the Netarhat plateau, the Agarias of Wadrafnagar (Chhattisagarh) and also of Ghatgaon in Maharashtra (for details, see Tripathi 2001). There are some minor changes in furnace design or size, but the basic nature of the process remains more or less the same in most of these cases.

A close observation of the metallurgical process emphasises some special features of pre-industrial metallurgy enumerated below:

- The size of the furnace, including those of the Sonbhadra-Sidhi area, is generally very small. The height of the furnace varies between 0.55 and 1.00 m and the diameter at the bottom is about 0.25–0.30 m.
- Charcoal is used as the fuel.
- Generally, no basic (CaO) flux is used, even when lime is easily available, which removes sulphur from the end product. Instead, calculations have shown the use of extra SiO_2 (sand) as acidic flux in the charge.[1]
- Despite these shortcomings, the quality of the end product is quite good. Its superior quality was recognised by local consumers, as stated earlier. Appreciating its advantages, European engineers had also considered the possibility of iron production with charcoal in certain parts of India, in particular Schwartz, a Swedish engineer, who attempted to work on this scheme in 1882. He had studied the ores of Lohra and Pipalgaon in central India with a view to establishing a smelting plant for large-scale production. In his opinion, 'it would be possible by proper conservation of forests to ensure the supply of enough charcoal for an output of 25,000 tonnes of iron per annum' (Schwartz 1882). Durgapur, on the Erai River, was suggested as a suitable site, however, the project never materialised for a variety of reasons including the lack of support from the British government.
- The temperature in the furnace is not uniform – it is low, around 600 °C, in the upper portion, increasing to 1300–1400 °C, or sometimes almost as high as 1500 °C, near the tuyère zone. A close study of the entire production process and temperature mechanism has led metallurgists to believe that 'the traditional smelter is constrained by the low temperature in the furnace and the process is finely tuned to his requirements, any rise in the slag melting temperature can ruin the product' (Ballal 1993).
- The tribal smelters are very careful about the specific type of material they use and have no wish to experiment with their working process in any way. Magnetite-bearing sand from rivers is used by several such groups because of its purity. Iron objects produced from it are in great demand locally because of their durability and long-lasting working edges.
- Despite the elementary-looking furnace, tools and technique, 'small blast furnaces operated with charcoal give superior metal products', according to an Indian Institute of Technology study (Ballal 1993).
- The smelters were capable of producing different types of iron, such as the famous *wootz* steel of

south India, of course using a different technique. Apparently, a similar white iron or *charka loha* is produced in the Sarguja district of Madhya Pradesh using magnetite sand. The Parsa people of the Mahuli Agarias are adept in this art and the product is used for making weapons. It is of a superior quality and is much more expensive. The metallurgical skill, however, is restricted exclusively to only a small ethnic group of artisans who have perfected the process over the ages.

- The special feature of pre-industrial iron is that it is a high purity wrought iron. It has high phosphorous and low sulphur contents, making it corrosion resistant. This feature is common to pre-industrial and ancient Indian iron.

Discussion: continuity of a tradition

Having described the basic metallurgical process of traditional ironworking and its study by metallurgists, it may be worthwhile examining the question of whether the metallurgy of iron has undergone changes in the course of time. The excavations of remains of ironworking areas reveal only small pits filled with charcoal and slag in association with iron objects. Both the furnace remains and the ancient objects need to be studied closely to be able to reconstruct the actual working of iron which must have been prevalent in ancient times. Some ancient objects have been analysed as have some pre-industrial iron objects. We propose to discuss these briefly in order to enable us to compare ancient and pre-industrial iron. Some important results of the analysis of objects from different periods are given in Tables 1 and 2.

Our energy-dispersive spectroscopy (EDAX) analysis conducted of an iron pickaxe fashioned by Pipara ironworkers indicates 99.48% iron, 0.52% silicon, and 0.2% carbon. Its hardness on the Vicker's scale is 125. This is comparable with the composition shown in Table 1. In this context, the remarks of Hadfield (1913–14) are important:

> ... the very low sulphur content in the two specimens analyzed is in accordance with our experience with practically all previous Indian specimens. The phosphorus content is rather low compared with many of these previous specimens, one of which ran as high as 0.303. At the same time there have been specimens (no. 2441-Khambaba) as low as 0.020. The chief feature of the present specimens is the high carbon percentage, 1.23 in no. 1 in Table 1. The Brinell hardness and the micro-structure of nos. 1 and 2 also indicate they are of similar high carbon. We have never come across any ancient specimens as high as this before.

A comparison with the ancient sample could be made with iron specimens produced by pre-industrial society. McWilliam (1920) studied the traditional iron-making process at Mirjati in the Chhotanagpur plateau (district of Manbhum). He reported: 'the smelters used small clay furnaces, brown hematitic ore (found in the vicinity) and charcoal as fuel. No flux was added. On analysis, it was found to be wrought iron of high purity comparable to Delhi iron pillar as reported by Hadfield (1912).' McWilliam further observed that the 'ore (was) very much alike' and also that by careful manipulation, especially of the proportion of charcoal to ore, a steely iron containing up to 0.9% carbon could be made. It is noteworthy that iron objects unearthed from the Iron Age site of Mahurjari (dated to *c.*900–800 BC) were found to contain 0.9% carbon. A comparison between objects from different cultural periods reveals an interesting pattern, as is evident from Table 2. On the basis of such studies, one finds a unique similarity in the basic composition of iron objects manufactured over a period of nearly 2,000

Table 1 Results of analysis of Taxila iron (after Marshall 1975).

S. no	Description	Brinell hardness		Analysis acc. no.	C (wt%)	Si (wt%)	S (wt%)	P (wt%)	Mn (wt%)
1	Adze for carpenters 5.75	–	–	5. 116	1.23	0.28	0.004	0.024	0.01
	Away from edge	220	240						
		240							
		259							
	Neat cutting edge	236	high carbon						
2	Axe 5.75	103	125 iron	6. 109					
		106							
		167							

Table 2 Analysis of ancient and pre-industrial iron produced in traditional furnaces (after Prakash 1997).

S. no.	Elements	C (wt%)	Si (wt%)	Mn (wt%)	S (wt%)	P (wt%)	Fe (wt%)
1	Delhi iron	0.08	0.05	0.00	0.01	0.11	99.72
2	Iron from Jabalpur, (pre-industrial) recent	0.59	0.110	0.41	0.20	N.A.	rest
3	Pre-industrial iron (produced 1918)	0.06	0.22	trace	0.03	0.02	–
4	Jiragora furnace	0.4	0.2	–	–	–	rest
5	Konark beam	0.110	0.100	0.00	0.24	0.015	rest
6	Iron from Bastar (about 100 years old)	0.25–0.45	–	18.7			

years – from the first centuries of the Christian era (Table 1) to the pre-industrial stage of iron production.

This suggests that there is continuity in the basic metallurgical techniques employed. In this context, our recent investigation in the Sonbhadra-Sidhi region is indeed highly significant. While it shows the survival of the traditional iron technology for which India was famous throughout the ancient world, it also provides a model for understanding the ancient metallurgical processes with greater clarity and opens up the possibility of a revival of a nearly lost technology, a heritage from the past.

Note

1. B. Prakash, pers. comm.

References

Alam, I. (2002) 'Iron manufactures in Golkonda in the 17th century', in G. Pande and J. af Geijerstam (eds), *Tradition and Innovation in the History of Iron Making: An Indo European Perspective*, 98–111. Nanital: PAHAR Parikrama.

Anantharaman, T.R. (1996) *The Rustless Wonder: A Study of the Iron Pillar at Delhi*. New Delhi: Vigyan Prasar.

Balasubramaniam, R. (2002) *Delhi Iron Pillar: New Insights*. New Delhi: Aryan Books International.

Ballal, N.B. (1993) *Interim Report on Indigenous Iron and Steel Technology of India*. Bombay: Department of Metallurgy, I.I.T.

Banerjee, N.R. (1965) *The Iron Age in India*. Delhi: Munshiram Manoharlal Publishers Pvt. Ltd.

Chakrabarti, D.K. (1985) 'The issue of the Indian Iron Age', in S.B. Deo and K. Paddayya (eds), *Recent Advances in Indian Archaeology*, 74–88. Poona: Deccan College Post-Graduate and Research Institute.

Chakrabarti, D.K. (1992) *The Early Use of Iron in India*. New Delhi: Oxford University Press.

Elwin, V. (1942) *The Agaria*. London: Oxford University Press.

Hadfield, R. (1912) 'Sinhalese iron and steel of ancient origin', *Journal of Iron and Steel Institute* 85: 134–74.

Hadfield, R. (1913–14) 'Archaeological survey report', in J. Marshall (1975) *Taxila*, 534–35. New Delhi: Motilal Banarasidass.

Krishnan, M.S. (1952) 'The iron ores of India', *Indian Minerals* 6(3): 113–30.

La Touche, T.H.D. (1918) *A Bibliography of Indian Geology and Physical Geography with an Annotated Index of Minerals of Economic Value. Part I-B*. Calcutta: Geological Survey of India.

Leuva, K.K. (1963) *The Asur*. New Delhi: Bhartiya Adimjati Sevak Sangh.

Marshall, J. (1975) *Taxila*. New Delhi: Motilal Banarisidass.

McWilliam, A. (1920) 'Indian iron-making at Mirjati, Chhotanagpur', *Journal of the Iron and Steel Institute* 102: 159–70.

Prakash, B. (1997) 'Metal and metallurgy', in A.K. Bag (ed.), *History of Technology in India. Volume 1: From Antiquity to c.1200 AD*, 80–174. New Delhi: Indian National Science Academy.

Sahi, M.D.N. (1979) 'Iron at Ahar', in D.P. Agrawal and D.K. Chakrabarti (eds), *Essays in Indian Protohistory*, 365–8. New Delhi: B.R. Publishing Corporation.

Sahi, M.D.N. (1994) *Aspects of Indian Archaeology*. Jaipur: Publication Scheme.

Schwartz, C.R. (1882) 'Report on the principal prospects of ironworking in the Chanda district, Shimla', quoted in M.S. Krishnan (1952) *Bulletin of the Geological Survey of India* 9: 102.

Srinivasan, S. and Ranganathan, S. (2004) *India's Legendary Wootz Steel: An Advanced Material of the Ancient World*. Bangalore: National Institute of Advanced Studies and Indian Institute of Science.

Tripathi, V. (2001) *The Age of Iron in South Asia: Legacy and Tradition*. New Delhi: Aryan Books International.

Tripathi, V. (2007) *History of Iron Technology in India: From the Beginning to Pre-modern Times*. New Delhi: Rupa and Company.

Tripathi, V. and Mishra, A.K. (1997) 'Understanding iron technology: an ethnoarchaeological model', *Man and Environment* 22(1): 59–67.

Tripathi, V. and Tripathi, A. (1994) 'Iron working in ancient India: an ethno-archaeological study', in J.M. Kenoyer (ed.), *From Sumer to Meluhha: Contributions to the Archaeology of South and West Asia in Memory of George F. Dales, Jr*, 241–51. Madison, WI: Department of Anthropology, University of Wisconsin.

Tylecote, R.F., Austin J.N. and Wraith A.E. (1971) 'The mechanism of the bloomery process in the shaft furnace', *Journal of Iron and Steel Institute* 209: 342.

Fine structures: mechanical properties and origin of iron of an ancient steel sword excavated from an old mound in Japan

Masahiro Kitada

ABSTRACT The microstructures and mechanical properties of a Japanese ancient sword excavated from an old mound (AD 200–600) have been investigated. The sword was excavated in 1885 from an old tomb called the 'General's Mound', near Tokyo. Fine structures of the specimen were examined with a scanning electron and transmission electron microscope. The elements of the non-metallic inclusions were analysed by an energy-dispersive X-ray (EDX) method. The mechanical properties were evaluated by tension and hardness tests. A banded structure consisting of soft and semi-hard steels was observed in the cross-sectional micrographs. The banded structure is the result of repeated forging. The carbon content of the soft steel (α-Fe) is 0.05 to 0.1 mass% carbon. The semi-hard steel contains 0.35 to 0.45 mass% carbon. Non-metallic inclusions are arranged along the bands. Transmission electron micrographs (TEM) of the non-metallic inclusions indicate a very complex pattern containing nanometer-size structures which contain iron, silicon, potassium, calcium and other trace elements. The analysed phases are FeO (ferrite), Fe_2SiO_4 and a mixture of FeO and silicate glass. Ferrite is the primary crystallised grain and the mixture of FeO and Fe_2SiO_4 represents secondary eutectic grains. The tensile strength and the elongation are from 450 to 700 MPa (44–69 kg/mm^2) and 15–30%, respectively.

Introduction

Japanese swords have been classified according to their fabrication era and the shape. Usually, swords excavated from an old mound (the Yayoi era: about AD 200-600) are called 'ancient' or 'antique' swords. Since the swords are straight in shape, they are also referred to as 'straight swords'. Two kinds of 'ancient' swords, some having one edge and some with two edges, have been found in excavations. Although swords were sometimes used in combat, some, on the other hand, were also treated as sacred treasures; a similar view was held of the art of swordmaking from ancient times in Japan. Swords made from the Heian era (AD 794–1191) to the present day are called 'Japanese swords'. These swords have a curved shape and are classified as: 'old swords' made from approximately 800–1603, 'new swords' in the first half of the Edo era (1604–1750), 'new-new swords' in the second half of the Edo era (1751–1867) and 'modern swords' from the Meiji era (1868) to the present day.

Ancient swords excavated from old mounds raise many questions about their detailed microstructure, fabrication technique, the materials from which they were made and other features. To clarify these aspects, the microstructures, non-metallic inclusions and mechanical properties of a Japanese ancient sword have been investigated systematically.

Experiment

The swords used for analysis[1] were excavated in 1885 from an old tomb called the 'General's Mound' near Tokyo, built in the Yayoi era (Fig. 1). Although the surfaces of the swords were completely consumed by rust, there was some steel remaining in the interior. The three sectional macrostructures were observed with an optical microscope. Carbon content was analysed by electron probe (50–200 μm ϕ) micro-analysis. The finer structure of the specimen was examined with a scanning electron and transmission electron microscope. The elements of the non-metallic inclusions were analysed using an energy-dispersive X-ray (EDX) method. The mechanical properties were evaluated by tension and hardness tests and by fractography.

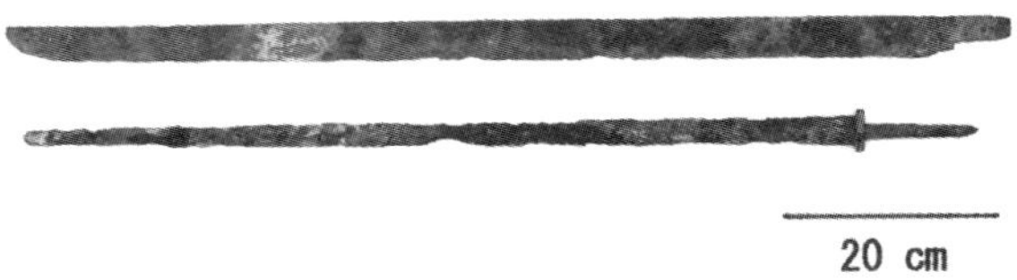

Figure 1 Straight sword excavated from an old mound.

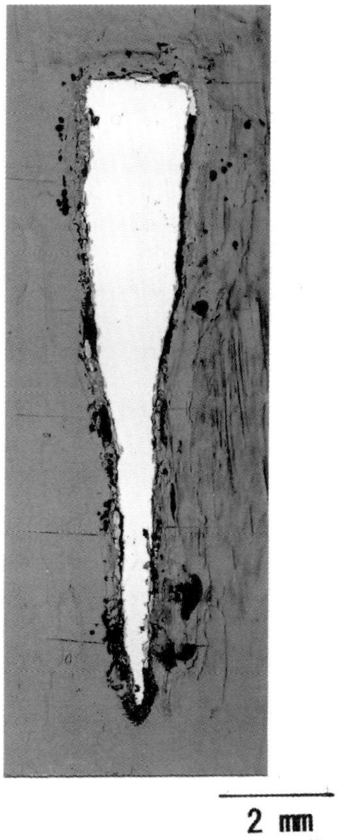

Figure 2 Cross-sectional shape.

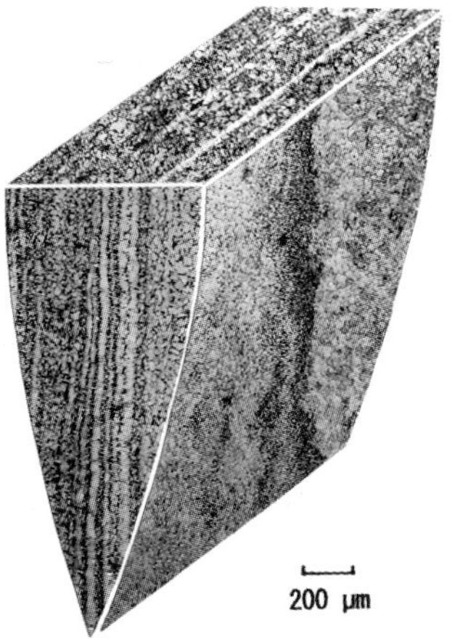

Figure 3 Three-sectional optical micrograph.

Results and discussion

Macro- and microstructures

The cross-sectional macrostructure of the as-polished sword is shown in Figure 2. The surface was consumed by rust, so that the height and width of the sword were smaller than the original size. No quenched area (martensite) was observed in the macrograph after etching.

A three-sectional optical micrograph of the specimen is shown in Figure 3. A banded structure consisting of soft and semi-hard steels was observed in the cross-sectional micrographs. The magnified micrograph of the cross-section is shown in Figure 4. The soft and the semi-hard layers were α-Fe and medium carbon steel, respectively. The thicknesses of the soft and semi-hard layers in the banded structure are approximately 50 μm and 150 μm, respectively. The semi-hard steel layer consists of α-Fe and pearlite grains. Since the crystal shape of α-Fe is polygonal, the structure is in a recrystallised and/or normalised condition. This indicates that the steel was air cooled from the forging temperature.

The carbon content of α-Fe is 0.05–0.1 mass% carbon. This falls under the category of the softest steel. The semi-hard steel contains 0.35–0.45 mass% carbon. The grain diameters are approximately 50 μm for the α-Fe and 20–60 μm for the medium carbon steel. The banded structure is comparatively uniform.

It was not possible under the optical microscope to distinguish a pearlitic structure in the semi-hard steel, but this was revealed by observations using an electron microscope. A transmission electron micrograph (TEM) of the pearlite structure is shown in Figure 5. The diameter of the cementite needles is approximately 0.05 μm. This means that the cooling rate from the forging temperature was comparatively high, although martensite was not produced.

To determine the origin of the banded structure, some possibilities concerning the original steel and the forging processes are shown in Figure 6: (a) a duplex process; (b) decarburised (or carburised) steels in layers and (c), two steels having different carbon contents. In modern steelmaking, when primary segregation in an ingot of low car-

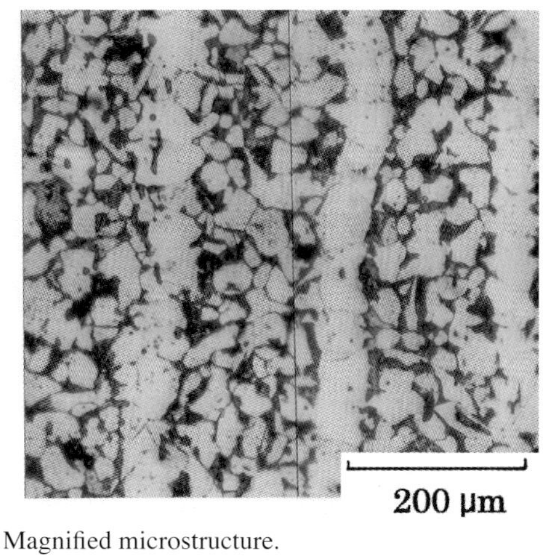

Figure 4 Magnified microstructure.

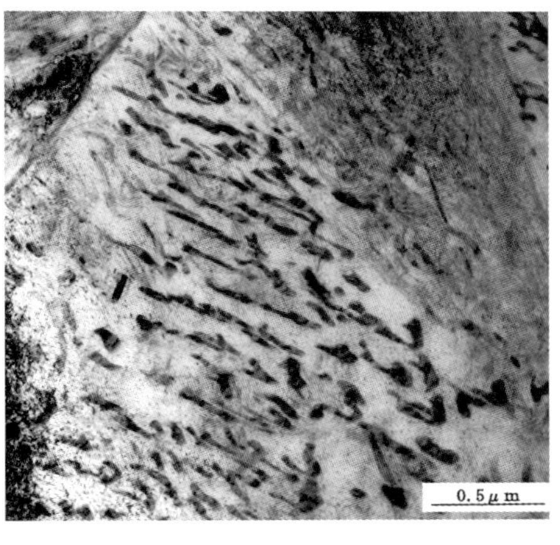

Figure 5 TEM of the pearlite structure.

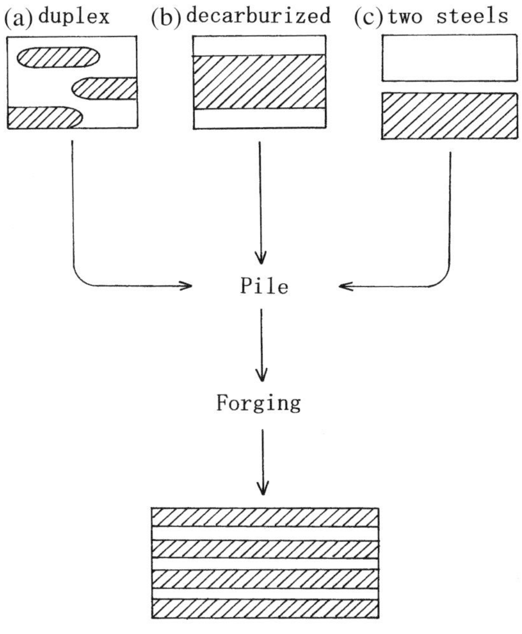

Figure 6 Possible fabrication techniques.

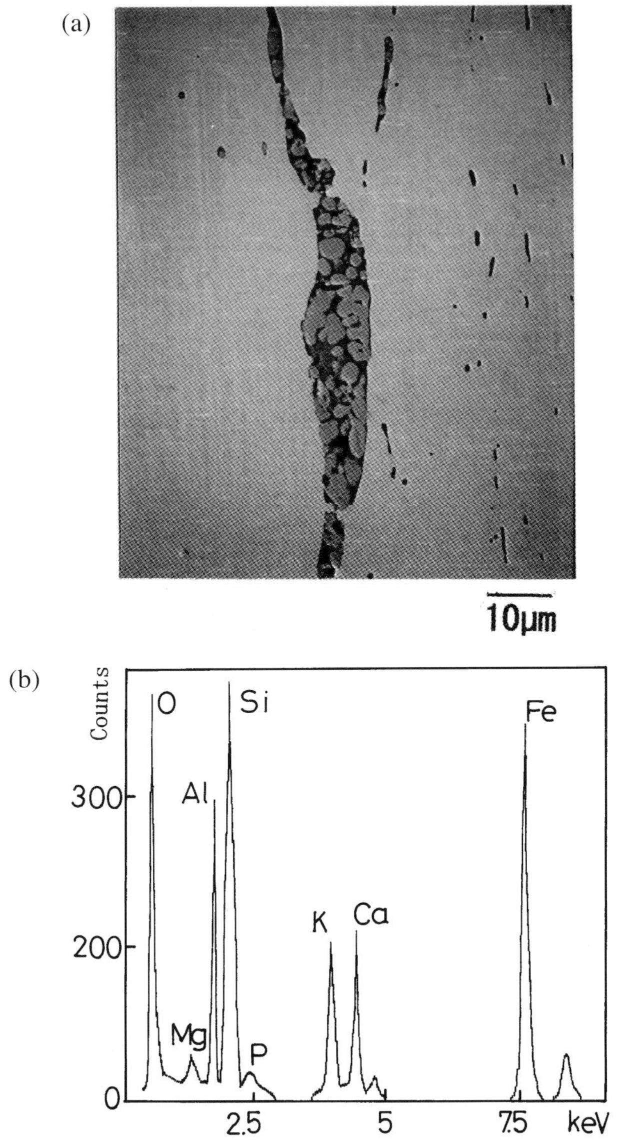

Figure 7 SEM image (a) and EDX spectrum (b) of inclusion.

bon steel is observed, a banded (striped) structure is often observed after hot forging. The techniques employed for (b) and (c) are more advanced than that of (a). In the case of (b) or (c), turned forging must have been used to produce the banded structure and the number of forgings must have been 5–6 (roughly estimated from the number of layers).

Non-metallic inclusions

Non-metallic inclusions are arranged along the banded structure and most of them are present near the semi-hard steel bands. Small and large size inclusions were observed. The typical microstructure and EDX spectrum of the inclusions are shown in Figure 7. The inclusions contain iron, aluminium and silicon, and other elements. The free energies ($-\Delta G$) of the impurities are higher than that of iron. The inclusions are a kind of segregated impurity and are the cause of the banded structure, as described above.

Titanium is usually detected in non-metallic inclusions in most Japanese swords because Japanese iron-sand contains ilmenite ($FeTiO_3$). However, no titanium was detected in the inclusions in the straight sword examined here. Therefore, it may be said that Japanese iron-sand containing titanium was not used to fabricate the straight sword. Iron made overseas, for example India, China and other Asian countries, does not usually contain titanium. It is possible that the iron materials were imported from those countries, either as bulky pig iron or as finished swords (ready made or made to order). On the other hand, if Japanese titanium-free iron ores, for example limonite ($2Fe_2O_33H_2O$), hematite (Fe_2O_3) or magnetite (Fe_3O_4), were used to make iron, titanium would not be detected in the specimen. This is unlikely to be a strong possibility because the smelting of bulky ores is difficult. One cannot exclude the remote possibility that titanium was removed by a refining process. In addition, it is noteworthy that phosphorus is detected.

From optical micrographs of the non-metallic inclusions, two kinds of inclusions whose size and microstructure are different were observed as described above. One category comprises the large inclusions which contain several different precipitates. Since the microstructures of these large inclusions clearly show that solidification has occurred, it proves that the inclusions were liquid at the elevated (forging) temperature. Typical fine structures of the large inclusions are shown in Figure 8a. The TEM indicates a highly complex pattern containing nanometer-size structures. The analysed phases are FeO (ferrite), Fe_2SiO_4 and a mixture of FeO and silicate glasses (amorphous). Ferrite is the primary crystallised grain and the mixture of FeO and Fe_2SiO_4 represents secondary eutectic grains. Saw-like crystals which have grown epitaxially from the residual liquid on the surfaces of primary solidified crystals are observed.

Furthermore, very fine structures containing nano-precipitates in an amorphous phase were observed to have solidified finally from the residual liquid. This result shows that the residual liquid separated into two phases: amorphous and crystalline. A typical TEM of the small inclusions is shown in Figure 8b. The inner structures are very simple, consisting of primary precipitate Fe_2SiO_4 crystal and

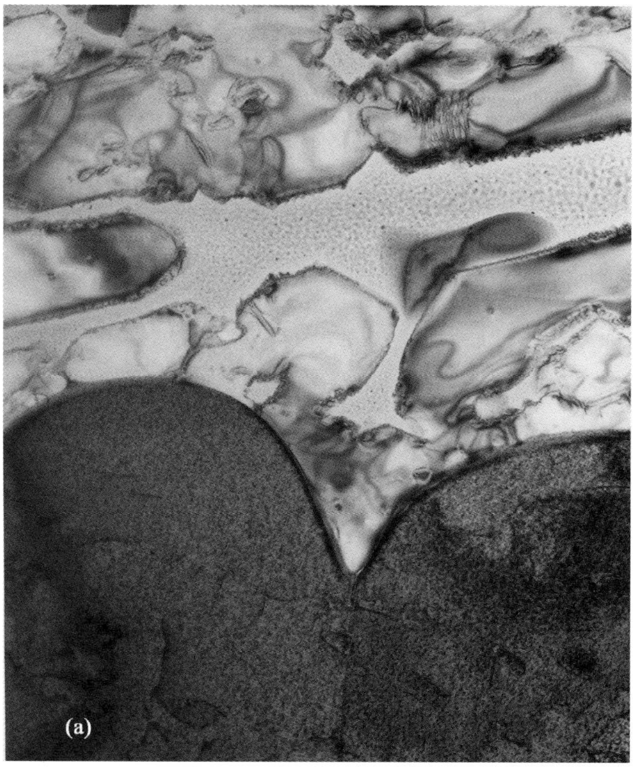

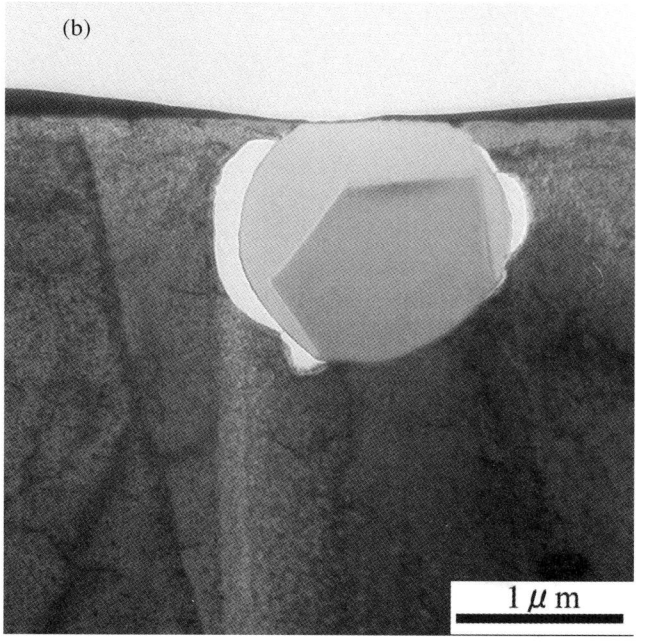

Figure 8 TEM of non-metallic inclusions: (a) large inclusion; (b) small inclusion.

secondary solidified amorphous glass. Where inclusions are surrounded by voids, this may have been a result of shrinkage when a molten inclusion solidified.

The structural difference between the large and small inclusions is thought to be due to the differences in the cooling rate from the forging temperature and differences in composition. The large inclusions could be broken up into the smaller inclusions by forging. The other possibility is that the large inclusions were used as a flux to weld the iron blocks together in the solid state (during forging), in the case of (b) and (c) in Figure 6. The flux removes the surface oxides which prevent solid state welding. Thus, the flux is liquid and shows high fluidity at the forging temperature.

On the other hand, the structures of the small inclusions, as shown in Fig. 8b, are relatively simple; only two phases, the primary crystallised Fe_2SiO_4 and a secondary solidified amorphous phase, are observed.

Mechanical properties

The test pieces were cut along the longitudinal direction of a sword. Both sides of the test pieces were twisted to tension the specimen (Fig. 9). The diameter of the test pieces was 2–3 mm. The typical stress–strain curves are shown in Figure 10. The tensile strength and the elongation are from 450 to 700 MPa (44–69 kg/mm^2) and 15–30%, respectively. The tensile strength and elongation are equivalent to those of modern steel, containing 0.2–0.5 mass% carbon. The difference of obtained values is due to a variation in microstructures.

From the observation of fracture surfaces, it may be said that the non-metallic inclusions and the banded structure

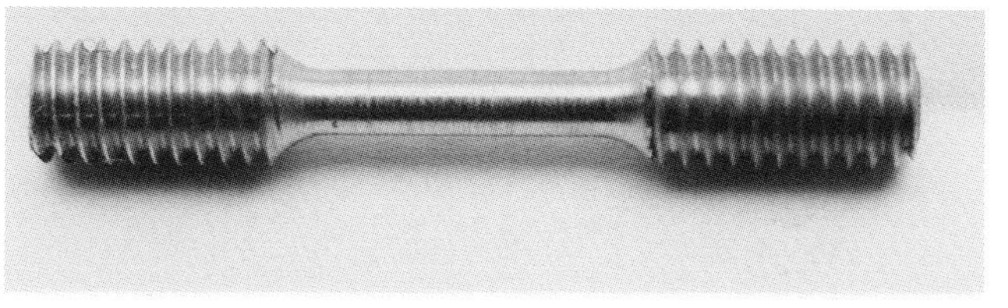

Figure 9 Test piece.

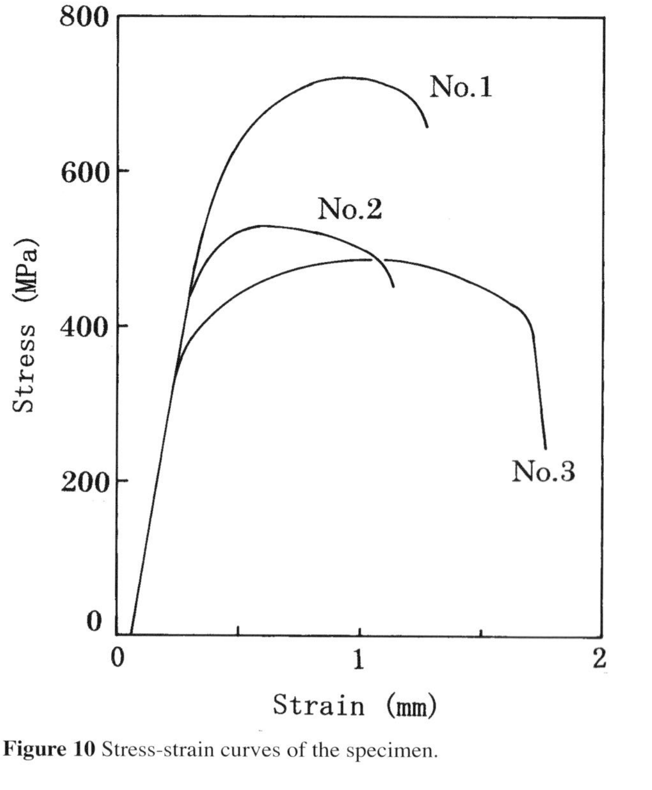

Figure 10 Stress-strain curves of the specimen.

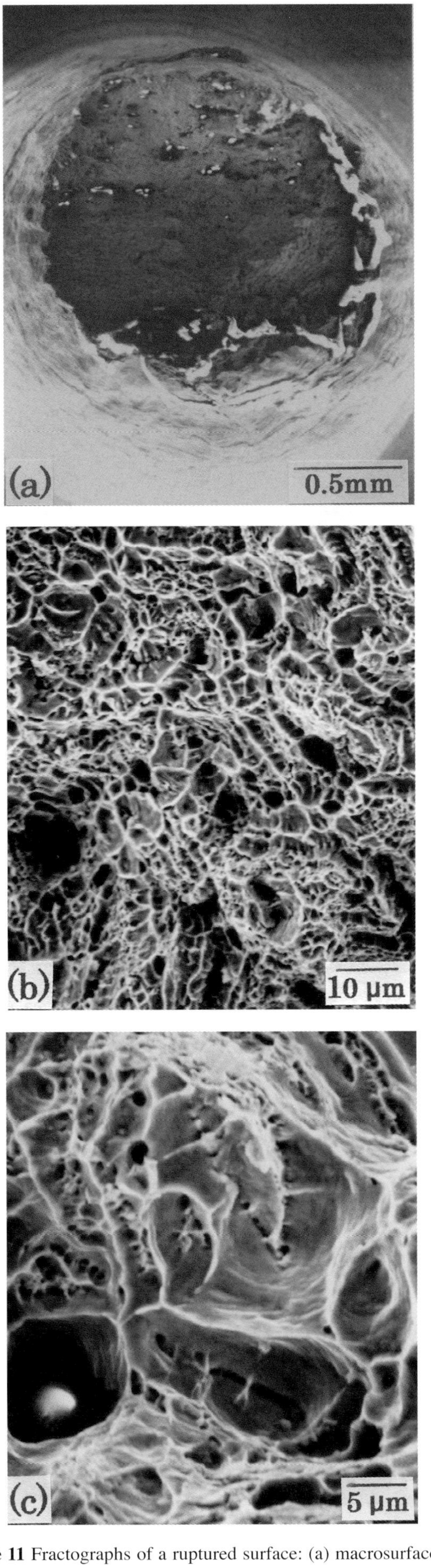

Figure 11 Fractographs of a ruptured surface: (a) macrosurface; (b) dimple pattern; (c) cave-like pattern.

strongly affect mechanical properties. The Vickers hardness of the cross-section containing the banded structure of α-Fe and a medium carbon steel did not depend on the point of measurement. This is because the semi-micro band structure is uniform. The Vickers hardness numbers (load:10g) for the α-Fe and pearlite grains are, on average, 218 and 345, respectively.

A typical fractography (fracture surface inspection) of the test pieces is shown in Figure 11. Many non-metallic inclusions are observed in the macrosurface (a). Most of the fracture surfaces showed a dimple pattern (b). This means that the fracture mode is ductile. However, the pattern size of the fracture surface around non-metallic inclusions is larger than that of ductile fractures (c).

Conclusions

Fine structures, non-metallic inclusions and mechanical properties of an ancient Japanese straight sword were investigated. The sword has the banded structure of soft and semi-hard steels and the pearlite is fine. Non-metallic inclusions contain iron, silicon, potassium, calcium and other metal elements. The inclusions show a solidified structure. The origin of the iron used in making the sword was discussed from the point of view of the detected elements.

Note

1. The specimens are the property of the author.

References

Kitada, M. (2005) 'Metallography of Japanese swords', in *Proceedings of the Symposium on the Arts and Science of Archaic Steel and Sword-making in Japan and Nordic Countries*, 34–55. Stockholm: UNESCO.

Kitada, M. (2006) 'Microstructure of Japanese ancient swords excavated from an old mound', in *Proceedings of the Spring Meeting of the Japan Institute of Metals*, 361.

Satou, T. (1960) *Iron and Steel*, Y. Imai (ed.), 59. Tokyo: Asakura-shoten [in Japanese].

Tawara, K. (1953) *Scientific Investigation of the Japanese Sword*, 378–419. Tokyo: Hitachi-hyoronsha (originally published in *Iron and Steel* 5 (1919) [in Japanese].

Specialisation in iron- and steel-making in the early Middle East and Central Asia: myths, assumptions and a reassessment of early manuscript evidence

Brian Gilmour

ABSTRACT It has not been firmly established how steel was made through much of the 1st millennium BCE and later. With the lack of detailed analytical work on which to draw, the theory still most commonly assumed is that iron was smelted in a largely carbon-free state then converted to steel by carburisation (or cementation). Written sources from about the 3rd century CE give details of the manufacture of crucible steel, and there is a considerable amount of written evidence from the 8th to the 11th century CE, although much of this has not been fully and systematically studied. The aim of this paper is to examine this evidence and determine what new insights can be gained into the way iron and steel was made and used over large parts of the Old World during the last half of the 1st millennium CE, and whether it provides a better framework around which to work out future research programmes in this field. The aim also is to examine how it enables us to recognise and assess just how scanty is the parallel archaeological evidence, and what can be done to improve it

Introduction

The detailed investigation of iron and steel artefacts, together with the wider study of ferrous metalwork, its technology and the cultural framework into which it fitted, has lagged behind comparable studies of objects made from copper alloys and the other main medieval and ancient metals. This must be due to the poor survival of ferrous metalwork and to the assumption that ironwork was, from early on, a utilitarian metal in contrast to gold, silver and copper alloys. Compared to other metals in antiquity, ferrous metalwork is also more difficult to study analytically, and this is rarely done despite the fact that the analytical study of metal can reveal not only its identity and type of iron or steel but also a great deal about its manufacturing history. Early written sources and their descriptions of the making and use of iron and steel are also very little known or recognised, although some highly informative early texts do exist.

The cultural framework within which this branch of metalworking developed is little considered, even though it can provide an understanding of why a particular technique or alloy was chosen. Perhaps least understood but most important of all is the development of the making of steel. But it is difficult to approach these problems until the scanty and uneven nature of the archaeological and other evidence relating to early ferrous metalwork is recognised, and how to make best sense of it. Instead of the rediscovery of a manufacturing and development sequence based on evidence, the existing interpretation about how and why iron and steel were made has come from a retrospective extrapolation of modern ideas and practices.

An additional problem is that most ironwork survives because it was deposited either in the ground or a watery place with no intention to recover it, although some deposits of ironwork were simply stored but never retrieved. The survival of iron is uneven both chronologically and geographically and, if we are to understand what is left – the types and status of artefacts as well as how representative they are likely to be – it is important to understand the circumstances surrounding the deposition of artefacts, as these will explain why the survival of ferrous metalwork is so variable in different parts of the world.

Unfortunately because so little is known about early iron and steel, various assumptions and myths about the metal have tended to serve as a replacement for actual evidence and an understanding of how and why it developed. This problem should be recognised by working towards a more consistent method of looking at the development of early iron and steel technology. The aim of this paper is to take an initial look at what is known of iron and steel in the wider Middle Eastern region, the generally accepted birthplace of the technology,[1] but an area where the very uneven or sparse nature of the evidence makes it difficult to examine and reveal the exact nature of the development of the technology.

Evidence for early ironworking

The biggest problem is that so little survives of the ferrous metalwork that once existed, and also that little of what has survived has been examined analytically, and hence identified. Steel may well have been known and exploited since the early stages of the Iron Age and might soon have followed the intentional exploitation of iron whenever and wherever that occurred. The skilled use of steel can be seen in swords which are typologically early, such as those with skeuomorphic hilts[2] from Luristan in western Iran, dated to the late 2nd millennium BCE (Rehder 1991: 13–14). Unfortunately because very little early iron and steel has been systematically analysed or even recovered, not much is known about how and when they were made, what they were used for, how production techniques became more specialised etc. The overall nature of the industry and the long timescales involved may only become apparent by combining scattered fragments of archaeological and textual evidence. For instance, the scale of the industry by the late 8th century BCE can be gauged from the enormous hoard of 163,000 kg of iron in the form of bars and the like (Pleiner and Bjorkman 1974) found in a strongroom at the short-lived Neo-Assyrian Palace of Sargon II at Khorsabad in the upper reaches of the Tigris River (Fig. 1).

Much of this is in the form of fish-shaped bars of trade iron known also from other Neo-Assyrian sites, but it seems to largely disappear from the archaeological record. However, a passing but detailed description of a bar of trade iron of exactly this size and form occurs some 1,500 years later in a treatise on the use of different types of iron and steel for making swords, written in Baghdad in the early 9th century by the Iraqi scientist and polymath Ya'qūb al-Kindī (discussed in more detail below).[3] Al-Kindī reports that these bars were known as *shabābīt*, a species of fish 'slender in the tail, wide in the middle part, soft to the feel, small in the tail' and were evidently still being made and traded in south-west Iran at the beginning of the 9th century,[4] some 1,500 years later (Hoyland and Gilmour 2006: 41 and note). The tradition of trade iron in the form of biconical bars continued in parts of Europe to at least the 14th or 15th century, when a batch of biconical bars formed part of the cargo of a ship that was wrecked near the mouth of the Ria Oria in the Basque country.[5]

Steel also represents a serious problem, as it has not yet been firmly established how this iron alloy was made through much of the 1st millennium BCE and later. Until recently, with little detailed analytical work on which to draw, the most common assumption was that iron was largely carbon-free when smelted, and then converted to steel by carburisa-

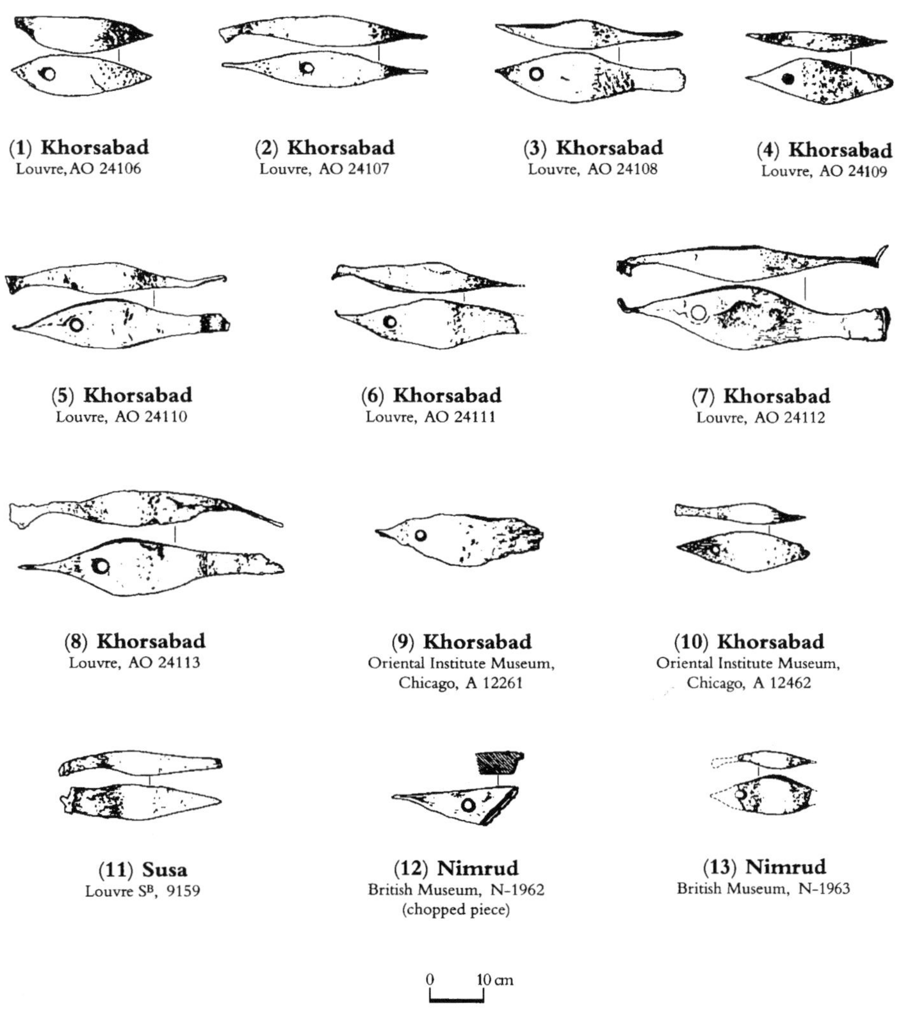

Figure 1 Fish-shaped bars of iron from later Assyrian sites.

tion rather than being made by direct (bloomery) smelting. It is known that written sources exist detailing the manufacture of crucible steel from at least as early as the 2nd century (Giumlia-Mair *et al.*, in this volume, pp. 35–43). The best and most detailed written evidence comes from Middle Eastern sources of the 8th to the 11th century CE, although much of this has not been fully and/or systematically studied. Most of this material is not well known or has not been studied in detail despite being often quoted. Previously unknown written sources or sources thought to be lost occasionally come to light, as has recently happened with the rediscovery of another 9th-century treatise relating to iron, steel, swords and heat-treatments by al-Kindī (Hoyland and Gilmour 2006: 10, n. 25).

The aim of this study is to look at the evidence and see what new insights can be gained into how iron and steel were produced and used over a large part of the Old World outside China (in this case the Middle East and Central Asia) during the second half of the 1st millennium. It is hoped to build up a better framework around which to work out future research programmes in this field and to examine how to recognise and assess how scanty the parallel archaeological evidence is and what can be done about it.

The finds of early iron are uneven in distribution and over time, and few of these iron finds have been properly analysed and identified. We also need to reassess the developments in iron manufacture, iron use and disposal in the past, how iron might have fitted into earlier cultures, and how important or valuable – economically and culturally – it might have been as a material. Finally, we need to work out how the metal actually looked, and if this can be deduced from archaeological and written sources. What appears to be a widely held view certainly needs to be confronted, namely that iron technology is likely to have been carried out at a rather rudimentary level until relatively recent times.

Swords are an exception to this view of iron, but unfortunately even here many modern myths persist about the qualities, composition, development and use of these weapons. 'Damascus steel' is perhaps the most entrenched and misleading myth among those relating to iron and steel. The manufacture of swords often tends to be regarded as a now lost secret art, only passed from generation to generation by a few select practitioners. These may be thought of as having almost supernatural skills and abilities, a view reflecting a merging of a modern idea of iron and its products with an earlier (Western epic) story-telling tradition left over from a pre-literate era. There are some Roman and pre-Roman European sources on ironworking, however it has seemed that little follows until a 1000 years or more later. This apparent state of affairs may, to a surprisingly large extent, simply reflect the lack of detailed research that has been carried out into the existing source material, both written and archaeological or artefactual, or the lack of realisation that it exists at all.

Specialisation in early iron production and use: what do we know?

Iron and steel technology may have ultimately derived from a common origin in Asia Minor in the early 2nd millennium BCE; however, it is clear that this developed along different lines in three wider regions: Europe, the Middle East

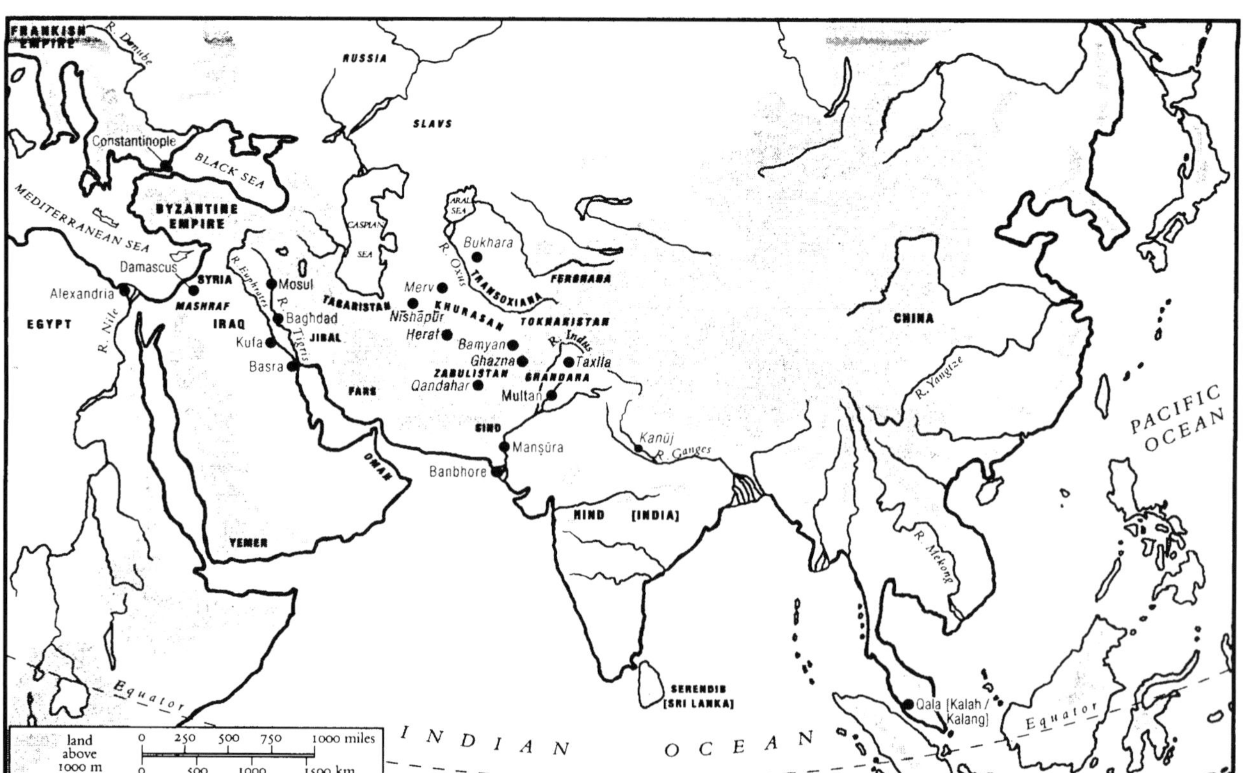

Figure 2 Map of the Old World showing all the places and regions mentioned in various written sources from the 2nd to the 11th century.

and central southern Asia, and China. It would appear that iron and steel technology developed separately in China, although it is possible that early developments in Central Asia were influenced by contemporary Chinese practices. The development of ferrous technology in China is being reconstructed from a wide variety of Chinese written sources, reinforced by analytical work, particularly on cast iron. This gives a good idea at least of how cast iron production and use progressed over the two millennia following the introduction of cast iron technology in this area, about 2,500 years ago.

Bloomery iron production reached China in about the 10th century BCE and, by some time in the 6th century BCE, cast iron was being put to a variety of uses, often on a large scale, such as the manufacture of blades to fit on wooden spades. Cast iron must have been known early on in other parts of the Old World, but it was only in China that it was used on any significant scale before the development of blast furnace technology in later medieval Europe, some 1,500 years later. And even here it was not really used much for castings before the 17th century.[6] The reason for the early and rapid development of cast iron technology in China was that the existing large-scale bronze-casting technology was readily adapted to the new material. In its turn, this technology developed as a consequence of the ready availability of copper and tin, especially from the vast deposits of high grade copper ores at Tonglüshan in Hubei Province in central China.[7]

In Europe a very different mass of evidence exists for reconstructing developments in iron and steel technology over this same 2,000-year period. Here there are a few useful written descriptions dating to before the 16th century, and a great deal of ironwork relating to this long time-span has been recovered from European sites, although the survival of ironwork is very uneven in terms of the different kinds of artefacts that survive at different times. Also very little systematic analytical work has been carried out on this material. The complexities of iron and steel production and use are only now beginning to emerge. However, in marked contrast to the contemporary developments in China, it is clear that the bloomery iron-making technology continued largely unchanged throughout this period, although we do not yet understand how the production technology was actually carried out in detail due to the lack of analytical information on the many surviving artefacts.

By far the most important development of all early ferrous technology is steel-making, but surprisingly this is still poorly understood both in Europe and China. In Europe there has long existed the widely held assumption that this was done by taking iron, produced by the bloomery process (assumed at this stage to be largely carbon-free), then getting this iron to absorb carbon by the process known as carburisation or cementation. However, there appears to be almost no real evidence to support this view, which seems to be the result of a retrospective extrapolation of post-medieval and later practices by archaeometallurgists who have performed investigatory work in the past century. This view, however, does not stand up to scrutiny and there is a small but growing body of evidence – archaeological and textual – indicating that steel was made as part of the bloomery process. This is a more logical interpretation as to how steel was made, and also leads to a better idea about how bloomery iron-making was actually carried out.

The most critical, but also the most elusive archaeological evidence for bloomery steel-making is represented by the blooms and especially the billets of consolidated metal made from these. Few iron blooms of any kind survive and small corroded pieces of billets are even less likely to make it through the archaeological process of being recovered and saved, then identified and analysed to reveal them for what they are, given that so little ironwork is analysed at all. Nevertheless, inhomogeneous steel blooms and billets have been identified in Britain. These are the 2nd-century Roman material from a smithy at Carmarthen in South Wales (Salter 1997: 100–101) and the remaining part of an ultra high carbon (hypereutectoid) steel billet from a Middle Saxon (approximately 8th century) context at West Heslerton in North Yorkshire (Fig. 3).[8]

As might be expected, this surviving material bears a remarkably close resemblance to the hypereutectoid *tamahagane* steel from one of the giant blooms still made by the *tatara* process in Japan, the one remaining example of the early bloomery process that has continued to operate up to the present day (Fig. 3). The *tatara* blooms are very heterogeneous. They were broken up and the iron was divided into

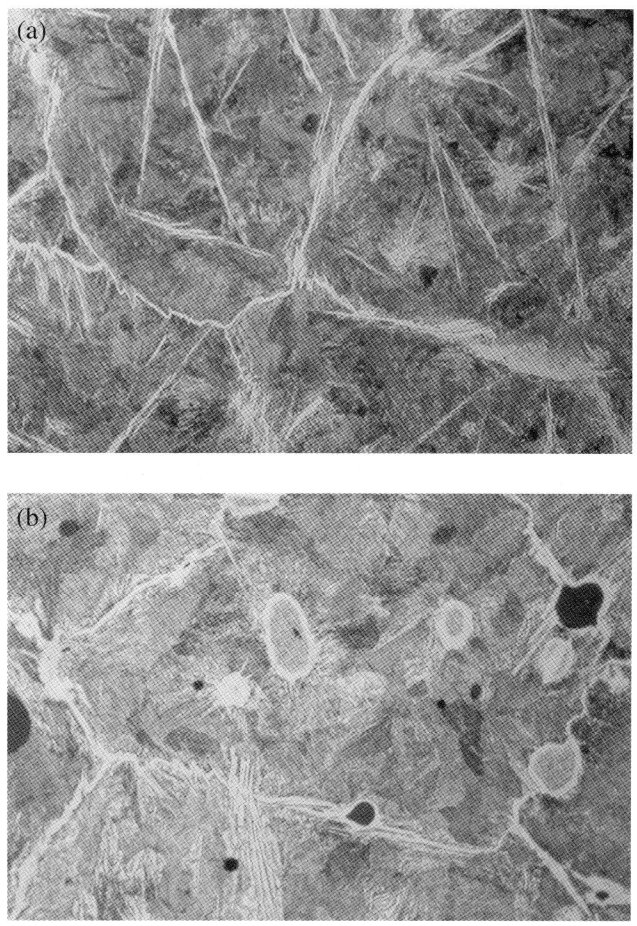

Figure 3 Ultra-high carbon microstructures of the Carmarthen 2nd-century steel billet (a) and of a piece of *tamahagane* from the steel part of a *tatara* bloom (b).

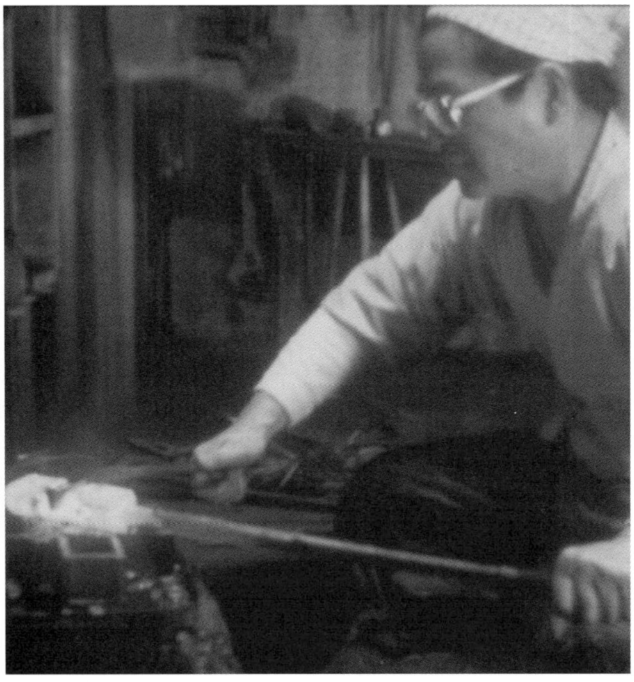

Figure 4 Controlled decarburisation and homogenisation of a piece of ultra-high carbon *tamahagane* (approximately 1.5% carbon but variable), prior to this being used as a piece of high carbon steel (approximately 0.7–0.8% carbon) in a traditional Japanese sword.

different grades depending on its carbon content, *tamahagane* being the highest grade. In its raw state, as extracted from the bloom, its carbon content is very high but variable, typically 1.2–1.7%. It is in this state that it is bought by the swordsmiths, who then process this ultra-high carbon steel – far too brittle in its raw state – before it is suitable to be incorporated into a sword blade.[9]

This processing takes the form of controlled decarburisation in which a consolidated bar of raw steel is coated with a thin slurry of clay mixed with rice-straw ashes to prevent excessive decarburisation (Fig. 4). The bar is then heated and forged out, folded in half along its length, and the process is repeated several times until the swordsmith judges that the steel has the proper carbon content (approx. 0.7–0.8%), and is also homogenised. The similarity of the 2nd-century Roman Carmarthen steel bloom and the 8th-century steel billet from West Heslerton to the ultra-high carbon *tamahagane* steel suggests that exactly the same process was carried out on the lumps or blocks of raw bloomery steel, to produce a usable high carbon (but not hypereutectoid) steel.

In China the process known variously as the 'thirty-fold', 'fifty-fold' and 'a hundred-fold' refining, known from written sources from the 1st century, may refer to a similar process for the homogenisation of steel. Some swords are actually inscribed as having been made using this material or process. This would have been done by forging out an inhomogeneous steel bar, then folding it and repeating the process until there were approximately 30 (or 50 or 100) layers.[10] Thus far the process is very similar to that used in Japan, and (with a fair degree of certainty) also in Roman and early medieval Britain for the manufacture of usable homogenised high carbon steel, made by and large for use with iron in composite welded artefacts such as edged weapons (especially swords) and tools, coin dies and the like. More research on suitable material is needed before it can be determined whether the steel that was being put through the 'thirty-fold' (etc.) refining process really was bloomery steel, as would seem likely by analogy with the evidence from both Britain and Japan.

Crucible steel: ferrous specialisation in Central Asia and the Middle East

Early ferrous technology is poorly understood in both Europe and China, but until very recently it has been a virtual blank in the Middle East and Central Asia. What is clear is that at some point bloomery technology gave way to crucible processing technology for the production of steel in the Middle Eastern region. We do not yet know when this happened or why, but we can be certain from two key pieces of evidence that it had happened in the early 1st millennium, and most probably began several centuries earlier.

First, a sword and an adze were recovered from 1st-century Parthian contexts during archaeological excavations at Taxila in the 1930s, and subsequent analytical investigation indicated these to have been made of ultra-high carbon crucible steel, by far the earliest to have been identified. Secondly, the earliest surviving description of crucible steel-making in this region is by Zosimos of Panopolis[11] who gives one crucible recipe and some associated description possibly deriving from a source several centuries earlier than his time of writing in the 2nd century.[12]

Several significant written sources of the later 1st millennium provide a great deal of information about iron and most particularly steel-making in the Middle Eastern region, especially when linked to information gained from a large body of Arab poetry from the late pre-Islamic and early Islamic period.[13]

Regrettably there is almost no archaeological or analytical database to set against the written sources, since very few specialised artefacts such as swords appear to survive, and little medieval and earlier ironwork of any kind has been recovered, much less scientifically investigated.

Myths and assumptions: why do these matter?

The discussion of myths and assumptions relating to early iron- and steel-making could make a study all of its own, but it should suffice here to draw attention to two of the most persistent and misleading examples. As mentioned already, one of the most persistent myths is that concerning 'Damascus steel'. The idea that Damascus was a great steel and sword-making centre in the medieval and pre-medieval period, stretching back into the 1st millennium BCE, has become ingrained and has found its way into many articles and books written on iron and steel, especially those relating to the Middle Eastern region. Like so many myths, once they get repeated and written down, they tend to become accepted as fact, even if the basis for this has been forgotten or was non-existent in the first place. Just as misleading is the persistent but baseless myth that specialised iron- and steel-making was a secret craft only practised and passed on by a few chosen practitioners.

Why should myths like these matter to us? They matter a great deal because they act as a kind of smokescreen to distract us from what really happened. Even worse, they also distract us from the need to question and undertake the research necessary to determine what really happened and lead us to accept background ideas or a 'given' framework into which to fit the results of bits of research, especially analytical results for iron.

Without the necessary questioning followed by research, we cannot hope to reconstruct the main aspects of the complex development and specialisation of iron manufacture and use across the world, and to understand something of the prevailing cultures and events which had a bearing on this. Research must be targeted and we must think very carefully about what individual researches have actually contributed, and where the gaps in our knowledge still lie. As long as we do not succeed in this aim, iron will continue to languish in the doldrums in most research agendas and our understanding of both the material and its place in society for the past 4,000 years from the Late Bronze Age will remain limited.

As early as the 17th century, a well-known scientist at the Royal Society of London, Robert Boyle, sent an associate to Damascus to find evidence of the famous steel that was supposedly made there. Needless to say, he failed to find any trace of it, but in response to his enquiries he was directed away from the city, ever further east, on a search that proved fruitless.[14] A more recent search for medieval references to this claimed industry also found no evidence for it (Elgood 1994: 103–9). That is not to say that early crucible steel was never made in or near Damascus, as the city was one of many places where it may have been made at some stage. This much we can say from a study of the known manuscript evidence which has recently been critically examined.

Early treatises on iron and steel and other manuscript evidence

Very little ironwork has so far been recovered and studied in the Middle East, Central Asia and the Indian subcontinent. But at least for the period from *c.*500 to 1500 BCE, there is some extremely useful written evidence particularly that of the later 1st millennium, mostly written in Arabic, which provides a framework for understanding this industry. By far the earliest of these surviving manuscript descriptions of early iron- and steel-making is that of Zosimos of Panopolis, a Greek-speaking Egyptian who lived in Egypt around the 2nd century (see Giumlia-Mair *et al.*, in this volume, pp. 35–43). One crucible steel recipe and some additional description from this were transcribed and translated into French in the late 19th century.[15] In this recipe, the steel is made by the carburisation of small pieces of plain or low carbon 'soft' iron mixed with carbonaceous matter – dried date skins and husks of palm fruit – plus powdered glass-workers' *magnesia* (probably manganese dioxide to offset the deleterious effects of any sulphur present) in the clay crucible.

Al-Kindī's two treatises relating to iron and steel and their use for swords

Many scholars working on iron and steel, particularly those interested in the early crucible steel industry of the Middle East and Central Asia, have long been aware of the existence of a relevant body of early manuscript material written in Arabic. Surprisingly little detailed work has been done on this material, despite its obvious potential not only for facilitating understanding of the development and specialisation of this industry and its applications, but also providing information on who wrote down these early technological descriptions and why.

The key document that provides a framework for early iron- and steel-making is the longer and better known of two treatises on this subject written by al-Kindī (Fig. 5). This was written in the form of a letter addressed to his patron Mu'tasim, the Abbasid caliph in Baghdad (832–841). The long and very detailed document describes, compares and discusses swords and the metal used to make them across the Middle Eastern region. Although long known, this treatise has only very recently been fully translated and studied in detail (Hoyland and Gilmour 2006).

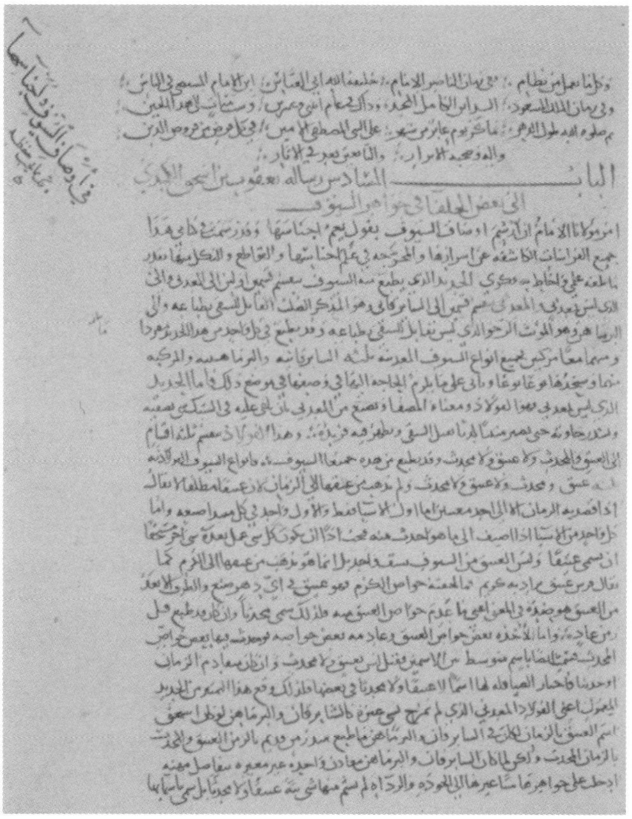

Figure 5 Page 1 of the Leiden manuscript copy of al-Kindī's sword treatise.

In this treatise, al-Kindī outlines the main varieties of iron in use in his day, explains how they are used for sword-making, and compares and contrasts the quality of the main varieties. The different main categories of iron are divided according to whether they are 'mined iron' (i.e. directly smelted bloomery of one sort or another or cast iron), or 'unmined' iron (i.e. indirectly processed iron, by which he means crucible steel (*fuladh*)). Each main form of iron is divided into the principal subcategories, and these are further subdivided in the form of a family tree for iron products (Fig. 6). The different types of swords are thus grouped under the different varieties of iron from which they are made, and these are divided into subcategories mainly according to origin (Hoyland and Gilmour 2006: fig. 8). He explains that different types of bloomery iron and steel are used in different parts of Europe for both composite and non-composite swords, thus providing more information about how iron and steel are made and used in Europe than any other medieval source. However, the main bulk of the text is concerned with the various types of swords made from crucible steel, and how to discern the differences. Even different testing procedures for this are described.

The emphasis on the different varieties of swords and on the ways to test them suggests that the treatise may have been inspired by a wider concern and perhaps a need or wish by the caliph to look at the problem of fraud in the manufacture and selling of swords across the Abbasid world. Clearly the fraudulent practice of making less expensive swords and passing them off as some of the more sought after and expensive varieties was rife at that time. Al-Kindī's sword treatise is certainly outstanding in the information it gives and framework it provides in helping us to understand early iron and steel use, particularly when compared with evidence from other written sources. Essentially, it is a detailed contemporary account and therefore a snapshot of what was happening in his day.

Given the overall detail and depth of discussion in al-Kindī's sword treatise, one might also expect to see some description of the manufacture of crucible steel and of heat treatments for steel, which he mentions in passing. This odd-seeming gap is explained by a second, virtually unknown treatise, which deals with these very aspects of steel-making, with nine crucible steel recipes plus many different quenching solutions and procedures.[16] This forgotten text was listed by a 10th-century bibliographical compiler, Ibn al-Nadīm (d.990), as one of two miscellaneous treatises on swords and iron by al-Kindī, and only came to light very recently (Hoyland and Gilmour 2006: 5, 9–10, n.25–26). Curiously we now know that it survives in at least two different volumes of manuscript compilations. One is in Turin and a more complete copy, unrecognised until now, is preserved in a well-known 14th-century Mamluk book on horsemanship.[17]

As with all early crucible steel recipes, the basic starting ingredient takes the form of small pieces of 'soft' or plain (low carbon) iron and this is mixed with a series of other

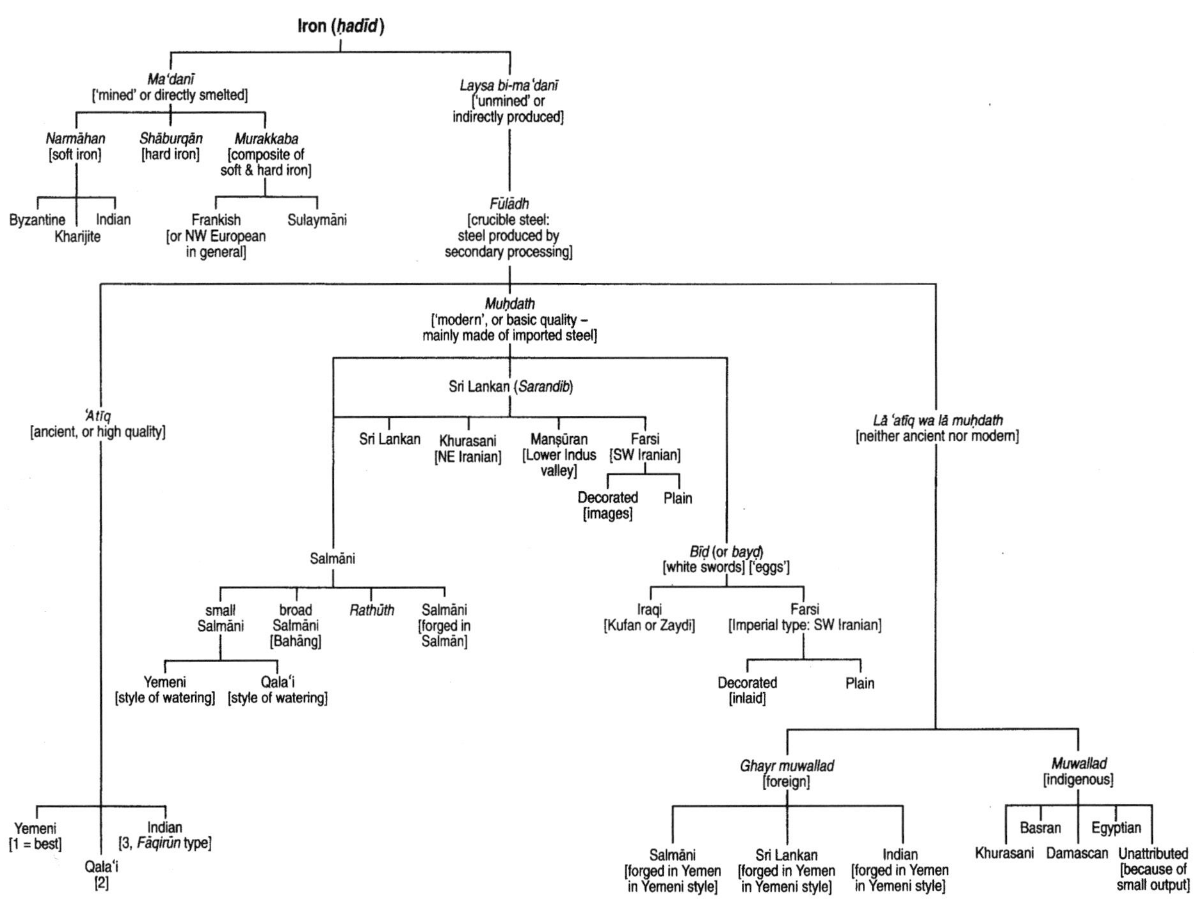

Figure 6 Diagrammatic representation of al-Kindī's classification of swords and the types of iron and/or steel from which they are made.

ingredients in a clay crucible with a (not always mentioned) lid. A source of carbon was needed for conversion of the soft iron to steel and this took the form either of carbonaceous organic matter, or of cast iron pieces. At about 900 °C absorption of carbon atoms at the surface of the iron (carburisation) took place and this continued until a pool of liquid steel was formed in the crucible, resulting in an ultra-high carbon steel cake or ingot (with somewhere around 1.5% carbon) when it cooled and solidified.

Jābir ibn Ḥayyān's 'Book on Iron'

Slightly earlier (probably later 8th century) is the fragment of the 'Book on Iron' by Jābir ibn Ḥayyān,[18] which survives as part of a much later commentary on this book by 'Izz al-dīn al-Jildakī (d.1342).[19] This fragment is especially valuable, as it takes the form of a two-part description of crucible steel-making. First of all, there is a description of the making of (white) cast iron (the only medieval example so far known to exist) and then of the conversion or processing of bars of iron. These are partly bars of the cast iron made in the first part of the description and partly the plain/soft iron necessary for the crucible processing. Unlike the writings quoted by al-Kindī and other medieval observers from this region, neither part of Jābir's description takes the form of a recipe, but both parts read more like a description of carrying out of the process.

Al-Bīrūnī on iron

The other surviving long medieval description of iron and especially crucible steel is that of Muḥammad ibn Aḥmad al-Bīrūnī, whose 11th-century book on mineralogy and related subjects ('The Sum of Knowledge about Precious Stones') contains a chapter on iron.[20] When compared with other early manuscript evidence, particularly an extensive body of Arab poetry of the later pre-Islamic and earlier Islamic period, this turns out to be a rather curious document, lacking the crisp scientific clarity and discussion of al-Kindī's better-known treatise 'On Swords and their Kinds'. It turns out to be a slightly odd and anachronistic compilation of contemporary oral information that reached al-Bīrūnī in Ghazna in central southern Afghanistan where he was based, and information from a variety of written sources, some much earlier, to which he had access in Ghazna.

Fortunately this area was on the eastern fringe of the region covered so well by al-Kindī two centuries earlier, and much of the oral information comes from the north and north-west of India, an area about which we otherwise have very little information concerning iron and steel. Some of the best steel was made from iron smelted near Kanūj, an ancient Indian regional capital city near the Ganges, already partly ruined when al-Bīrūnī visited it, probably in *c.*1020, following the Ghaznavid Sultan Mahmūd's campaigns in this area. Kanūj is also mentioned in the *Shahnameh* in a famous reference in which a series of precious gifts is said to have been sent to the mid-6th-century Sasanian Persian emperor, Anushirvan, by the Raja of Kanūj. These gifts appear to have included watered crucible steel swords and the earliest known chess set, carried along the southern silk route from north India through Afghanistan to Persia (Hoyland and Gilmour 2006: 65, n. 63). Crucible steel from this north Indian region would also seem a very good potential source for the best quality steel mentioned by al-Kindī.

Al-Bīrūnī mentions several earlier written sources, including that of al-Kindī, but he seems to rely most heavily on earlier sources about places well known for the making of crucible steel. The principal places or areas he mentions are Yemen and Mashraf, but the context is difficult to date until his references are compared to names and descriptions in well-dated early poetry. From this we can work out with a fair degree of certainty that the Yemen was already renowned in the 6th century for swords and the crucible steel from which they were made. Various references to Mashraf also mark it out as a contemporary centre of steel production at this time (Hoyland and Gilmour 2006: 95–6, 152, 165–6). Yemen continued to be famed for its swords in al-Kindī's day, nearly three centuries later, unlike Mashraf, which seems to have disappeared by then and was unknown to al-Kindī. Mashraf can be placed approximately in a mountainous region in southern Syria, near the Jordanian border, some 80 or 90 km south-east of Damascus. The area was then controlled by the 6th-century Ghassanid dynasty, although it was still run as part of the Byzantine Empire. Thus, unwittingly, al-Bīrūnī has highlighted an early crucible iron production centre that might otherwise have remained unknown.

In al-Bīrūnī's work there is a single reference to al-Kindī, which says that al-Kindī has incorporated into his treatise material from an earlier book by Mazyad ibn 'Alī, the Damascan smith. Al-Bīrūnī states that this book covered the entire process of watered crucible steel sword manufacture, beginning with the building of the furnace and the design of the crucibles, including the specific types of clay to be used, and continuing with the recipes and procedures for the making of the steel ingots used in sword production. He quotes one of the recipes he attributes to Mazyad ibn 'Alī, in which old soft iron horseshoes and nails are the source of the iron placed in the crucibles, in this case with dried organic matter as the source of the carbon, and with powdered *magnesia* (probably manganese dioxide) (Hoyland and Gilmour 2006: 154–6, 172–3). Unfortunately, Mazyad ibn 'Alī's description does not appear to survive in or with the versions of al-Kindī's treatises that we now have, although research on the second treatise is not yet complete. This earlier description is difficult to date, but, from the context, it seems most likely to belong either to the period when Damascus was the capital of the Umayyad dynasty (660–750), or to the 6th and early 7th century Ghassanid period when Mashraf was still famous as the production centre for crucible sword steel in this region when it was still under overall Byzantine control.

Until now al-Bīrūnī's report that al-Kindī incorporated the earlier written work of Mazyad ibn 'Alī into his own treatise has seemed slightly odd for a different reason. Al-

Kindī does not give crucible recipes or heat treatments (apart from a brief description of tempering) in his sword treatise, although the information on the different places of origin that he does give is so good for describing the wider picture of what happened and where, that it would seem inconceivable that al-Bīrūnī would not use it, instead of the much less complete and much earlier sources that he does use. The answer would seem to be that al-Bīrūnī only possessed al-Kindī's second treatise – on crucible recipes, heat treatments and the like – and that this had Mazyad ibn 'Alī 's earlier treatise (dealing with the same material) incorporated within it. Luckily for us, al-Bīrūnī, being the good scholar that he was, identified both these earlier writers, giving us yet another piece of this jig-saw puzzle of early source material.

Other references to crucible steel-making and the heat treatment of steel in medieval sources

At the beginning of his long treatise, written for the famous Ṣalāh al-Dīn (Saladin), entitled 'How to stay alive in battles' (Hoyland and Gilmour 2006: 7, n. 17), Murḍā ibn 'Ali al-Ṭarsūsī (late 12th century) gives four recipes for crucible steel followed by several heat treatments to apply to the swords made from this steel (see Cahen 1947–48: 106-7, 127–8; Allan and Gilmour 2000: 63–4). At least two of the recipes are taken from al-Kindī's second treatise and the wording of the other two suggests they might either come from a longer version of al-Kindī's second treatise or from a similar early compilation of crucible steel recipes. Likewise, the heat treatments appear to come either directly from a version of al-Kindī's second treatise or from a similar manuscript source, possibly a much earlier one. Similarly, a crucible steel recipe relating to the making of Indian swords which comprises an anonymous fragment of a written source – referred to as the Berlin manuscript (Ragib and Fluzin 1997: 68) – is actually a recipe in al-Kindī's second treatise.

Where do we go from here?

If we are to work towards a better understanding of how iron use began, how it developed subsequently in various parts of the world, and how this influenced or was influenced by contemporary culture, or other factors of a particular area or time, then the available evidence must be confronted in a much more detailed, consistent and systematic way than has hitherto been the case. We must recognise where the gaps are in our knowledge and we must not fall into the trap of substituting popular myth for real research and interpretation based on evidence. We must also assess how different factors have interacted to initiate change or to inhibit it. We need to look for, examine and assess the evidence for how iron use developed, how it changed, when this happened and why.

The manufacture and use of iron and steel developed differently in different regions, but the picture is indistinct at present, largely because the base of reliable evidence with which we are working is far too narrow. As discussed in the introduction, there are big gaps in our knowledge of early iron, partly because there are big and very uneven gaps in the available evidence, but also because the examination and assessment of the existing evidence is very patchy, uneven and inconsistent. More metallographic and compositional analysis of iron and steel artefacts needs to be undertaken in order to compile a database of analytical results that can be used to identify and track the ways in which ferrous technology varied and changed both chronologically and geographically. Both the metallographic and compositional analysis needs to be done systematically and to a high standard. Except for carbon content determination, which can be estimated during metallography, compositional analysis needs to be done by using electron probe microanalysis (EPMA) to determine the proportions of elements present in both the metal and in any non-metallic inclusions present.

In general, we need to put the study of iron and steel onto a much better footing. Sometimes it is simply a matter of looking at the problem from the right vantage point, as in the question of Damascus. Once a Middle Eastern perspective is taken, and once the available evidence has been scrutinised, it becomes clear that there is hardly any evidence to support the Damascus hypothesis, and plenty to suggest that it started as a late medieval misconception and grew from there.

Virtually the only existing mention related to this hypothesis, possibly not just of local significance (as for instance the presence of foundries – Al-Hassan 1979: 38), is the shadowy figure of Mazyad ibn 'Alī, the Damascan smith mentioned by al-Bīrūnī. However, it is difficult to gauge the significance of this, and it may not amount to very much. Al-Kindī does mention Damascus, but clearly could find very little hard evidence of local steel-making, as he says that it was carried out there 'in olden times'. This is the only place that he mentions it in this context. It may be that the Mashrafi industry, which was famous for steel in the 6th century (and possibly a little later), has been confused with Damascus given that it was the nearest city of significant size. Further research may clarify this.

The current project, which aims to look systematically at all the available written evidence from the Middle East, has been making good progress and is ongoing.[21] Taking all the evidence into account, the one steel-making centre that lasted a long time and had the best reputation, certainly from the 6th to the 9th century, was the Yemen. However, this does not mean that we should change the terminology and call this crucible steel, Yemen steel. We now know from the work on the available medieval manuscript sources that the crucible steel industry was very widespread and well developed by the mid-to-late 1st millennium. The wide availability of written descriptions indicates that there was nothing secret about the material or the process – quite the reverse.

One very interesting possibility concerns the nature of the written evidence for crucible steel-making and its possible antiquity. Al-Kindī's sword treatise is outstanding as a contemporary description and discussion of the state of the sword-making industry. This text was known throughout the Abbasid world and beyond, and without it a viable framework for the early iron and steel industry and many detailed aspects of it would be very difficult to construct. It is very fortunate that this text has survived, and it seems perfectly possible and even quite likely that al-Kindī's sword treatise always was the best of its kind. But this only appears to be true of al-Kindī's first sword treatise, not of the second treatise, which is mainly a series of crucible steel recipes, heat treatments and associated descriptions. The one crucible steel recipe we have (so far) from the 2nd century writer, Zosimos of Panopolis, is so similar in the way it is set out as to suggest that all the recipes quoted by al-Kindī and others may even have derived from earlier sources, possibly lost works of Zosimos, perhaps only surviving as fragments translated into Arabic.

At last, we are making better sense of the surviving manuscript evidence for iron- and steel-making, particularly relating to crucible steel. A better interpretation should give us a framework for working out the development of iron- and steel-making and use across the Old World. However, this will involve a great deal of investigative research on all the different kinds of iron that were used before we can reconstruct a more complete picture of how the metal began, and how it developed in its manufacture and use over the following three millennia.

Notes

1. So far the earliest consistent dates for iron smelting and its subsequent development come from a series of furnace sites in the Supsa-Gubazeuli river system in western Georgia, where a series of calibrated radiocarbon dates indicated that the industry in this area had begun by *c.*1800 BCE and that it had spread to the rest of western Georgia by *c.*1200 BCE (Khakhutaishvili 2009).
2. These hilts were made of iron to resemble specific cast bronze examples or styles; they were made either using small strips of iron pinned together, or forged to shape; see for example Maxwell-Hyslop and Hodges 1966: 168–9.
3. The description is slightly corrupt, probably through copyists' errors.
4. All dates given thus are to be taken as CE (Common Era) whereas earlier dates are stated as BCE (Before Common Era); these are of course equivalent to BC/AD dates.
5. The bars, some 80 of which were found, were approximately 1m long, twice the length of the ancestral fish-shaped form, but the iron 'blooms' from which they were made – flattened and cut versions of which were also found in the shipwreck – were also correspondingly bigger, reflecting the more industrial scale of iron-making, the use of water-powered hammers and so on; Crew *et al.* 1997.
6. At Rievaulx in 1591–92, approximately 3.5% of the blast furnace iron output was for making miscellaneous castings, the rest was converted to wrought iron; see Schubert 1957, appendix 10.
7. For Tonglüshan and the adaptation of bronze technology for cast iron in this region see Wagner 1993: 42–9.
8. Powlesland, pers. comm.
9. This process is illustrated and described in detail in Kapp *et al.* 1987: 68–78.
10. As pointed out by Dr Donald Wagner, archaeometallurgical research in China relating to iron has mainly focused on cast iron, and little material of possible bloomery origin, iron or steel, has been studied in any detail. What little material exists has been described and discussed recently by him and he has given extensive translations of some of the technical reports that have been carried out – see Wagner 1993: 267–95. These and the many photomicrographs that illustrate the results of these analyses (296–332) show that (much as elsewhere) research here has yet to concentrate on sorting out the technological sequence and development of the bloomery process in China from its beginning, and to work out for how long and to what extent it was practised alongside cast iron technology, particularly for specialist uses such as steel-making.
11. Not to be confused with Zosimus, the 5th-century Byzantine historian.
12. Alessandra Giumlia-Mair, who is working on the Zosimos manuscript material, has discussed this possibility with me. The original source can be dated between 323 and 168 BCE.
13. These sources, especially al-Kindī's sword treatise, also form almost the only surviving written record of early iron and steel exploitation in Europe. See Hoyland and Gilmour 2006.
14. Moxon 1703: 59; for this description and a discussion of the overall problem, also see Allan and Gilmour 2000: 76–9.
15. Berthelot 1888: 332; see also Allan and Gilmour 2000 (46, n. 34) for an English translation of this French passage.
16. Research by Robert Hoyland and Brian Gilmour on al-Kindī's second sword treatise is in progress.
17. The *Nihāyat al-su'l wa-l-umniyya fī ta'līm a'māl al-furūsiyya* (The Last Word on Horsemanship) by Muḥammad ibn 'Īsā al-Aqsarā'ī (d.1348), with the best and earliest known copies being those in the Chester Beatty Library, Dublin (dated 1366) and the British Library (dated 1371) – see Hoyland and Gilmour 2006: n. 25 and also Smith 1979: 28; but al-Kindī's second treatise is not mentioned in this booklet.
18. Also known in the West as Geber.
19. For a new translation and commentary on this see Hoyland and Gilmour 2006: 144–7.
20. For many years al-Bīrūnī worked in Ghazna for his patron Mahmūd, the Sultan of Ghazna, and he accompanied Mahmūd on his campaigns in northern India, following his invasion of this region in 1017. It may have been during this time that al-Bīrūnī gathered his information about crucible steel-making in the region. However, al-Bīrūnī wrote his book on mineralogy late in his life, probably during the reign of Mahmūd's grandson Mawdüd (r.1041–1050). For a new translation and commentary of this, see Hoyland and Gilmour 2006: 148–74.
21. The first result of this is the publication of *Medieval Islamic Swords* (Hoyland and Gilmour 2006), this being an analysis of the most readily available primary sources.

References

Al-Hassan, A. (1979) 'Iron and steel technology in medieval Arab sources', *Journal for the History of Arab Science* 2: 31–52.

Allan, J. and Gilmour, B. (2000) *Persian Steel: The Tanavoli Collection*. Oxford: Oxford University Press (for the Board of the Faculty of Oriental Studies, University of Oxford and the British Institute of Persian Studies).

Berthelot, M. (1888) *Collection des Anciens Alchemistes Grecs, Vol. 3*. Paris: G. Steinheil.

Cahen, C. (1947–48) 'Un traité du armurie compose pour Saladin', *Bulletin d'Etudes Orientales* 12: 103–63.

Crew, P., Crew, S., Gilmour, B. and Urteaga, M. (1997) 'Split blooms and pointed bars from the *Ria Oria* wreck: trade iron from the Basque country', in P. Crew and S. Crew (eds), *Early Ironwork-*

ing in Europe: Archaeology and Experiment, 101–2. Blaenau Ffestiniog: Snowdonia National Park Study Centre.

Elgood, R. (1994) *Arms and Armour of Arabia*. Aldershot: Scholar Press.

Hoyland, R. and Gilmour, B. (2006) *Medieval Islamic Swords: Kindī's Treatise 'On Swords and Their Kinds'*. Oxford: Gibb Memorial Trust.

Kapp, L., Kapp, H. and Yoshihara, Y. (1987) *The Craft of the Japanese Sword*. Tokyo/New York/London: Kodansha.

Khakhutaishvili, D. (2009) *The Manufacture of Iron in Ancient Cholchis*. BAR International Report 1905. Oxford: Archaeopress.

Maxwell-Hyslop, K. and Hodges, H. (1966) 'Three iron swords from Luristan', *Iraq* 28: 164–76.

Moxon, J. (1703) *Mechanick Exercises: Or the Doctrine of Handy-Works*. London: Printed for Dan. Midwinter and Tho. Leigh.

Pleiner, R. and Bjorkman, J. (1974) 'The Assyrian Iron Age: the history of iron in the Assyrian civilisation', *Proceedings of the American Philosophical Society* 118(3): 283–313.

Ragib, Y. and Fluzin, P. (1997) 'La fabrication des lames damassées en Orient', *Journal for the Economic and Social History of the Orient* 40: 30–71.

Rehder, J. (1991) 'The decorated iron swords from Luristan: their material and manufacture', *Iran* 29: 13–20.

Salter, C. (1997) 'Bloomery steel: an overlooked material', in P. Crew and S. Crew (eds), *Early Ironworking in Europe: Archaeology and Experiment*, 100–101. Blaenau Ffestiniog: Snowdonia National Park Study Centre.

Schubert, H. (1957) *History of the British Iron and Steel Industry: from 450 BC to AD 1775*. London: Routledge and Kegan Paul.

Smith, G.R. (1979) *Medieval Muslim Horsemanship: A Fourteenth Century Cavalry Manual*. London: British Library.

Wagner, D. (1993) *Iron and Steel in Ancient China*. Leiden: E.J. Brill.

Ancient metallurgical and manufacturing processes

The early history of lost-wax casting

Christopher J. Davey

ABSTRACT The different processes of lost-wax casting are described according to the technique used to form the wax model. The current assessments of the early history of the technique are discussed. The evidence from early literature, the moulds and metal artefacts are described and lead to the conclusion that lost-wax casting was a common process in much of the ancient world from the time that copper metallurgy was first developed. The development of the process begins with solid wax models in the late 5th millennium and ends with skin of wax models for life-size statuary formed by the 'sloshing' method in the classical period. Its origins are therefore much older than previously thought and its geographical spread is broader. The moulds show that lost-wax casting was often used for simple prosaic tools and ornaments and complexity of shape is not a determining feature of the process. In the ceramic cultures of the ancient world, lost-wax casting was not an exotic process used for complicated and prestigious items; in fact, the opposite seems to be the case. The construction of heterogeneous moulds with a gradation of clay coarseness aimed at preserving details on the wax model begins during the 3rd millennium BC and mould fragments displaying this variable fabric may be expected in the archaeological record.

Introduction

Lost-wax casting is acknowledged to have a long and continuous history, the origins of which have generally been traced to 3rd-millennium Mesopotamia (Penniman 1975: 144). However, recent discoveries and research have revealed that this view was premature.

Unequivocal evidence for casting practice comes from the discovery of the moulds used but, as lost-wax moulds are shattered as part of the process, they are not common. Broken unused moulds currently known are from Tell edh-Dhiba'i (Davey 1983, 1988; Al-Gailani 1965) and Poliochni (Bernabò-Brea 1964: 591, pl. 85d; de Jesus 1980: 41). The large number of used lost-wax mould fragments found at Gussage All Saints, Dorset, England from the pre-Roman Iron Age represents the only example of such material with published analysis (Foster 1980).

This lack of evidence therefore necessitates the physical examination of objects to determine the possible means of fabrication. The discovery of the Nahal Mishmar 'Cave of the Treasure' and the more recent examination of the objects has added another chapter to the history of lost-wax casting (Shalev 1999), as has material from Mehrgarh in Pakistan (Mille 2006).

This paper aims to provide a framework for further research into lost-wax casting by drawing on ancient literature and evidence from the moulds and objects associated with the process. Complexity of shape has often been assumed to be the determinant of the process, but the evidence discussed below would appear to question this approach. Also questioned is the exotic nature of the process itself.

The process

There are many descriptions of lost-wax casting in the literature. The techniques described vary depending on the circumstances, as they are defined by the available materials, traditions inherited and the size and form of the object to be cast. It is common for historical comment to focus on one particular method and to ignore other related techniques. It is therefore necessary to outline the range of lost-wax casting systems. One approach is to classify the nature of the wax model.

Wax models may be fashioned in a number of ways and may take a variety of forms:

- solid wax made by hand – suitable for a pin;
- solid wax encompassing a simple core – suitable for a mace-head;
- solid wax formed in a piece-mould possibly incorporating a simple core – suitable for a shafted axe-head;
- skin of wax around a directly securable core – suitable for a figurine;
- skin of wax around a complex core – suitable for statuary;
- skin of wax formed in a piece-mould that is swished around (the 'sloshing' method) – suitable for a life-size bust.

Simple cores are those that produce a hole in the model so that the object may be hafted. Complex cores are those which carry much of the shape of the object, thereby reducing the volume of metal required for the casting and hence its weight. These cores may be keyed into the outer (investment) mould through holes in the wax and ultimately the casting itself. Alternatively, they may be secured with metal pins (chaplets) that become part of the final casting. The core material is often removed after casting so that the object is hollow. The skin of wax models produced by the 'sloshing' method must have a core of sand or clay poured in and baked solid when the mould is heated.

All these techniques seem to have been used in antiquity. Before the investment mould is formed over the wax models, sprues, gates and runners will be added to facilitate the introduction of molten metal and the escape of air at the time of casting. The design of these features embraces much of the skill associated with the technique, especially where delicate or large objects are to be cast.

The investment mould formed around the wax model may take a number of forms depending on the shape, size and complexity of the model:

- simple – one piece;
- one piece with a core secured directly to the investment mould;
- complex, where the core is secured to the investment mould by metal pins or chaplets.

The investment moulds can be made in a single phase where a refractory clay material is laid directly over the wax model. They may also be built up in successive layers from a number of applications of different materials. When the mould is heated, the wax runs out. The temperature to which the mould is heated will depend on the materials used; today 600 °C is the common temperature.

General historical assessments

Moorey (1994: 271f.) claims that lost-wax casting was practised in Mesopotamia from at least Early Dynastic I (2900 BC) on the basis of the grave goods from Kish. However, all of Moorey's evidence is derived from artefacts, none of which seem to have been subject to expert metallurgical examination.

Ogden argues that lost-wax casting was practised in Egypt from the Old Kingdom, but does not cite any analyses that would support the claim. Indeed he is not able to reference an analysis of Egyptian metalwork for any period that would conclusively identify lost-wax casting and he relies heavily on the 1927 work by Garland and Bannister (Ogden 2000: 157–60).

Bannister, a professor of metallurgy at the University of Liverpool, provided reliable comments on fabrication processes. Garland and Bannister (1927) note that most Egyptian bronze statues have a hole through which cores could be secured. Iron nails were also used to secure cores, but not bronze chaplets. Cores are reported to be made from blackened sand with some organic material. Large statues from the New Kingdom period have limbs cast separately and joined to the body by mechanical means (Garland and Bannister 1927: 39–54). Lucas (1989: 221f.) provides some detail about Egyptian lost-wax casting, but his examples are New Kingdom or later.

Indus River practice is attested by the well-known 'dancing girl' from Mohenjo-daro dating to the early 2nd millennium BC (Penniman 1975: 144: Marshall 1931: 345, pl. 94).

Intricately decorated Chinese Shang and Zhou bronzes are now known to be the products of piece-moulds and not lost-wax casting (Smith 1981: 128). Earlier commentators proposed that the fine detail of these objects must have been produced by lost-wax casting but, as will be argued below, complexity is not an indication of lost-wax techniques. There is a general view in China that lost-wax casting was not used before the Eastern Zhou period. This is also questioned below.

Branigan has assembled other evidence for lost-wax casting in the Aegean (1974: 83). This includes the use of bronze moulds, which he believes were for casting models that may have been made of lead instead of wax, as evidenced by an Early Bronze Age shaft-hole axe-head from Samos (Branigan 1974: pl. 13, 587).

Classical lost-wax casting has been discussed by Mattusch (1996) in the context of detailed X-ray examination of bronze sculpture held in North America. The X-radiographs reveal the use of bronze chaplets and a range of means of assembling the complete statues which could not be cast in a single pour.

Ancient and more recent sources

Classical sources on metallurgy include Pliny the Elder's *Natural History*, Books 34–35, written during the 1st century AD. Pliny provides a commentary on famous Greek and Roman bronze statues and their fabrication (Mattusch 1996: 29ff.). He refers to lost-wax casting, stating that:

> The first person to make a plaster likeness of a human being from the actual face, and having poured wax on to [into] this plaster mould, to make final corrections to the wax cast, was Lysistratus of Sicyon, the brother of Lysippus I have mentioned ... Lysistratus also invented the taking of casts from statues, and this practice became so widespread that no statues or figures were made without a clay model (Pliny 2004: 336f).

This is a description of the 'sloshing' method technique referred to above. The technique relies on the production of durable plaster piece-moulds made over a model which, according to Pliny, may be an actual body part, a statue or a clay model. This enables the sculptor to make multiple wax models which, as Pliny said, may be modified.

In the absence of any evidence to the contrary, Pliny's comment may reasonably be taken to attest to the origin of this method. Lysippus was a contemporary of Alexander the Great and is said to have worked for him, although there is no record of him travelling to the East (Mattusch 1996: 98). The development of this technique partly explains the prolific nature of classical bronze sculpture. Pliny's comment about making corrections is significant. Prior to the development of modern fettling equipment, it was preferable to produce a clean cast to one with mould joints and imperfections that required laborious grinding and polishing.

Of medieval and Renaissance texts, Vannoccio Biringuccio's *Pirotechnia* (1540) has a detailed description of lost-wax casting, as does Benvenuto Cellini's *Treatise on Sculpture*, first published in abbreviated form in 1568. The full manuscript was not published until 1857 and gives an account of the casting of *Perseus and the Gorgon's Head* (Cellini 1967: 114–26). All the accounts of mould-making by these writers describe the use of fine clays coating the wax model and coarser refractory clays enclosing the mould.

The technical aspects of the lost-wax technique have also been discussed in three early Sanskrit texts, the *Manasollasa*, the *Silparatna* and the *Manasara*, in the context of solid or hollow casting of idols. The 12th-century *Manasollasa* describes the making of lost-wax moulds with a smooth paste made by grinding a mix of clay, rice husks, 'cotton severed a hundred times' and salt. The paste was smeared over the wax (thin enough to be transparent). After drying for two days a second then a third coat were applied, each being thicker than the previous one (Krishnan 1976: 2). The 16th-century *Silparatna* also describes the making of lost-wax moulds by applying very soft wet clay to the wax followed by a coat of soft clay once it had dried. Finally, hard clay was applied. Hard clay was made from material procured from an ant hill mixed with brick powder and arecanut husk juice. Soft clay was made from four parts of pot clay and ten of ordinary clay. Very soft clay was soft clay mixed with cow dung (Krishnan 1976: 4).

While these descriptions postdate the period of interest, they do provide an insight into traditional practices of lost-wax casting that predate the use of modern materials, and they illuminate the issues that are important for the study of the history of the process. Some of these practices still exist in the developing world (Feinberg 1983: 29–31).

The Tell edh-Dhiba'i mould

Tell edh-Dhiba'i, located in the suburbs of Baghdad, was excavated by the Directorate of Antiquities, Iraq, in 1965 under the direction of Lamia al-Gailani (Al-Gailani 1965). A coppersmith's workshop was found in the course of the excavation. Its date is consistent with that of the site, i.e. the Isin-Larsa and Old Babylonian periods, the first half of the 2nd millennium BC (Al-Gailani 1965: 33, 35).

A preliminary report of the site contained comments by W. Winton of the Science Museum, London, who studied photographs of the finds and identified the object with Field No. 614/19 as a bellows funnel (Al-Gailani 1965: 38). The object was not considered significant enough to receive an Iraq Museum number (Fig. 1).

In 1977 the author was able to study the Tell edh-Dhiba'i collection in the course of which a number of discoveries were made, including the fact that 614/19 was a lost-wax mould (Davey 1983). Moorey describes and discusses the material in some detail (1999: 265–72).

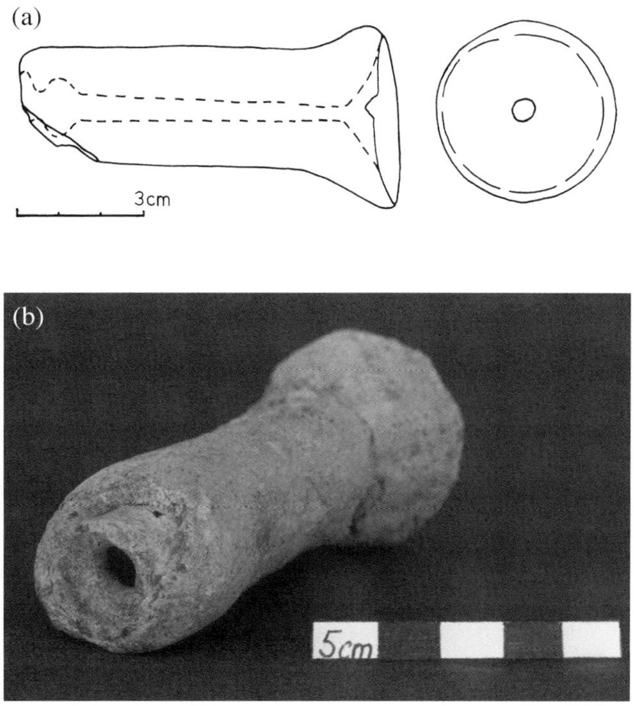

Figure 1 A drawing (a) and photograph (b) of the lost-wax mould from Tell edh-Dhiba'i. (Photo: C.J. Davey.)

One end of the mould was funnel-shaped and the other was broken. The broken end revealed that it was made from at least two distinct types of ceramic: a fine cream-coloured inner layer and a coarser buff-coloured refractory outer layer. The inner surface was smooth and had the shape of a short pin with a knob and probably a spherical top. The extent of breakage was not sufficient to allow the extraction of the casting and so it may be concluded that the mould suffered an accident prior to use and was therefore unused.

Pins with round heads have been found in Early Dynastic graves. The 'A' cemetery at Kish, for example, has a number of pins, the round heads of which were deemed by the excavators to be cast (Mackay 1929: 171, pl. 40). This cemetery has been redated to Early Dynastic III (2500–2600 BC) by Whelan (1978: 96) that is about contemporary with the royal cemetery of Ur. This reliably attests to this process of lost-wax casting in Mesopotamia at about 2500 BC.

The copper dagger from Kish cited by Moorey (1999: 271) for an Early Dynastic I use of lost-wax casting is not entirely convincing. However, the rein-guides of a slightly later date, also from Kish, do seem to have characteristics in common with lost-wax casting (Watelin 1934: 33, pl. 24). These objects are similar to those found at Ur in the royal cemetery and depicted on the 'War' panel of the 'Standard' (Woolley 1934: pl. 92). From this evidence it can reasonably be concluded that the lost-wax tradition represented by the Tell edh-Dhiba'i mould was well established in Mesopotamia at least 900 years earlier.

The capacity of modern technology to create wire-like objects can mask the difficulty of doing this by ancient mechanical means. The ease with which a round cross-section can be formed using a soft wax material explains much of the attractiveness of lost-wax casting. The shape thus formed may not be complex, but it has smooth curved surfaces and it is this feature that is the hallmark of objects produced by lost-wax casting. Objects with delicate round sections such as the Mohenjo-daro dancing girl are classical lost-wax casting shapes.

The second important feature of the Tell edh-Dhiba'i mould is its heterogeneous composition (Fig. 2). The use of fine and coarse clays reveal that the practices described by the Sanskrit and medieval texts were established by 1700 BC in Mesopotamia. The attention to the quality of the casting's surface represents an advance in the technology of lost-wax casting.

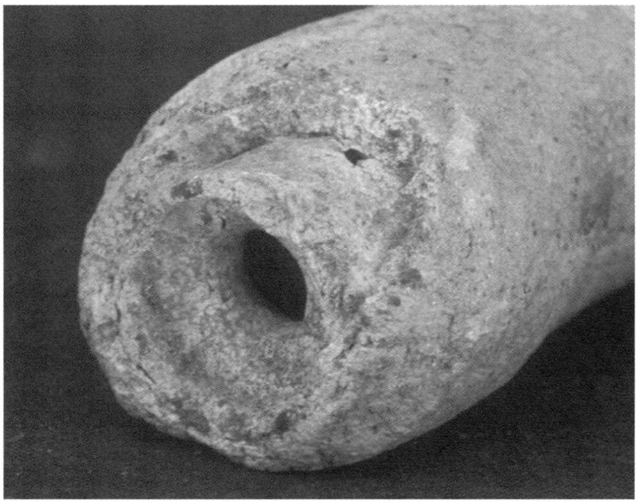

Figure 2 Detail of the lost-wax mould from Tell edh-Dhiba'i showing its layered construction. (Photo: C.J. Davey.)

The Akkadian copper head

The Akkadian copper head of a king from Nineveh (Strommenger 1964: 405, pls 22–3) (Fig. 3) and the fragmentary copper casting of a standard bearer (Al-Fouadi 1976) reveal that lost-wax casting of large-scale statuary was well developed in Mesopotamia in the second half of the 3rd millennium BC. Both objects are made of copper, and thus casting temperatures were comparatively high. A consequence of this may have been the greater thickness of copper when compared to statues of later times.

Strommenger's (1985) analysis of the Akkadian head revealed chaplets protruding 2.5 cm on the inside. The ears were made separately and are solid. X-radiographs confirm that the hair lines were chased onto the object after casting. These objects reveal that in Mesopotamia, 3rd-millennium lost-wax casting technology had progressed to the point where only the last stage, the 'sloshing' method, was yet to be developed.

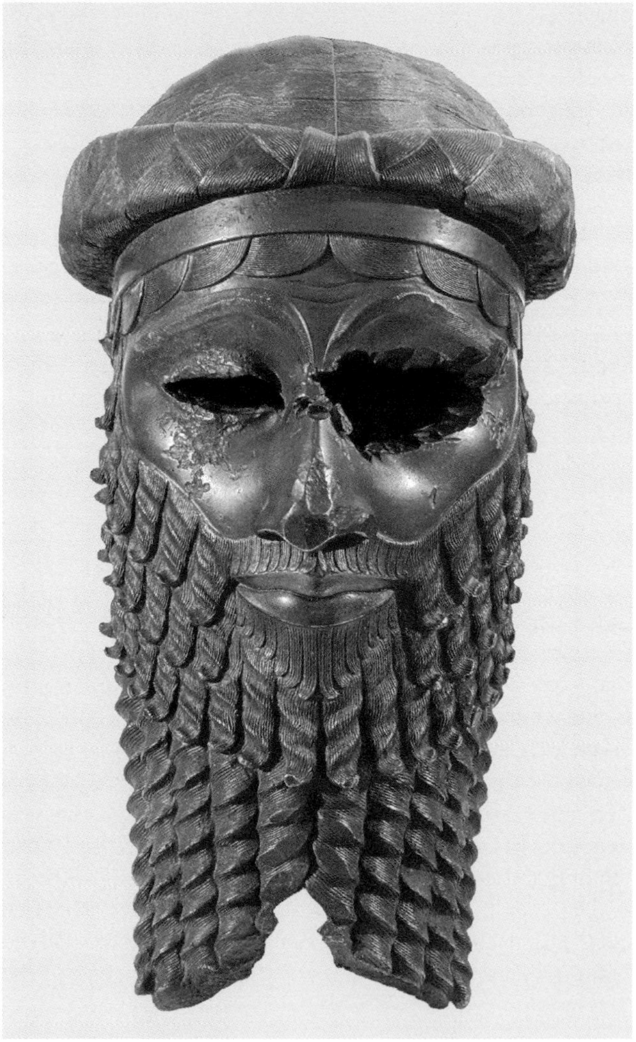

Figure 3 The Akkadian head from Nineveh, 2300–2159 BC (from *Iraq* 3 pl. 6, used with the permission of the British School of Archaeology in Iraq).

Nahal Mishmar objects

In 1961 a hoard of copper objects dating to the Chalcolithic period was found in a cave above Nahal Mishmar, a wadi on the western side of the Dead Sea (Bar-Adon 1980). In total there were 429 objects: 416 of metal, six of ivory and six hematite mace-heads and one mace-head of limestone.

A technical analysis has revealed that the objects are made from antimony-arsenic-rich copper, pure copper or nickel-arsenical copper. The source of the ores remained largely problematic (Tadmor *et al.* 1995: 143), however, high nickel coppers are reported to be found in artefacts from Ur and this leads to some speculation about the Iranian plateau as an origin.

The ceremonial objects are reported to be cast using lost-wax techniques over cores in single castings. There was no evidence of bivalve mould casting. Only two of the pure copper tools were cast solid, but the assumption that these were also the products of a lost-wax technique was not explained. Where repairs were necessary, they were done by casting, not welding (Tadmor *et al.* 1995: 144).

Shalev (1999) reports on a casting experiment of the Nahal Mishmar hoard at Pangolin foundry. The mace-heads were made over a wooden stick, the standards had clay cores and the crowns were formed from slabs of wax. The clay used by Shalev contained dung and grog. Shalev makes no comment about the quality of the metal surface achieved by this approach, but he does comment that the final grinding and polishing could be done by unskilled workers, such as women and children, because it is the most time-consuming aspect of the entire process (1999: 298). He also reports that a surface analysis of the ceremonial Nahal Mishmar objects detected extensive grinding and polishing. It appears that the homogeneous moulds used by Shalev produced objects in need of further treatment.

The Poliochni lost-wax mould

An Early Bronze Age lost-wax mould was found at Poliochni (Bernabò-Brea 1964: 64, 591, pl. 85(d); de Jesus 1980: 41, fig. 15; Branigan 1974: 82, fig. 4). The mould is for an axe-head and was broken before it was used. It is made from coarse clay and is reddish on the exterior and black on the interior. The blackness may be the result of the firing of the mould and the melting of the wax model.

The mould appears to have been found in a metalworking context. The description and the photograph of the mould indicate that it is homogeneous in composition and that it would have produced a casting with fairly rough surfaces.

No discussion of the context of the shape of the axe-head contained in the mould has been offered by the excavators or Branigan. Such an analysis would assist in establishing the history of lost-wax casting in the Aegean. The mould is about 1,000 years earlier than that of Tell edh-Dhiba'i and it seems to represent an earlier lost-wax casting tradition.

Mehrgarh

Benoit Mille has drawn attention to copper alloy 'amulets' discovered in the early Chalcolithic (late 5th millennium) levels of Mehrgarh in Baluchistan, Pakistan. He reported that metallographic examination established that the ornaments were cast by the lost-wax method (Mille 2006). The amulets were made from copper alloyed with lead. Mehrgarh is well recognised as a centre for early pyrotechnologies. The wax models of the amulets would have been solid and may have had a simple core inserted. This is understandably the first stage in the technology. Mille also draws attention to the 'Leopard Weights' from Baluchistan, dating to about 3000 BC, which were made using a complex core keyed into the investment mould.

China

With over 100,000 piece-mould fragments from the Shang dynasty capital of Anyang, it can be asserted with some conviction that lost-wax casting technology was not part of Anyang metallurgy. However, the inspection of a mask dating to the late Shang period (about 1300 BC), now in the National Museum, Beijing, indicates that lost-wax casting may have been practised elsewhere in China (Fig. 4).

The mask was part of a sacrificial deposit discovered in Sanxingdui Pit 2, Sichuan (Institute of Archaeology of Sichuan Province 1999: 197, pl 57; Bagley 1999: 213f). The rear side of the mask still has parts of the investment mould attached to it. The wax model of the mask would have been made from sheets of wax. The openings would have been cut with a knife and in some places on the rear of the mask what may be false cuts of the wax are preserved in the bronze casting (Fig. 5). There was no evidence of chaplets, but the holes in the mask may have provided the necessary support.

(a)

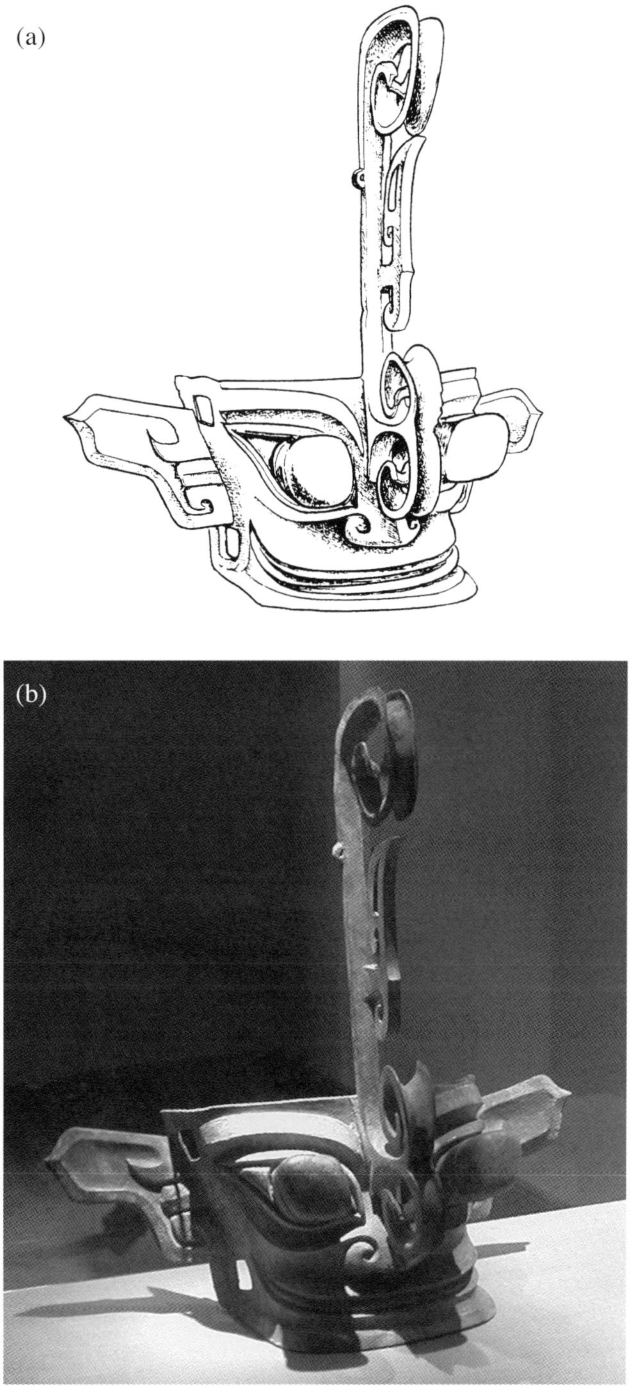

Figure 4 A drawing (a) and photograph (b) of the mask from Sanxingdui. (Photo: C.J. Davey.)

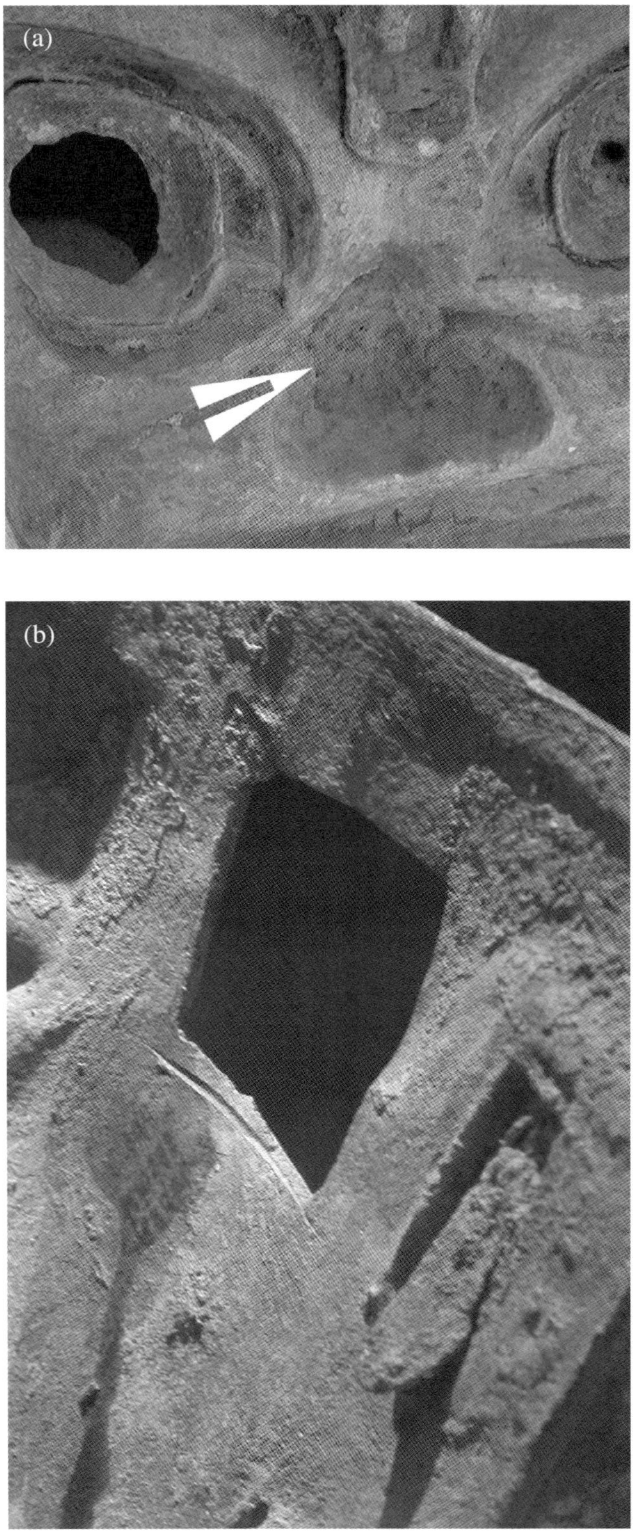

Figure 5 Detail of the rear of the mask from Sanxingdui showing (a) mould remains in the nose cavity and (b) possible 'false cuts' originally made in the wax model. (Photos: C.J. Davey.)

Also found in the Sanxingdui Pits were 4,600 cowrie shells (Anon. 1999: 419). Possibly of an Indian Ocean origin, these may have travelled along the 'south-western silk road' to Sichuan. It is therefore logical that technologies such as lost wax could have travelled the same path. An examination of the objects from this location in relation to lost-wax casting is warranted.

Gussage All Saints, Dorset

About 7,318 pieces of lost-wax mould were found at Gussage All Saints dating to the pre-Roman Iron Age. Radiocarbon dating confirmed a date of about 100 BC (Wainwright 1979: 125–53; Foster 1980). According to Tylecote (1986: 35), this is the first evidence in the British Isles of lost-wax casting. The lost-wax moulds were exclusively for the fabrication of horses' bridles. A petrological analysis of the mould fragments by Hilary Howard is reported in Foster (1980: 39–45) and establishes that two basic clays were used. The discovery of such a volume of mould fragments holds out the prospect that similar discoveries may be made elsewhere in earlier contexts.

Other evidences

At Byblos, a series of piece-moulds for fenestrated axe-heads was found (Dunand 1954: 98, pl. 184). These items showed no signs of being used for the casting of bronze and it is therefore proposed that they were used for the casting of wax models. The shape of the objects indicates a date at the end of the 3rd millennium BC.

Unidentified moulds and fragments have been found in the literature. There are two objects at Thermi that may be unused lost-wax moulds (Lamb 1936: 121, fig. 37). The purpose of the objects was not determined, but they are said to be made from 'coarse red ware', which would be consistent with being a lost-wax mould. One object was unstratified and the other was in an Early Bronze Age context.

Two clay fragments found at Hazor, dating from the Middle Bronze Age, may have been used for casting something like the horns of a bull and may have been part of a larger mould (Yadin 1965: 130, pl. 96 no. 30, pl. 98 nos. 30–31). The dimensions of the objects are not clear. The attribution of 'crucible' seems unlikely, but the object does indicate a metalworking context and the description of the fabric as coarse, red and with traces of straw is consistent with lost-wax moulds.

Conclusions

Lost-wax casting began in the Chalcolithic period before 4000 BC. The attestation of the technique at Mehrgarh points to an eastern origin, possibly on the Iranian plateau. The Nahal Mishmar material some 500 years later reveals the existence of a large-scale industry using lost-wax casting. Traditional theories that the technique began in Mesopotamia are no longer viable as it appears that lost-wax casting began soon after the start of copper metallurgy itself and well before the development of the Sumerian civilisation.

The Poliochni Early Bronze Age lost-wax mould shows that wax models were solid or made over a core that was secured directly to the outer mould. The moulds themselves were homogeneous in composition and made from coarse refractory clay that would have produced objects requiring surface enhancement by polishing and grinding. It also shows that by the beginning of the 3rd millennium BC lost-wax casting techniques were known from India to Europe.

At some stage during the 3rd millennium BC, more care was taken in constructing the moulds so that the surface of the casting would not require attention. A fine slip-like material, probably mixed with cow dung, was applied repeatedly to the model until a reasonable thickness was established; a coarse refractory clay was then applied to complete the mould. This technique is demonstrated by the Tell edh-Dhiba'i mould.

Lost-wax casting does not appear in the Eurasian steppe before the late Shang period in China, or even later if the above analysis is not accepted. Elsewhere it was ubiquitous and was used for prosaic objects with rounded surfaces as well as more intricate designs. The ease with which wax models could be perfected and accurately reproduced, thus obviating the need for polishing and grinding, made it the method of choice for many metalworkers.

Cores secured by chaplets were used in the casting of the Akkadian head, indicating that complex core systems were known by about 2400 BC in Mesopotamia. Egyptian statuary of the period had the cores secured directly to the outer mould. Chaplets are known in New Kingdom Egypt.

Pliny testifies to the classical use of forming wax skin models in piece-moulds, and the analysis of statuary by Mattusch indicates that bronze chaplets were the normal means of securing the cores by that time. The only aspect of lost-wax casting that was not practised by the Roman period was the single casting of large statues. Cellini's description of the casting of the three and a half tonne statue of *Perseus and the Gorgon's Head* implies that this was rare even in his day.

The modern perspective that the technique was sophisticated ignores the fact that the ancient world was a society based on ceramics and that the production of a pottery mould was straightforward. Smith is at least partly right to view the metalworker as related to the potter (1981: 127). Modelling in a soft material and then reproducing the shape accurately in metal circumvented the need to mechanically shape the metal, something that in the ancient world was difficult and time consuming. Lost-wax casting was not only used for intricate shapes but also for those of a reasonably routine nature that had curved surfaces. In an era when systems such as wire pulling and extruding are normal, it is easy to forget that objects with a circular cross-section were difficult to produce by primitive mechanical means.

The importance of achieving a smooth surface of the casting led to the construction of moulds with heterogeneous sections from almost the origin of the technique. The variable ceramic cross-section of the Tell edh-Dhiba'i mould and the descriptions of most of the ancient authorities lead to the conclusion that lost-wax moulds from at least the Middle Bronze Age will normally not have a homogeneous ceramic composition.

As no other pottery is formed in this manner, the discovery of heterogeneous pottery fabrics may well attest to lost-wax casting. The shapes of such material will be nondescript and in the past would have escaped the attention of most archaeologists who seek diagnostic forms and not significant fabrics. Until metallurgical workshops are excavated with the precision of prehistoric archaeology, many of the precise details of the smith's techniques will remain obscure.

Postscript

After this paper was submitted, Yuval Goren published a paper in 2008 entitled 'The location of specialized copper production by the lost wax technique in the Chalcolithic Southern Levant' (*Geoarchaeology: An International Journal* 23(3): 374–97) in which he documents the different clays used in the moulds associated with the copper objects from Nahal Mishmar and other Chalcolithic sites. I am indebted to Professor Goren for a copy of his paper and the news that a copper standard recently found at Qiryat Gat has a multilayered mould remaining inside the shaft. There is a layer of clay and grass and a layer of lime and quartz sand. This demonstrates that there was discrimination of clay types for moulding prior to the 3rd millennium, the purpose of which is uncertain.

Acknowledgements

The author acknowledges the helpful discussions with Cameron McIndoe and Sean Elliot of Fundêre Fine Art Foundry and Studios, Melbourne, and Richard Stringer who taught him bronze sculpture.

References

Al-Fouadi, A-H. (1976) 'Bassetki statue with an Old Akkadian royal inscription of Naram-Sin of Agade (2291–2255 BC)', *Sumer* 32: 63–76.

Al-Gailani, L. (1965) 'Tell edh-Dhiba'i', *Sumer* 21: 33–40.

Bagley, R. (1999) 'Shang archaeology', in M. Lowe and E. Shaughnessy (eds), *The Cambridge History of Ancient China: From the Origins of Civilisation to 221 BC*, 124–231. Cambridge: Cambridge University Press.

Bar-Adon, P. (1980) *The Cave of the Treasure*. Jerusalem: Israel Exploration Society.

Bernabò-Brea, L. (1964) *Poliochni, citta preistorica nell' isola di Lemnos, Vol. 1*. Rome: L'Erma di Bretschneider.

Biringuccio, V. (1966) *Pirotechnia* (trans. by C.S. Smith and M.T. Gnudi from the 1st edn, 1540). Cambridge, MA: MIT Press.

Branigan, K. (1974) *Aegean Metalwork of the Early and Middle Bronze Age*. Oxford: Clarendon Press.

Cellini, B. (1967) *The Treatises of Benvenuto Cellini on Goldsmithing and Sculpture* (trans. by C.R. Ashbee from the Marcian Codex of *c*.1565). New York: Dover.

Davey, C.J. (1983) 'The metalworkers' tools from Tell edh-Dhiba'i', *Bulletin of the Institute of Archaeology University of London* 20: 169–85.

Davey, C.J. (1988) 'Tell edh-Dhiba'i and the southern Near Eastern metalworking tradition', in R. Maddin (ed.), *The Beginning of the Use of Metals and Alloys*, 63–8. Cambridge, MA: MIT Press.

de Jesus, P.S. (1980) *The Development of Prehistoric Mining and Metallurgy in Anatolia*. BAR International Series 74. Oxford: BAR.

Dunand, M. (1954) *Fouilles de Byblos Tome II 1933–1938*. Paris: Librairie d'Amerique et d'Oriente Adrien Maisonneuvre.

Feinberg, W. (1983) *Lost-wax Casting: A Practitioner's Manual*. London: International Technology Publications.

Foster, J. (1980) *The Iron Age Moulds from Gussage All Saints*. British Museum Occasional Paper 12. London: British Museum.

Garland, H. and Bannister, C.O. (1927) *Ancient Egyptian Metallurgy*. London: Charles Griffiths & Co Ltd.

Institute of Archaeology of Sichuan Province (ed.) (1999) *Excavation of the Sacrificial Pits at Sanxingdui*. Beijing: Cultural Relics Publishing House.

Krishnan, M.V. (1976) *Cire perdue Casting in India*. New Delhi: Kanak.

Lamb, W. (1936) *Excavations in Thermi Lesbos*. Cambridge: Cambridge University Press.

Lucas, A. (1989) *Ancient Egyptian Materials and Industries*, 4th edn, rev. by J.R. Harris. London: Histories & Mysteries of Man Ltd.

Mackay, E. (1929) *A Sumerian Palace and the 'A' Cemetery at Kish, Mesopotamia*. Chicago: Field Museum Press.

Marshall, J. (1931) *Mohenjo-daro and the Indus Civilization, Vol. 1*. London: Arthur Probsthain.

Mattusch, C.C. (1996) *The Fire of Hephaistos: Large Classical Bronzes from North American Collections*. Cambridge, MA: Harvard University Art Museums.

Mille, B. (2006) 'On the origin of lost-wax casting and alloying in the Indo-Iranian world', in *Metallurgy and Civilisation: 6th International Conference on the Beginnings of the Use of Metals and Alloys, University of Science and Technology, Beijing, BUMA VI*.

Moorey, P.R.S. (1994) *Ancient Mesopotamian Materials and Industries: The Archaeological Evidence*. Winona Lake, IN: Eisenbrauns.

Ogden, J. (2000) 'Metals', in P.T. Nicholson and I. Shaw (eds), *Ancient Egyptian Materials and Technology*, 148–76. Cambridge: Cambridge University Press.

Penniman, T.K. (1975) 'A note on specimens in the Pitt Rivers Museum illustrating *cire perdue* casting', in H.H. Coghlan (ed.), *Notes on the Prehistoric Metallurgy of Copper and Bronze in the Old World*, 143–5. Oxford: Oxford University Press.

Pliny the Elder (2004) *Natural History: A Selection*, J.F. Healy (trans.). London: Penguin Books.

Shalev, S. (1999) 'Recasting the Nahal Mishmar hoard: experimental archaeology and metallurgy', in A. Hauptmann, E. Pernicka, Th. Rehren and U. Yalçin (eds), *The Beginnings of Metallurgy*, Der Anschnitt Beiheft 9, 295–9. Bochum: VFKK.

Smith, C.S. (1981) *A Search for Structure: Selected Essays on Science Art, and History*. Cambridge, MA: MIT Press.

Strommenger, E. (1964) *Art of Mesopotamia*. New York: Harry N. Abrams.

Strommenger, E. (1985) 'Early metal figures from Assur and the technology of metal casting', *Sumer* 42: 114–15.

Tadmor, M., Kedem, D., Bergemann, F., Hauptmann, A., Pernicka, E. and Schmitt-Strecker, S. (1995) 'The Nahal Mishmar hoard from the Judean desert: technology, composition and provenance', *'Atiqot* 27: 95–148.

Tylecote, R.F. (1986) *The Prehistory of Metallurgy in the British Isles*. London: Institute of Metals.

Wainwright, G. (1979) *Excavations at Gussage All Saints*. Archaeological Report Series 10. London: HMSO.

Watelin, L.C. (1934) *Excavations at Kish*. Paris: Paul Geuthner.

Whelan, E. (1978) 'Dating the A Cemetery at Kish: a reconsideration', *Journal of Field Archaeology* 5: 79–96.

Woolley, C.L. (1934) *Ur Excavations II: The Royal Cemetery*. London and Philadelphia: British Museum and University Museum Philadelphia.

Yadin, Y. (1965) *Hazor III–IV*. Jerusalem: Magnes Press.

A natural draught furnace for bronze casting

Bastian Asmus

ABSTRACT This paper discusses the design and use of natural draught furnaces for non-ferrous casting purposes. A series of experiments was conducted to demonstrate the possibility of using such a furnace without artificial oxygen induction. For comparative reasons the use of bellows was also tested. This paper explores problems of crucible handling, fuel charge and gas flow in the suggested furnace design, and illustrates the thermodynamic principles of such a design and explains why this furnace works.

Introduction

This paper aims to re-evaluate our concept of reconstructing the metallurgical production process of bronze casting, or rather the preceding step of bronze melting. The use of bellows seems all too often implicitly accepted and this experimental case study on low shaft, natural draught furnaces[1] offers an alternative way for interpreting features related to metal casting traces at archaeological sites. Natural draught furnaces are not to be confused with wind-powered furnaces, where wind above the furnace is the main factor for producing sufficient updraught (e.g. Juleff 1996; Tabor *et al.* 2005).

The use of natural draught furnaces for metallurgical processes has been noted and published for several decades, mostly in relation to smelting processes (e.g. Percy 1864; Cline 1937; Tylecote 1987). Despite this and the fact that these furnaces are still in use in present-day bronze art foundries, they seem not to be recognised as viable alternatives for the production of non-ferrous metal artefacts in antiquity. In the archaeological literature they are typically mentioned in connection with extractive metallurgy (e.g. Rehder 1987), but never with the casting of bronze. This is all the more intriguing as for the high-medieval period Theophilus Presbyter reports the use of a natural draught furnace for the melting and production of brass (Hawthorne and Smith 1979; Brepohl 1999) and in the Renaissance, Benvenuto Cellini (Brepohl 2005) reports the use of natural draught for casting furnaces.

Why it is believed that furnaces need forced air supply

The bias in favour of bellows and tuyères,[2] i.e. forced draught furnaces, seems deeply rooted in archaeologists' perception of high temperature processes and might have misled our interpretations, or at least distorted the image of metal foundry in antiquity. The lack of clear-cut archaeological evidence for the design of bronze casting furnaces certainly constrains our interpretation. Tuyères are frequent, but their number bears no relation to the amount of archaeological bronze objects – it would be expected that a highly fired or vitrified artefact such as a tuyère should survive in the archaeological record. In contrast, bellows – probably manufactured from organic materials – would not be expected to survive and are accordingly very rare in the archaeological record. We notice that there are neither enough tuyères nor bellows in the record to explain why high temperature processes are usually reconstructed in a way that assumes their presence. As a rule, the archaeological record provides furnace bases at best, whereas the once existing superstructure is almost always lost and the reconstruction has to be based on a set of assumptions.

The following aspects may have biased these assumptions: first, the tradition in the metal crafts, maintained by present-day craftsmen. However, only a few archaeologists have had formal training as craftsmen, and thus evaluating information retrieved from such a source might prove difficult and lead to misinterpretations. Secondly, historical sources such as Georgius Agricola (2003), Vannoccio Biringuccio (Smith and Gnudi 1990) and Lazarus Ercker (1598) report the use of bellows and emphasise their usage. Further back in human history, Roman and Greek depictions of metalworking scenes exhibiting the use of bellows can be found. Even some Egyptian wall paintings, for example those from Rekh-Mi-Re (Davies 1973), Mereruka (Duell 1938) or Meir (Blackman 1914), give evidence for the use of bellows in high temperature processes, including the casting of metals. Archaeologically, Davey (1979) has shown the existence of pot bellows for the Near East, at least from the 2nd millennium BC onwards, which tallies well with the above-mentioned pictorial evidence in Egyptian tombs. Looking closely at this evidence it can be seen that there is one instance from Egypt and one from the Greek world in which casting of metals is depicted (Oddy and Swaddling 1985; Weisgerber and Roden 1986; Tylecote 1987; Pusch 1990, 1994). In Roman times, depictions of bellows are associated predominantly with blacksmithing (Weisgerber and Roden 1985). It might not yet be fully appreciated in the archaeological world that there is a clear distinction between foundry work and metal forging, and subconsciously the terms 'metal-founder' and 'metalsmith' might have been used synonymously. It can be argued that this ambiguity might well be partly responsible for the bias in favour of forced draught furnaces.

It is generally believed that, in Europe, separate chimneys have been in existence only from the 12th century onwards; this obscures the fact that even an ordinary campfire produces an updraught above the fire, and in doing so creates a low pressure zone at the base of the fire. The difference between ambient air pressure and the lower pressure zone at the campfire's base will result in an air flow from the higher pressure zone towards the lower pressure zone; thus, air is transported towards the fire and supplies it with oxygen. In summary, there is ample reason to assume that natural draught furnaces may have played a much larger role in metal casting than hitherto acknowledged.

Aims and methods

A series of experiments with low shaft furnaces was conducted to evaluate the potential of natural draught shaft furnaces for melting, superheating and casting non-ferrous metal. The experiments draw upon more than 10 years of professional experience in bronze casting, both with modern and reconstructed casting tools and equipment. Questions regarding the functionality and practicability of this design are:

- Will the furnace work at all?
- What temperature will the furnace reach?
- Will the temperature be consistent enough to melt and superheat the metal?
- How does charging affect the furnace's performance?
- Does fuel lump size have an effect on performance?
- Is the handling of crucibles without modern steel or iron tongs possible?
- How much time does the furnace need to dry? Can it be used immediately?
- Is there a difference in performance between a cold furnace and a continuously used hot furnace?
- Is this mode of furnace operation practicable for a production environment?
- How much investment of labour and material does a furnace like this require? Is it valuable? Does it need sheltering?
- Can an experiment answer these questions?

Experimental set-up

Furnace

The furnace type that was tested is a simple shaft furnace circular in plan and slightly conical in side view (Figs 1 and 9). The external diameter is 42 cm at the bottom, 27 cm at the top and the height is 45 cm. The inner diameter, which is 20 cm at the bottom and 15 cm at the top, is of importance. The bottom of the furnace is slightly bowl-shaped. At ground level there are three 10 × 6 cm air inlets. The building material is clay with 60–70% quartz temper, and wall thickness varies from 11 cm at the bottom to 6 cm at the top. The furnace was built entirely from fresh clay and dried within the day by firing it with wood and charcoal for several hours preceding the experiment.

The furnace works with a crucible placed in the charcoal charge, whereby the bottom of the crucible is approximately 6 cm above ground level. This ensures that the crucible receives the maximum heat during charcoal combustion.

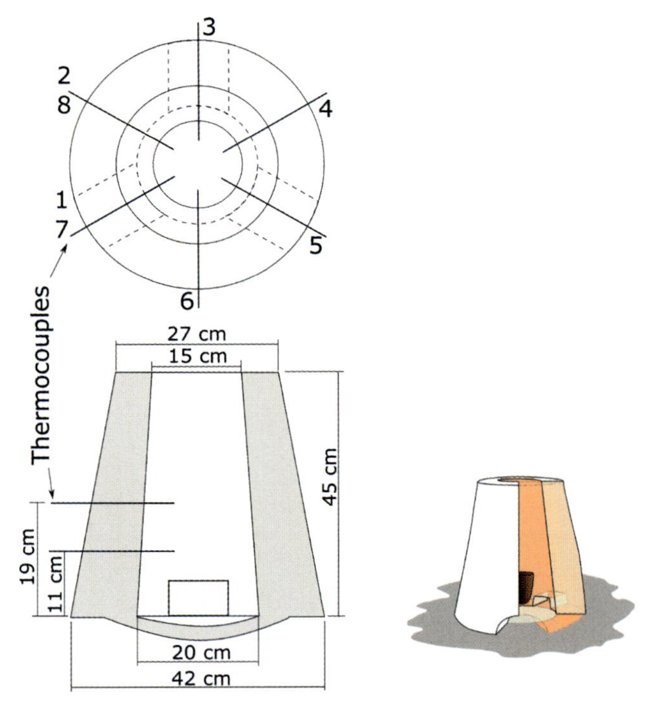

Figure 1 Plan and side view of the experimental shaft furnace. The numbers 1–8 denominate the two principal sets of positions of the thermocouples; the side view shows the thermocouples' elevation above ground.

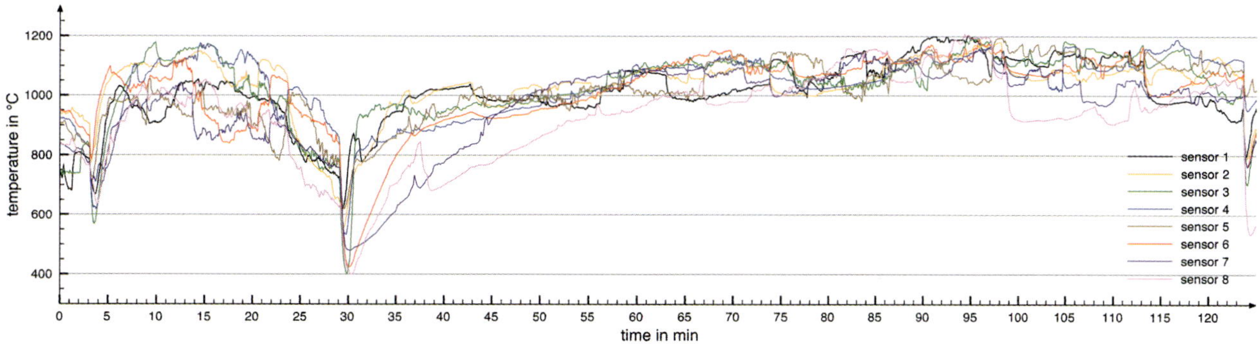

Figure 2 Plot 1: an episode with the shaft fully charged with charcoal. Note that the y-axis starts only at 300 °C.

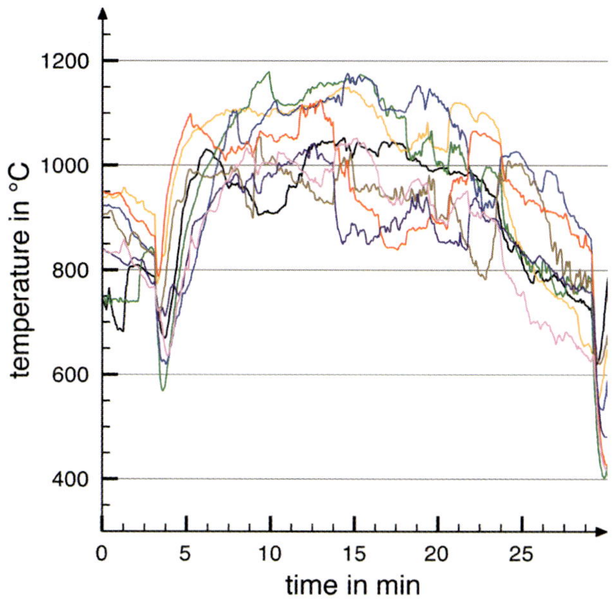

Figure 3 Plot 2: shaft half-filled with charcoal. Note that the y-axis only starts at 300 °C.

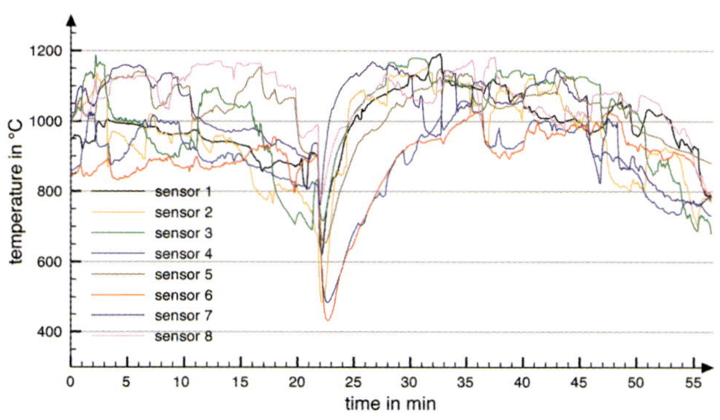

Figure 4 Plot 3: two consecutive cycles with the shaft half-filled. Note that the y-axis only starts at 300 °C.

The hottest zone in the furnace is easily identified by the yellow-orange colour of the embers; additionally, the embers will exhibit white edges if they burn hot. The crucible itself is filled with metal and completely immersed in charcoal. No lid or cover for the crucible is necessary – the charcoal on the metal prevents oxidation while producing additional heat from the top.

Temperature measurements

Temperatures within the furnace were taken with type N thermocouples and data logging equipment, which could handle up to eight thermocouples simultaneously. Previous experience led to the observation that the highest temperatures are to be expected immediately above the air inlets. Therefore, two distinct sets of positions for thermocouples were chosen: positions 1, 3, 5 and 7 above air inlets and positions 2, 4, 6 and 8 between air inlets. The thermocouples protruded 5 cm into the furnace. Six thermocouples were placed 11 cm above ground and two others 20 cm above ground (Fig. 8).

Bellows were available for the experiments and their impact on the melting process was monitored in one single cycle. The bellow nozzle[3] did not protrude into the furnace. Only one air inlet was supplied with a forced draught.

The charcoal used for the experiments was obtained from local char-burners and the main wood species used were oak and beech.

Results

Plot 1 in Figure 2 represents temperature data of a furnace cycle with a shaft completely filled with charcoal. Figure 3 represents one cycle with the shaft half-filled and Figure 4 shows the temperature in two consecutive cycles with the shaft half-filled. The drop in temperature in the plots represents recharging events with fresh charcoal. It can clearly be seen that the temperature decreases by 500 °C after each charging event. This loss of temperature also affects the metal in the crucible and in doing so prolongs the time that is required to liquefy the metal.

A maximum temperature of just over 1200 °C was reached, however the average operating temperature of a continuously used furnace was found to range between 1000 °C and 1200 °C. The temperature was maintained sufficiently long to melt a 90/10 tin bronze with a liquidus temperature of 990 °C (Shim *et al.* 1996). The time the high temperature range lasted depended on the amount of charcoal – from 15 minutes with a half-filled shaft (Figs 3 and 4) to up to an hour with a completely filled shaft (see Fig. 2). The degree of variability in the operating temperature is less if the furnace is filled completely with charcoal and the melting of bronze was accomplished within this cycle. However, it was found that melting and superheating the metal was achieved considerably faster with two cycles of charging the shaft halfway.

The wet furnace immediately after building did not reach the necessary temperature. No measurements were taken, but the bronze in the crucible remained solid and even after several hours of firing it had only a dark red glow colour which attests a temperature of approximately 700 °C.

The performance of the furnace increased with continuous use. A cold but dry furnace did work to melt the bronze, but took considerably longer and consumed a much larger amount of charcoal than a hot furnace, which was in continuous use. In a cold furnace, the melting of bronze took 60–75 minutes whereas it only needed 20–25 minutes in a hot furnace.

It was found that the charcoal lump size plays a major role in the melting process. During combustion the charcoal fragmented into numerous smaller lumps. Once the gross charcoal lump size is below 15 mm in the area of the air inlets, the air flow within the furnace is impeded to such an extent that temperatures above 1000 °C are not achievable.

Because the crucible moves downwards (and away from the hottest zone in the furnace) in the shaft as the combustion of the charcoal progresses, its position has to be monitored and corrected regularly. The crucible was handled with tongs made from green wood, which is more difficult than using steel tongs, but nevertheless it was possible and casting could proceed without any noticeable delay (Fig. 5). The tongs also proved sufficient for the repeated positional corrections of the crucible during melting. Because wood has a lower thermal conductivity than metal, wooden tongs kept heat loss to a bare minimum at the most crucial part of the casting sequence, i.e. retrieving the crucibles and casting the metal – metal tongs immediately cool down the crucible (as evidenced by the crucible's change of colour) whereas the crucibles retrieved with wooden tongs maintained their temperature.

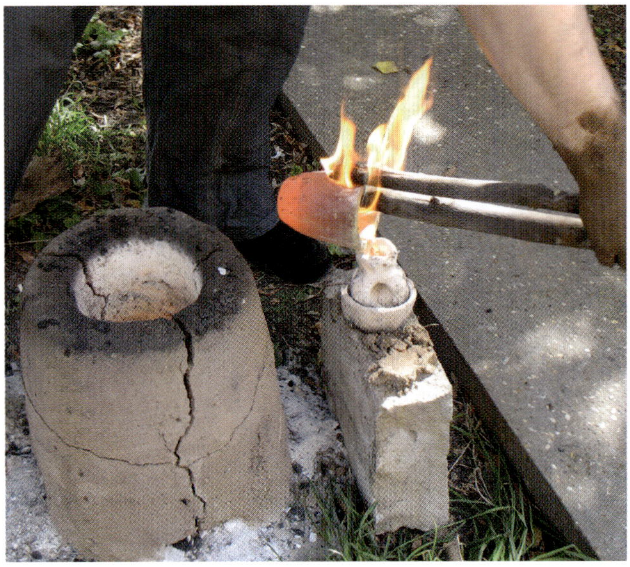

Figure 5 Casting with wooden tongs.

During experimentation, a simple and effective heat protection for hands was also found: clay slips applied on hands and arms protect the skin adequately when pulling the crucible and pouring metal into the moulds, even though a considerable amount of time is spent above the shaft of the furnace.

Observations on the use of bellows

During the experimentation with bellows, the temperature in the furnace rose immediately above the range of the thermocouples and they had to be withdrawn from the furnace. Simultaneously, charcoal consumption increased noticeably. It appeared, however, that the melting speed did not increase significantly, despite the increased oxygen supply: the charcoal burned hotter, but also much faster, and the necessity to recharge the furnace with charcoal more often seemed to counterbalance the effect of the hotter furnace. This touches on a general problem with melting metal in solid fuel furnaces, i.e. that the operator has to constantly evaluate the temperature versus the availability of fuel. In this respect it was found that the use of bellows enabled the operator to achieve better furnace temperature control. Towards the end of a charcoal cycle, instead of recharging it with charcoal, there is the opportunity to raise the temperature very quickly with the use of bellows and use this additional energy to superheat the metal.

In summary, the furnace operated in a very satisfactory fashion and several kilograms of bronze were cast into shape during the experiments.

Discussion

If we strive to reconstruct how metal is molten in a furnace, we need to understand the physical processes that govern how metal melts, what happens during combustion of a fuel, and the behaviour of gas and flue in our furnace. With regard to the melting of metal we have to consider the following three physical properties of matter: (1) the melting temperature, (2) the specific heat capacity and (3) the standard enthalpy of fusion.

Table 1 Physical properties of metals.

	Melting temperature in °C	Specific heat capacity (c_p) in kJ $kg^{-1}K^{-1}$	Standard enthalpy of fusion ($\Delta_{fus}H$) in kJ kg^{-1}
Copper	1083	0.385	205
Tin bronze	1020	0.38	190
Silver	962	0.232	105
Brass 63/37	920	0.377	166
Gold	1064	0.129	63
Lead	327	0.129	25

Melting temperature is the temperature at which a material is fully liquefied, also referred to as the liquidus temperature (see Table 1 for melting temperatures of various metals). However, a heat source that exceeds the melting temperature of a metal does not necessarily melt it. Consider the following: a candle flame has a temperature of 1300 °C but it cannot melt bronze. This is why we need an additional concept to that of temperature: energy. Energy and temperature together can explain the physical reactions during the melting of metals. The specific heat capacity (c_p) is a measure of how many units of energy are necessary to increase the temperature of 1 kg of matter by 1 °C (or 1 K = Kelvin). In order to heat 1 kg of bronze from room temperature to its melting temperature, we can calculate the energy Q with the equation:

$$Q = m_{bronze} * c_p * \Delta t$$

where:

c_p specific heat capacity for 90/10 tin bronze: 0.38 kJ kg^{-1} (see Table 1)

Melting temperature 90/10 bronze: 990 °C

m_{bronze} mass of bronze 1 kg

$$\Delta t = 990\ °C–25\ °C = 965\ °C = 965\ K$$

$$Q = 1\ kg * 0.38\ kJ\ kg^{-1}\ K^{-1} * 965\ K = 367\ kJ$$

Although the metal is at melting temperature, after this amount of energy is applied it is still in a solid state due to the microstructure of metal – because the metal is held together in a crystal lattice, extra energy has to be applied to dissolve the metallic bond. This amount of energy is called the standard enthalpy of fusion ($\Delta_{fus}H$) and the amount of energy Q_{fus} required can be calculated as follows:

$$Q_{fus} = m_{bronze} * \Delta_{fus}H$$

$$Q_{fus} = 1\ kg * 190\ kJ\ kg^{-1} = 190\ kJ$$

One kilogram of bronze needs 190 kJ to dissolve its crystal lattice. The total amount of energy Q_{melt} is:

$$Q_{melt} = Q + Q_{fus} = 367\ kJ + 190\ kJ = 557\ kJ$$

But in order to be able to retrieve the crucible from the furnace and pour the metal into the mould, its temperature needs to be raised by at least another 150 °C – this is called superheating. Again, the specific heat capacity is used to calculate the energy Q_{super} required to superheat the metal:

$$Q_{super} = 1\ kg * 150\ K * 0.38\ kJ\ kg^{-1}\ K^{-1} = 57\ kJ$$

$$Q_{tot} = Q_{melt} + Q_{super} = 567\ kJ + 57\ kJ = 624\ kJ$$

The additional energy requirement is 57 kJ. Altogether 624 kJ is required to melt and superheat 1 kg of tin bronze.

Let us now examine the heating value of charcoal to understand how much energy it provides. Because charcoal is produced under non-standardised conditions, it is difficult to quantify its heating value. This is further complicated by the fact that wood species can vary widely in their chemical composition and their calorific values. For the purposes of this paper, an average value for the higher heating value (HHV) of charcoal is assumed to be 28 MJ kg^{-1}, taken from the *Kirk-Othmer Encyclopedia of Chemical Technology* (Durbak *et al.* 1999). However, the amount of energy that can be used in our scenario is the lower heating value (LHV), which takes the heat loss of water evaporation (i.e. water that is present in the fuel) into account. As a rough approximation, this value is 90–95% of the HHV (Rehder 2000: 35), so that a conservative value of 25 MJ kg^{-1} is used. This means that 1 kg of charcoal generates 25 MJ of energy in the form of heat during combustion.

Although this value seems high at first glance, providing 40 times the energy that is required to melt 1 kg of bronze, in practice 1 kg of charcoal is not necessarily enough fuel to melt the bronze. The main obstacle to overcome is that of heat loss. In order to melt the metal successfully, the sum of the heat losses should be smaller than the lower heating value plus the energy required for the melting of the metal. Heat losses include losses through heat conduction, i.e. heat convection and heat radiation (Rehder 2000: 10); in other words, the heating of the furnace wall, the heating of the crucible, the heating of ambient air around the furnace and of course the evaporation of water, which is abundantly present in all the structural elements built from clay. This is the reason why a wet furnace does not work properly. In order to evaporate water, tremendous amounts of energy are required. The most substantial 'heat loss', however, is caused by the large upper diameter of the shaft through which the flue escapes and which creates the updraught. Looking at the gas flow in more detail, the thermodynamic principles at work in the shaft furnace are (1) Bernoulli's theorem and (2) the hydrodynamic paradox. These two phenomena can be used to explain how and why the air supply of this furnace works.

Bernoulli's theorem

In fluid flow an increase in velocity occurs simultaneously with decrease in pressure. Bernoulli's theorem can be used to explain why any fire is sufficiently supplied with oxygen. In thermodynamics, the term fluid is applied to gases and liquids alike. We are concerned with the flue of the furnace. The flue consists mainly of carbon monoxide and by-products of the furnace operation, but also other components of air, such as nitrogen, which are of no concern here. Importantly, the gas directly above the charcoal bed is more than 1000 °C hotter than the air outside the furnace, which means that the velocity of its molecules is increased notably. This increase in velocity results in a decrease of pressure above and around the furnace, the updraught. This lack of pressure is immediately counterbalanced by the atmospheric pressure, which pushes cold air towards the low pressure zone and supplies the furnace with air. This air is used for combustion and thereby heated, maintaining the low pressure zone above and around the furnace as long as it is hot (see Fig. 6).

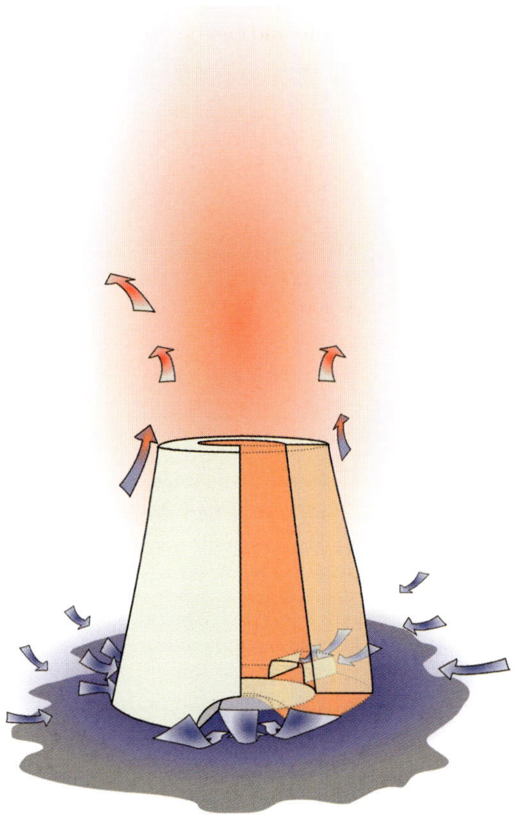

Figure 6 Bernoulli's theorem: hot air rises, leaving a zone of lower pressure at air inlets. Atmospheric pressure compensates the lower pressure zone by pushing air into the inlets.

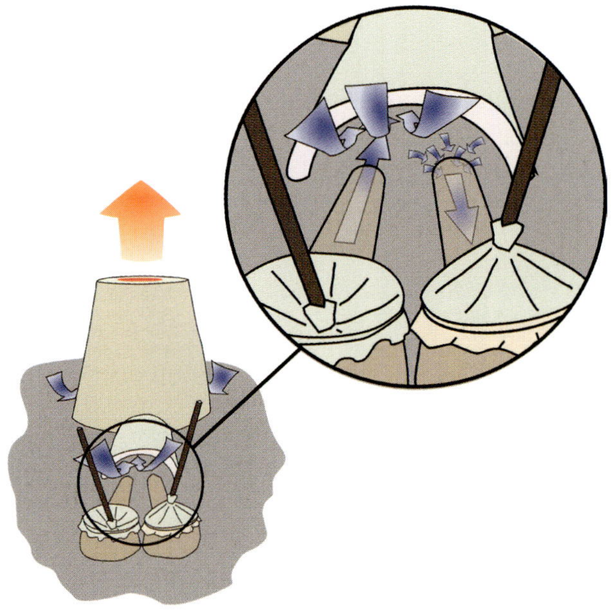

Figure 7 The hydrodynamic paradox: objects bordering streaming fluids are pulled towards them and not pushed away.

The hydrodynamic paradox

Objects bordering a streaming fluid are pulled towards it and not pushed away (see Fig. 7). This is especially important for the use of pot bellows without valves. Because these bellows release the air via the same nozzles through which they aspirate, the bellows must not protrude into the furnace in case they suck in hot fumes or embers, alighting the bellows. Nevertheless, such bellows (see Fig. 10) were and are used successfully for metallurgical processes. Because we are able to push an air column for a few centimetres, it is possible to induce air into a furnace through air inlets. Additionally, ambient air is swept along the air column into the furnace, inducing more air into the furnace than that supplied by the bellows. Fortunately it is impossible to draw a column of air into the bellows – as soon as the leather skin is pulled upwards, an under pressure is produced and the atmospheric pressure pushes air into the bellow.

The value of a furnace

Obviously it is very difficult to assign values to material objects of past societies, especially if they are to be understood solely through their material remains. Nevertheless, an understanding of the value of such a device can be facilitated by examining the investment of labour in order to produce such a structure. 'Value' refers to the willingness of protecting this installation against atmospheric conditions or other forces that might lead to its destruction.

The tempering and kneading of the clay took about an hour. The time to build the furnace was approximately 40 minutes. It seems unlikely that a shelter for furnaces of such a design would have been built if we compare the amount of labour and raw material procurement to build even the simplest shelter with the labour that went into the building of the furnace.

One of the furnaces was left unprotected throughout the winter on the experimentation field in London. The structure proved comparatively resilient against weathering and could be used in the spring without requiring major repairs. There is no doubt however that rain in combination with frost would have led to its destruction within the year, with hardly any trace being left in the archaeological record apart from a few droplets of metal.

Conclusions

The experiments clearly demonstrated that the use of natural draught is sufficient to melt and superheat a tin bronze with 10% tin content. After continuous use of several hours the furnace maintained a temperature above 1100 °C for periods of up to 15 minutes with the shaft half-stacked and up to an hour with the shaft full of charcoal. A maximum temperature of 1208 °C was reached. At each charging episode the temperature decreased by approximately 500 °C for a

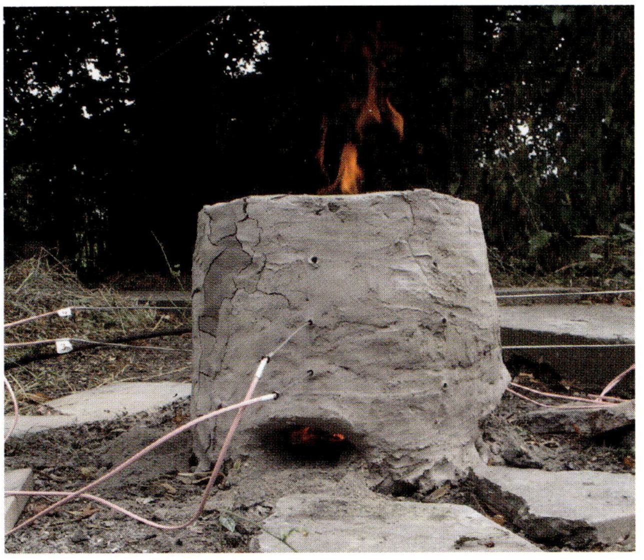

Figure 8 Experimental set-up in one of the experiments.

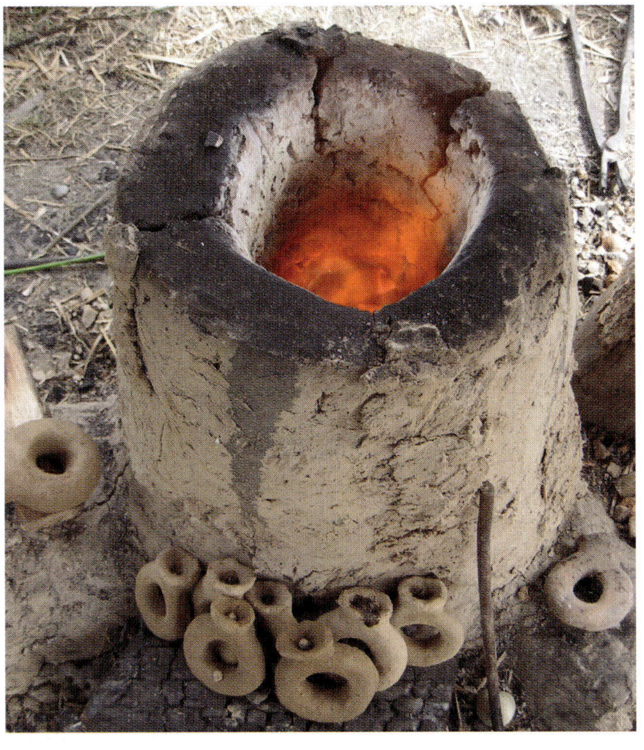

Figure 9 Furnace after one week of continuous operation. Cracks may easily be repaired with fresh clay. Because of the conical shape, the furnace is unlikely to break apart.

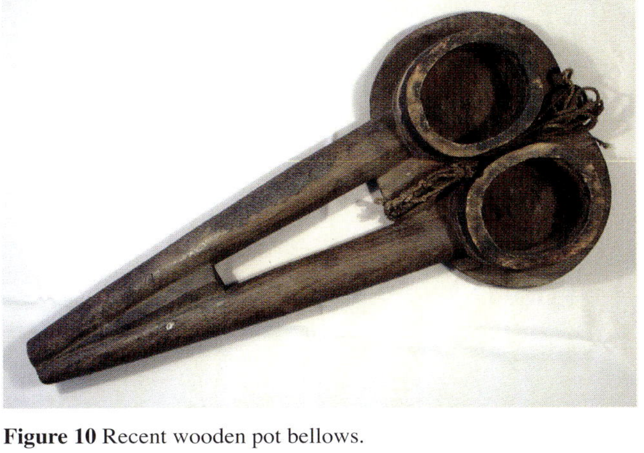

Figure 10 Recent wooden pot bellows.

period of 7–15 minutes, depending on the amount of charcoal charge in the shaft. It was observed that charcoal lump sizes below 15 mm impede the gas flow to such an extent that temperatures above 1000 °C are not achievable. The use of wooden tongs for managing the crucible in the furnace as well as handling the hot crucible proved very successful, and helped to minimise heat losses of the melt due to lower thermal conductivity of wood compared to metal.

In summary, it can be stated that low shaft natural draught furnaces are easy to build and can be used very successfully for the production of the majority of bronze artefacts of a mass ranging from 1 to 2 kg.

Notes

1. A furnace is a structure used for high temperature processes such as casting, glassmaking or smelting.
2. Tuyères are discrete implements, and usually not structural features of a furnace. They are generally more or less conical tubes of refractory material and taper towards the end that points (and may protrude) into the furnace. It is their function to induce air into the furnace.
3. The definition of nozzles is somewhat imprecise: on the one hand, they are often seen used in conjunction with blowpipes in order to prevent the tip of the organic blowpipe from alighting, but they are smaller in dimensions than tuyères. On the other hand, nozzles are structural parts of bellows where the air leaves the bellows (e.g. Davey 1979), and because of this they are sometimes confused with tuyères.

References

Agricola, G. (2003) *De Re Metallica Libri XII – Zwölf Bücher vom Berg- und Hüttenwesen.* Berlin: Fourier Verlag.

Blackman, A.M. (1914) *The Rock Tombs of Meir.* London: Egypt Exploration Fund.

Brepohl, E. (1999) *Theophilus Presbyter und das mittelalterliche Kunsthandwerk. Band 2 Goldschmiedekunst.* Cologne: Böhlau Verlag.

Brepohl, E. (ed.) (2005) *Benvenuto Cellini. Traktate über die Goldschmiedekunst und die Bildhauerei.* Cologne: Böhlau Verlag.

Cline, W. (1937) *Mining and Metallurgy in Negro Africa.* Menasha, WI: George Banta Publishing Company.

Davey, C. (1979) 'Some ancient Near Eastern pot bellows', *Levant* 11: 101–11.

Davies, N. de G. (1973) *The Tomb of Rekh-mi-Re at Thebes.* New York: Arno Press.

Duell, P. (1938) *The Mastaba of Mereruka.* Chicago: University of Chicago Press.

Durbak, I., Green, D.W., Highley, T.L., Howard, J.L., MacKeever, D.B., Miller, R.B., Petterson, R.C., Rowell, R.M., Simpson, W.T., Skog, K.E., White, R.H., Winandy, J.E. and Zerbre, J.I. (1999) 'Wood', in *Kirk-Othmer Encyclopedia of Chemical Technology*, 627–64. New York and Chichester: John Wiley & Sons, Ltd.

Ercker, L. (1598) *Beschreibung allerfürnemisten mineralischen Erzt und Bergwercksarten. Wie dieselbigen und eine jede in Sonderheit irer Natur und eigenschafft nach auf alle Metalla probirt und im kleinen Feur sollen versucht werden.* Frankfurt am Main: Johann Feyerabendt (electronic copy).

Hawthorne, J.G. and Smith, C.S. (1979) *Theophilus: On Divers Arts. The Foremost Medieval Treatise on Painting, Glassmaking and Metalwork.* New York: Dover.

Juleff, G. (1996) 'An ancient wind-powered iron smelting technology in Sri Lanka' *Nature* 379(4): 60–63.

Oddy, W.A. and Swaddling, J. (1985) 'Illustrations of metalworking furnaces on Greek vases', in P.T. Craddock and M.J. Hughes (eds), *Furnaces and Smelting Technology in Antiquity*, 43–58. British Museum Occasional Paper 48. London: British Museum.

Percy, J. (1864) *Metallurgy: The Art of Extracting Metals from their Ores, and Adapting them to Various Purposes of Manufacture*. London: J. Murray.

Pusch, E.B. (1990) 'Metallverarbeitende Werkstätten der frühen Ramessidenzeit in Qantir-Piramesse/Nord', *Ägypten und Levante* 1: 75–113.

Pusch, E.B. (1994) 'Divergierende Verfahren der Metallverarbeitung in Theben und Qantir? Bemerkungen zu Konstruktion und Technik', *Ägypten und Levante* 4: 145–69.

Rehder, J.E. (1987) 'Natural draft furnaces', *Archeomaterials* 2: 47–58.

Rehder, J.E. (2000) *The Mastery and Uses of Fire in Antiquity*. Montreal and Kingston: McGill-Queens University Press.

Shim, J., Oh, C., Lee, B. and Lee, D. (1996) 'Thermodynamic assessment of the Cu-Sn system', *Zeitschrift für Metallkunde* 87(3): 205–12.

Tabor, G.R., Molinari, D. and Juleff, G. (2005) 'Computational simulation of air flows through a Sri Lankan wind-driven furnace', *Journal of Archaeological Science* 32: 753–66.

Tylecote, R.F. (1987) *The Early History of Metallurgy in Europe*. London: Longman.

Smith, C.S. and Gnudi, M.T. (trans.) (1990) *The Pirotechnia of Vannoccio Biringuccio: A Classic Sixteenth-century Treatise on Metals and Metallurgy. New York:* Dover.

Weisgerber, G. and Roden, C. (1985) 'Römische Schmiedeszenen und ihre Gebläse', *Der Anschnitt* 37(1): 2–21.

Weisgerber, G. and Roden, C. (1986) 'Griechische Metallhandwerker und ihre Gebläse', *Der Anschnitt* 38(1): 2–26.

The liquation process utilised in silver production from copper ore: the transfer to and development in Japan

Eiji Izawa

ABSTRACT From ancient times, silver was derived mainly from argentiferous lead ores. After smelting lead ores, silver was extracted by cupellation of lead. Another important source of silver was represented by complex copper-silver ores, which ancient peoples smelted with lead. Silver-bearing lead was gravitationally separated from the copper-rich upper layer (dross) and silver was recovered from the lead by cupellation. The liquation method, the new silver extraction process for complex copper-silver ores, was invented in the middle of the 15th century in Nürnberg, Germany. The liquation process was adopted in several places in Germany during the 15th century and then spread all over Europe. A Japanese metallurgist learned about liquation in 1591 from a Portuguese, and the process was applied successfully for silver extraction from copper in Japan. The method diffused to many areas in Japan during the 17th and 18th centuries and was used until the end of the 19th century. In the early 17th century, liquation was adopted in smelting workshops to obtain silver from silver-copper ore. In those days, silver was the major product and copper was the by-product. In the later 17th century, copper became the more important export commodity. The aim of the liquation technique shifted to refining copper, and silver became a by-product.

Introduction

From ancient times, silver was derived mainly from argentiferous lead ores. After smelting lead ores, silver was extracted by cupellation of lead. In the case of simple silver ores containing native silver and argentite, the ores were smelted with lead. The resulting silver-bearing lead was cupellated to extract silver. Another important source of silver was represented by argentiferous copper ores. However, extracting silver from copper ores requires more complex processes as compared to those involved in the extraction from lead ore. In 1453, a newly invented liquation method for silver extraction was first documented in Nürnberg, Germany (Suhling 1976: 57; Braunstein 1983: 588). This paper discusses the transfer of the liquation technology to Japan and the spread of the method in early modern Japan.

Liquation: methods of separating lead from copper

Biringuccio (1540; English translation 2005: 156) describes two methods of separating lead from copper. The first method involves the gravitational separation of lower density copper-rich dross from higher density liquid lead. Argentiferous copper ore is smelted with lead or galena (PbS) in the blast furnace. The resulting melt (copper-lead-silver alloy) flows into the large forehearth and the molten alloy is allowed to separate into the copper-rich upper layer (dross) and the bottom lead layer. After removal of the upper layer, the silver-bearing lead at the bottom is recovered. The separation of lead from copper is necessarily incomplete and hence repeated smelting of the dross with lead is necessary. Finally, silver is separated from lead by cupellation.

Ancient people probably adopted this method to extract silver from argentiferous copper ore. Su briefly describes the procedure of silver extraction from copper ores as follows: 'If silver occurs along with copper in ores, then people must smelt the ores with lead repeatedly' (Su 1061: 27). 'Silver-copper ores occur in southern China. The ores are smelted with lead and silver comes out with lead' (Su 1061: 43–4). In *Tian-gong kai-wu* 天工开物 (Song 1637) a similar smelting process of copper-lead ore is described (Fig. 1). Although the text 'lead flows out from upper hole' (Song 1637; Japanese translation 1969: 267) is confusing, the figure illustrates that lead is flowing out from the bottom hole and copper is flowing out from the upper hole. This description seems to reflect the ancient smelting method of copper-lead ore.

The second method described by Biringuccio is the liquation process using the difference in melting temperatures of lead (327 °C) and copper (1083 °C). The liquation process for argentiferous copper was invented in the mid-15th century in Nürnberg. Liquation is an important technique for the exploitation of argentiferous copper ores during the late medieval to early modern period. The liquation method is described in detail in *De re metallica* (Agricola 1556: English translation 1950: 504–21) (Fig. 2).

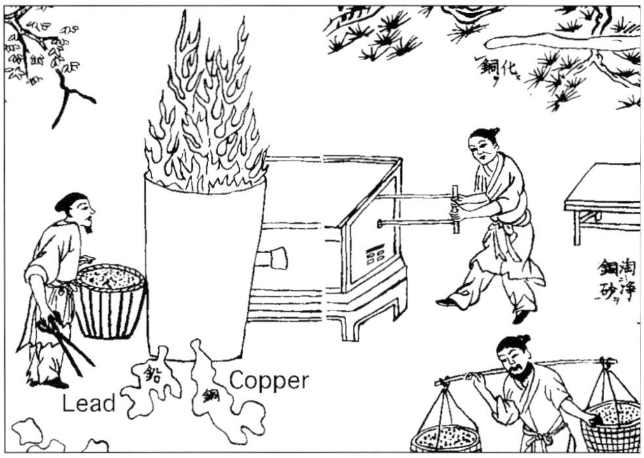

Figure 1 Gravitational separation of lead from copper. Charcoal and lead-copper ore are placed in a blast furnace, which has two tapping holes. The central part of the original figure. *Tian-gong kai-wu* (Song 1637, Kyushu University Museum Collection, Volume 3, leaf 13 (verso), leaf 14 (recto)).

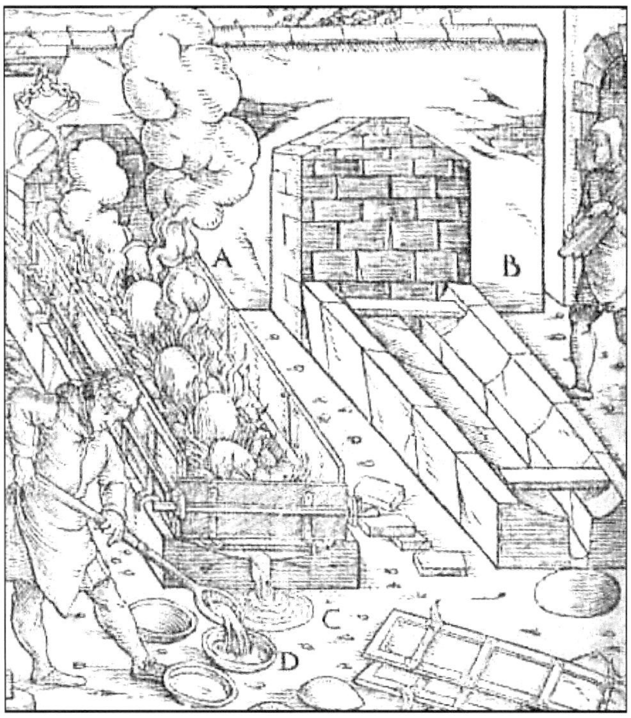

Figure 2 Liquation separating lead from copper (Agricola 1657: 41). Several cakes of silver-bearing lead-copper alloy are set with charcoal and heated in the liquation furnace. Silver-bearing lead liquates away from solid copper and flows down into the receiving pit (the lower part of the original figure is omitted).

From the 15th century to the early 16th century, the liquation process was operating in several places in and around Germany (Lynch 2002: 22–4). Liquation refineries (*Seigerhutten*) were constructed at Thüringen in 1462, Chemnitz in 1471, Schwaz (Tyrol) and Neusohl (Banska Bystrica, Slovakia) in the late 15th century (Suhling 1976: 61, 66–77), and Grunthal in 1537 (Kasper 1993: 12). The liquation method was adapted to the copper ore from Falun, Sweden, in the late 18th century. A new silver segregation plant was completed near Falun in 1789 (Lindroth 1955: 329) and the production of Falun silver was recorded from 1790 (Tegengren 1924: 46). By 1830 liquation (Seigern technique) appeared in the Russian Altai region (Suhling 1976: 163).

Namban-buki: liquation in Japan

In 1591 a Japanese metallurgist, Soga Ri'emon Jusai (1572–1636), learned about the process of liquation from a Portuguese in Port Sakai near Osaka. This statement is described in *Kodo-zuroku* 鼓铜图录 (Masuda *c.*1811; English translation 1983: 84; German translation 1984: 94). *Kodo-zuroku* is the well-known illustrated book of copper metallurgy in early modern Japan, which was first published in 1811–16 (Yasukuni 1999: 16). Soga Ri'emon, who was described as 'Sumitomo Jusai' in *Kod-ozuroku*, was the father of Izumiya Rihei (1607–1662), who founded the copper refinery using the liquation process in Osaka in 1623.

There is no reliable document about the origin of liquation in Japan and the year when the procedure was allegedly learned, 1591, is merely a legend from the Sumitomo family. However, it seems to have a grain of truth. This technology transfer took place in the context of the growing network of global trade. In 1543 a Portuguese ship came to a southern island of Japan, sailing from the south, the first encounter between the Japanese and Europeans. Soon after, Portuguese ships exported a large amount of silver from Japan. Jesuits also came to Japan and visited several cities, including Port Sakai. For example, Luis Frois stayed in Japan from 1563 to 1597 and Alessandro Valignano from 1579 to 1601. Around 1580, Portuguese ships exported 5000 or 6000 *kan* (about 20 tonnes) of silver every year from Japan (Kobata 1965: 253). It is conceivable that one of the merchants or Jesuits might have brought knowledge of European liquation to Japan.

There is some dispute about the relationship between the method described in *Tian-gong kai-wu* (Fig. 3) and *namban-buki* (smelting of the 'southern barbarians' style), the Japanese liquation technique. However, the description

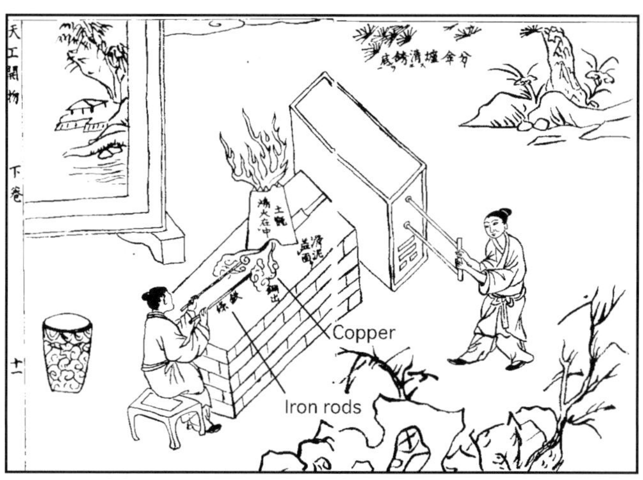

Figure 3 Furnace for metal separation from scrap copper-silver-lead. *Tian-gong kai-wu* (Song 1637, Kyushu University Museum Collection, Volume 3, leaf 13 (verso), leaf 14 (recto)).

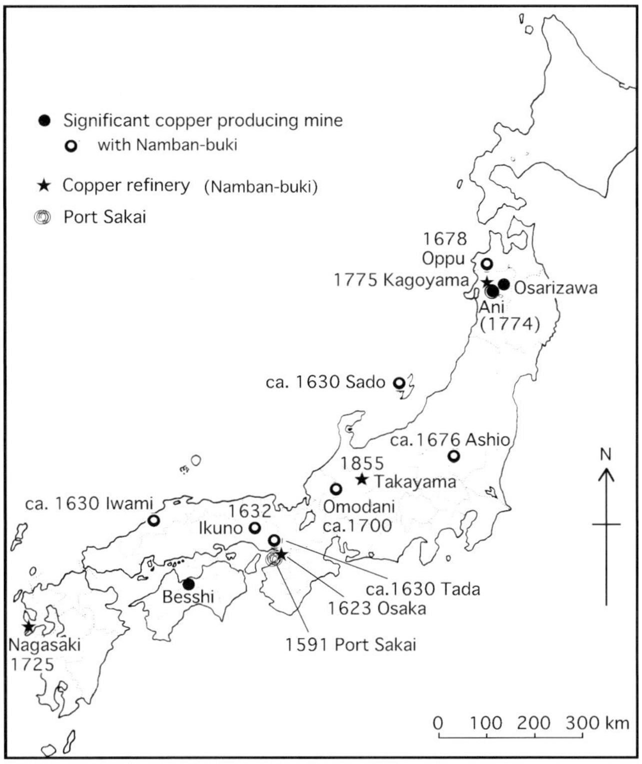

Figure 4 Location of significant copper-producing mines and central copper refineries in the 16th–18th centuries in Japan. The year when the liquation process was introduced is also shown.

(Song 1637; Japanese translation 1969: 265) is not clear enough to understand the process. The method is applied not to refining crude copper but to the recovery of silver from scrap metal.

The first liquation refinery was founded by Izumiya Rihei in Osaka in 1623. Later, a group of metallurgists opened refineries in other parts of Japan (Imai 1980: 3–4). The liquation process (*namban-buki*) was successfully applied to silver extraction from argentiferous copper.

In the early 17th century, the liquation technique spread to silver mines exploiting copper-bearing silver ores, such as Tada in *c.*1630 (Kobata 1968: 749) and Ikuno in 1632 (Kobata 1968: 178) (Fig. 4). Iwami and Sado also have similar silver-copper ores, and hence the liquation process was probably used at the same time (*c.*1630) as at Tada and Ikuno. An old map (*c.*1618) illustrates several working sites in the silver-copper ore area of Iwami (Shimane Prefectural Board of Education 1999: 45). This suggests the early use of liquation (Fig. 5) in that area, although the earliest document related to copper production from Iwami is dated to 1683 (Imai 1998a: 31). In the case of Sado, the production of liquated copper was documented in 1682 (Fumoto 1956: 174).

After the prohibition of the export of silver in 1668, the copper exports increased remarkably and the Osaka copper-refinery complex became the centre of copper refinement (Fig. 6). Liquation was also used in several copper mines. Ashio copper (probably liquated) was exported from Nagasaki from 1676. The liquation technique was used at Oppu in 1678 (Hasegawa 2004: 8) and Omodani in *c.*1700 (Kobata 1993: 512).

In 1725 a large-scale copper refinery was opened in Nagasaki and was in operation until 1738 (Iwasaki 1998: 19). In 1775 the central refinery in northern Honshu was constructed at Kagoyama after the experimental refining at Ani in 1774 (Ani-machi 1992: 697–8). A refinery in central Honshu was also constructed at Takayama in 1855 (Mitsui Mining and Smelting Co. Ltd. 1970: 147) (Fig. 7). However, Osaka remained the national centre of copper refinement through the Edo period.

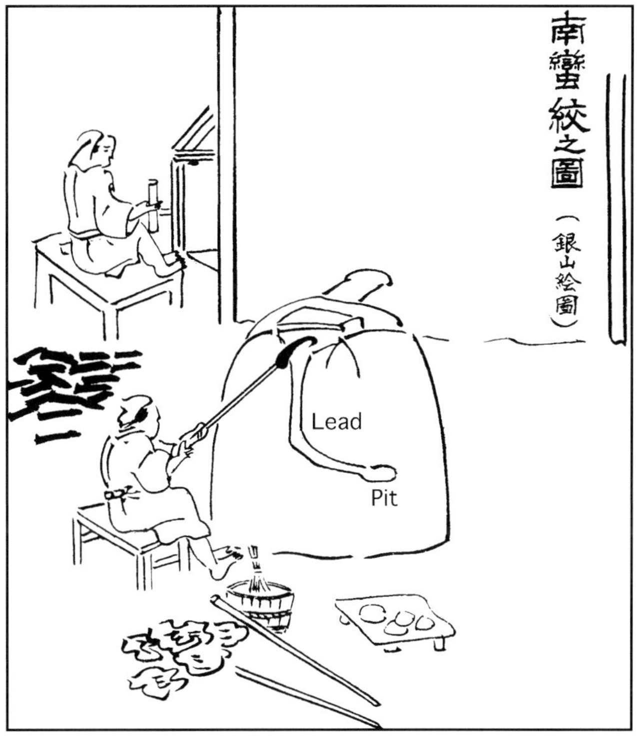

Figure 5 The Namban (liquation) furnace in the Iwami silver mine cited in Yamane 1932 (reprinted 1974 edition, figure 13 in the Appendices). The year of the original drawing is unknown.

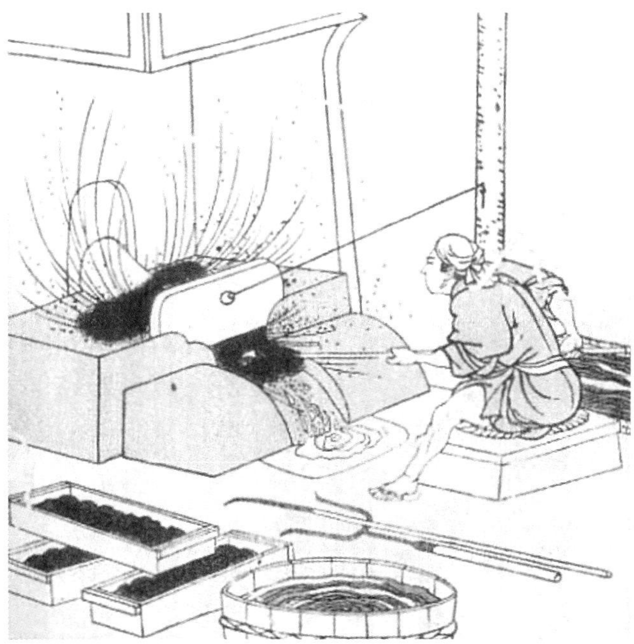

Figure 6 *Namban-buki* (liquation) at the Sumitomo refinery in *Kodo-zuroku* (Masuda *c.*1811, Kyushu University Museum Collection, leaf 6 (verso)).

Characteristics of *namban-buki*

Japanese liquation (*namban-buki* or *namban-shibori*) was modified greatly from the old German method (*Das Saigern*), although its principles were adapted from the European ones. In the early 17th century, liquation was adopted in the smelting workshops for extracting silver from copper-bearing silver ore. In those days, silver was the major product and copper was the by-product.

In the later 17th century, however, copper became the more important export commodity. The aim of the liquation technique shifted to refining copper, and silver became a by-product. As Gowland (1894: 464) stated, the Japanese metallurgists of the time developed the method and achieved complete separation of the two metals.

During the Edo period, copper ore was smelted in each mining area. The resulting crude copper was generally sent to the Osaka refineries and if the copper contained a certain amount of silver, it was refined to separate it from silver by liquation (*namban-buki*). The refining work, including liquation, consists of three sub-processes: (1) silver-bearing crude copper is melted with lead to form a copper-lead-silver alloy and cooled; (2) liquation process, i.e. the copper-lead-silver alloy is heated in a liquation furnace at a

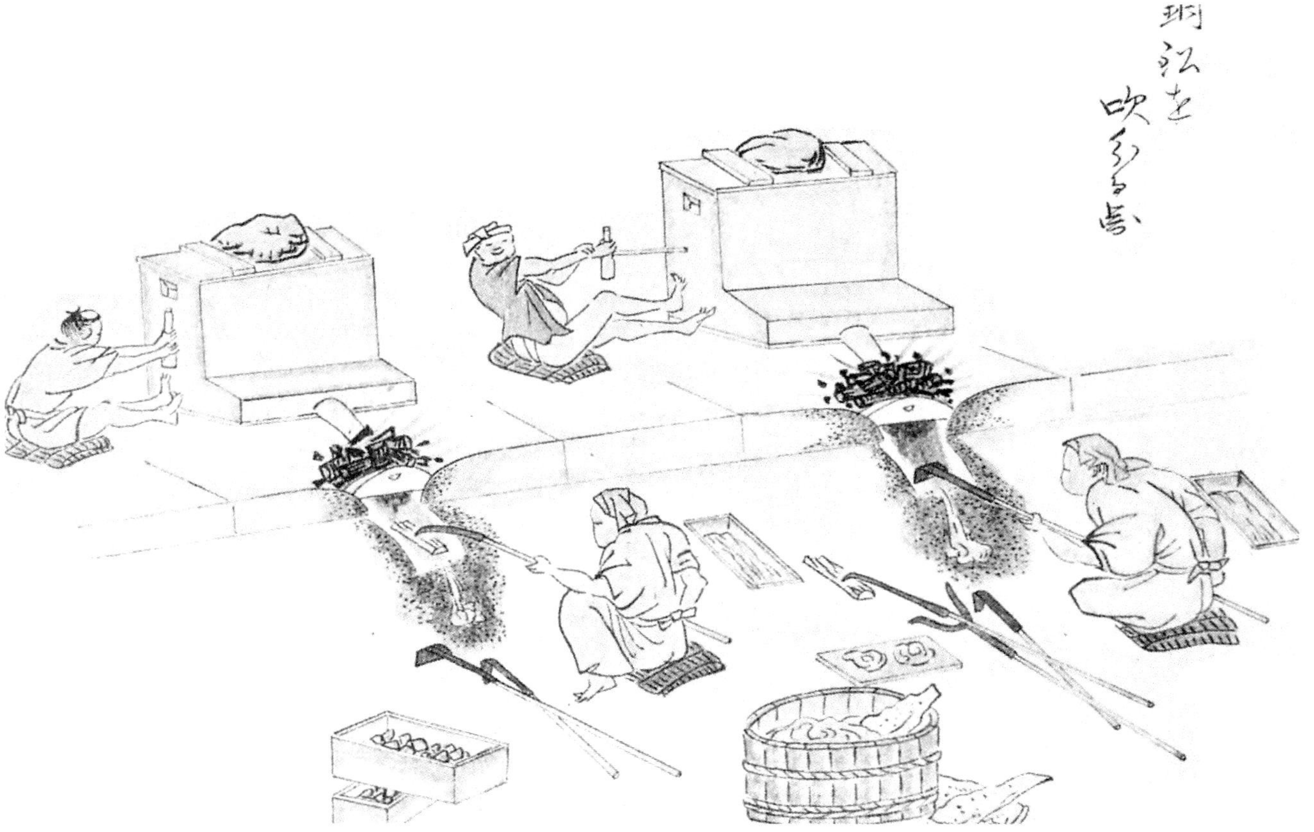

Figure 7 Liquation in the Takayama refinery drawn in *Hida Kanayama Ezu* (Sashida 1858; Kamioka-machi 1975: Historical Materials 2).

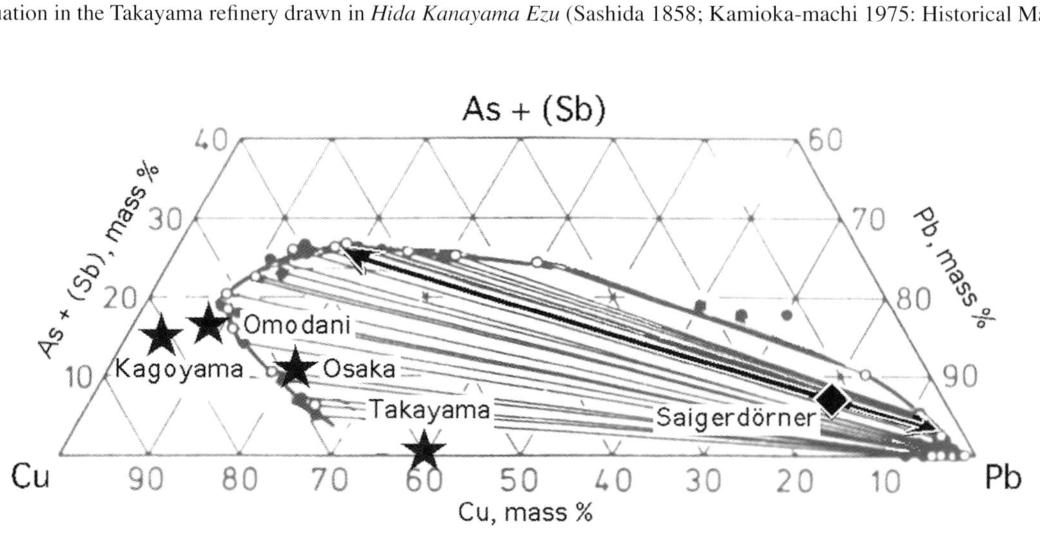

Figure 8 Compositions of alloys in the pseudo-ternary Cu-Pb-(As+Sb) system. The diagram shows the miscibility gap at 900 °C (Hino and Azakami 1978: 873; Espeleta *et al.* 1990: 827). Takayama (crude copper) (Gowland 1894: 464), Omodani (*shirome*) (Gowland 1894: 468), Kagoyama (*shirome*) (Goto 1998: 508), Osaka (*shirome*) (Osaka City Cultural Properties Association 1998: 519) and Saigerdorner (Percy 1880: 325).

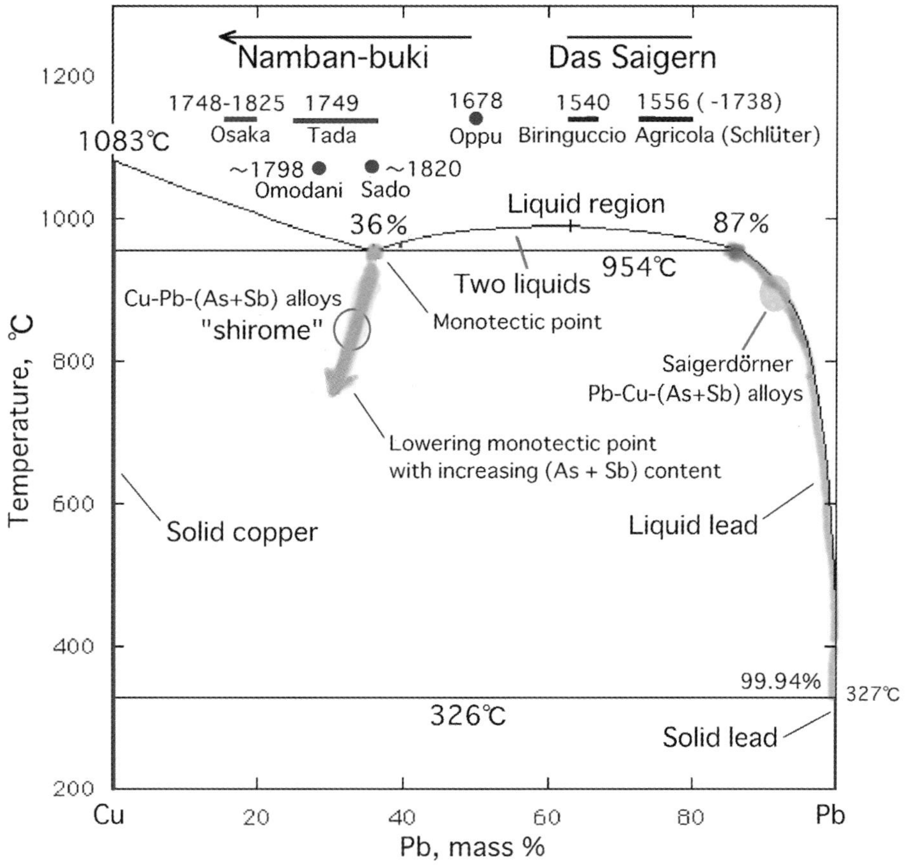

Figure 9 Compositions of liquation alloy and by-products *shirome* and Saigerdoerner on the phase diagram of binary copper-lead alloy (redrawn from Hansen and Anderko 1958: 610). The effect of As + Sb on the monotectic point is indicated.

temperature around 900 °C and silver-bearing liquid lead is squeezed out, leaving a pasty second liquid phase (*shirome*) and solid copper in the furnace; (3) cupellation, i.e. silver-rich lead is oxidised in the cupellation furnace and lead oxide flows out from the molten metal leaving pure silver.

The characteristics of *namban-buki* can be summarised as follows: squeezing of liquation alloy; a small-scale operation using many pieces of liquation alloy; small lead ratio in liquation alloy; and by-product *shirome* (copper-lead-arsenic-antimony alloy). The chemical compositions of various metals and liquation alloys are shown in Figures 8 and 9.

Das Saigern uses static heating and liquated metal (silver-bearing lead) flows down toward the front, leaving porous solid copper in the furnace. In the case of *namban-buki*, liquation is effected by squeezing the pasty alloy with a wooden stick fastened to an iron hook (Fig. 6).

In the liquation process, *Das Saigern* uses six or eight disk-shaped cakes of about 100 kg each (Biringuccio 1540; English translation 2005: 157). The cakes may be 60 cm in diameter, 8 cm in thickness and weigh 160 kg each (Schlüter 1738 in Percy 1880: 304). *Namban-buki* uses many pieces of alloy. Liquation alloys were unearthed from the Sumitomo copper refinery site in Osaka and they are curved plate-like fragments of 8–16 cm in diameter, 1–3 cm in thickness and weigh 460–690 g (Osaka City Cultural Properties Association 1998: 277–9). The fragments were made using the following treatment: 'a curious tool consisting of a rudely spherical-shaped iron head about 4 in. or more in diameter attached to a long handle. This is dipped into the molten alloy, lifted out, and the adhering crust of metal knocked off and thrown into water' (Gowland 1894: 465). The *namban-buki* worked on about 24 kg of alloy in each run at the Sumitomo refinery in Osaka (Imai 1980: 28) and 15 kg at Takayama (Mitsui Mining and Smelting Co. Ltd. 1970: 172).

The proportion of lead to copper in the liquation alloy is large in the European liquation process (Fig. 9). The lead content in the liquation alloy is 63–67% (copper:lead = 1:2 or 1:1.7; Biringuccio 1540; English translation 2005: 157), 73–80% (Agricola 1556; compiled by Suhling 1976: 116) and 79% (copper:lead = 3:11; Schlüter 1738 in Percy 1880: 304). However, Japanese liquation (*namban-buki*) uses a small amount of lead. During the period between 1748 and 1840, lead in the liquation alloy ranged between 15 and 20% at Osaka, and 16–37% at Tada and Kagoyama (Imai 1980: 23; 1998b: 464), an exception being 50% lead in the liquation alloy at Oppu in 1678 (Hasegawa 2004: 8).

By-product metal (copper-lead arsenic-antimony) *shirome* is different from *Saigerdorner* in its chemical composition (Fig. 8). Using a lowered monotectic temperature, the wringing out of impurities from copper is completed. The low-melting-point metal *shirome* is obtained as a by-product. It is a kind of speiss or dross, consisting of copper, lead, arsenic, antimony and bismuth. Thus, the products of the liquation process are silver-bearing lead and purified copper, and the by-product is *shirome*. Silver is recovered from the lead using cupellation. The copper is refined to high purity copper. *Shirome* is also useful as an alloying material for coin and bronze works. 'The objects of its use in these cases were: to give additional hardness to the alloy without impairing its fusibility at moderate temperatures, and to obtain, when cast, a sharper impression of the mould than was possible with the copper-lead alloy alone' (Gowland 1894: 469).

Conclusions

Liquation was invented in the early 15th century in Nürnberg, Germany. Knowledge of liquation reached Japan in 1591 in the context of the growing network of global trade.

At first, Japanese liquation was used for silver extraction from argentiferous copper, which was produced from copper-bearing silver ores. In the later 17th century, liquation became the core technology not only for silver production, but also for the production of high quality copper.

Acknowledgements

I would like to thank Mr Koichi Ueda (Japanese Study Group for the History of Mining and Metallurgy), Ms Noriko Imai (Sumitomo Shiryokan) and Dr Joerg Matschullat (Technische Universitaet Bergakademie Freiberg) for their help to access several important documents of medieval and early modern times. I also thank Dr Katsunori Miyazaki and Dr Tetsuya Nakanishi at Kyushu University Museum for their cooperative study. This study has been supported by Grant-in Aid for Scientific Research of Priority Areas at the Japanese Ministry of Education, Culture, Sports, Science and Technology (No. 14023212).

References

Agricola, G. (1556) *De re metallica*, H.C. Hoover and L.H. Hoover (trans). New York: Dover, 1950.

Agricola, G.. (1657) *De re metallica*. Basel: Emanuelis Koenig.

Ani-machi (1992) *Ani Choshi* (The History of Ani Town). Ani: Ani-machi [in Japanese].

Biringuccio, V. (1540) *De la pirotechnia*, C.S. Smith and M.T. Gnudi (trans.). New York: American Institute of Mining and Metallurgical Engineers, 1942. (Republished using the 1959 edition by Dover Publications, Inc., New York, 1990 and 2005.)

Braunstein, P. (1983) 'Innovations in mining and metal production in Europe in the late Middle Ages', *Journal of European Economic History* 12: 573–91.

Espeleta, A.N., Hino, M. and Azakami, T. (1990) 'Phase separation and silver distribution in the Pb-Fe-Sb and Pb-Cu-Sb systems at 1200°C', *Shigen-sozai gakkai-shi* (Journal of the Mining and Materials Processing Institute of Japan) 106: 825–30 [in Japanese with English abstract].

Fumoto, S. (1956) *Sado Kin'gin-zan Shiwa* (Chronicle of the Sado Gold-silver Mine). Tokyo: Mitsubishi Metal Mining Co. Ltd. [in Japanese].

Goto, S. (1998) 'Metallurgic considerations of *namban-buki*', in Osaka City Cultural Properties Association, *Archaeological Reports of the Sumitomo Copper Refinery Site*, 481–510. Osaka: Ōsaka-shi bunkazai kyôkai [in Japanese].

Gowland, W. (1894) 'A Japanese pseudo-speise (*shirome*) and its relation to the purity of Japanese copper and the presence of arsenic in Japanese bronze', *Journal of the Society of Chemical Industry* 13: 463–71.

Hansen, M. and Anderko, K. (1958) *Constitution of Binary Alloys*, 2nd edn. New York: McGraw-Hill.

Hasegawa, S. (2004) 'Enpo-Tenna-ki no Mutsu-koku Oppu gin-do-zan – Tsugaru-ryo oteyama no han'ei to suitai' ('The Oppu silver-copper mine of Mutsu-koku during Enpo-Tenna periods: prosperity and decline of the mine owned by Tsugaru-ryo'), *Bulletin of Humanity, Faculty of Humanity, Hirosaki University* 12: 1–32 [in Japanese].

Hino, M. and Azakami, T. (1978) 'The miscibility gap and the distribution of silver between metallic lead and speiss in Pb-Fe-As and Pb-Cu-As ternary systems: fundamental studies on speiss in lead smelting (1st report)', *Nippon Kogyokaishi* (*Journal of the Mining and Metallurgical Institute of Japan*) 94: 871–6 [in Japanese with English abstract].

Imai, N. (1980) 'Kinsei Sumitomo no fukisho no kenkyu' ('Study of the Sumitomo refinery during the early modern age'), *Sen'oku Soko* 19: 1–80 [in Japanese].

Imai, N. (1998a) 'Kinsei Iwami-ginzan no sando ni tsuite' ('On the copper production from the Iwami silver mine during the early modern age'), *Nihon Kogyo-shi Kenkyu* (*Journal of the Historical Study Group of Mining and Metallurgy, Japan*) 35: 30–35 [in Japanese].

Imai, N. (1998b) 'Management of the Sumitomo Nagahori refinery', in Osaka City Cultural Properties Association, *Archaeological Reports of the Sumitomo Copper Refinery Site*, 463–72. Osaka: Ōsaka-shi bunkazai kyôkai [in Japanese].

Iwasaki, Y. (1998) 'The study of Do-Fuki-Sho (factory for copper refining) in Nagasaki', *Shien* (*Journal of History*) 135: 19–42 [in Japanese].

Kamioka-machi (1975) *Kamioka Cho-shi, Shiryo-hen Chu-kan* (The History of Kamioka Town, Historical Materials, Volume 2). Kamioka: Kamioka-machi [in Japanese].

Kasper, H.H. (1993) *Von der Saigerhutte zum Kupferhammer Grunthal 1537–1873*. Olbernhau: Saigerhüttenverein e.V. Olbernhau-Grünthal.

Kobata A. (1993) *Nihon Dokogyoshi-no Kenkyu* (The Study on the History of Japanese Copper Mining). Kyoto: Shibun-kaku [in Japanese].

Kobata, A. (1965) 'The production and use of gold and silver in sixteenth- and seventeenth-century Japan', *Economic History Review*, 2nd ser. 18: 245–66.

Kobata, A. (1968) *Nihon Kozanshi-no Kenkyu* (The Study on the History of Japanese Mines). Tokyo: Iwanami-shoten [in Japanese].

Lindroth, S. (1955) *Gruvbrytning och Kopparhantering vid Stora Kopparberget II Kopparhanteringen*. Uppsala: Almqvist & Wiksells Boktryckeri Ab.

Lynch, M. (2002) *Mining in World History*. London: Reaktion Books Ltd.

Masuda, Tsuna *c*.1811. *Kodo-zuroku* (*Kodo-zuroku* Illustrated Book on the Smelting of Copper), C.S. Smith (trans.), Niwa, Tokei (illus.). Norwalk: Burndy Library, 1983 (German translation by A. Hauptmann, 1984, *Kodo-zuroku 'Illustrierte Darstellung der Kupfergewinnung in Japan'*, Bochum, Deutsches Bergbau-Museum).

Mitsui Mining and Smelting Co. Ltd. (1970) *Kamioka Kozan-shi* (History of the Kamioka Mine). Tokyo: Mitsui Mining and Smelting Co. Ltd. [in Japanese].

Osaka City Cultural Properties Association (1998) *Archaeological Reports of the Sumitomo Copper Refinery Site*. Osaka: Ōsaka-shi bunkazai kyôkai [in Japanese with English abstract].

Percy, J. (1880) *Metallurgy: Part 1 Silver and Gold*. London: John Murray.

Sashida, T. (1858) *Hida Kanayama Ezu*, reproduced in Kamio-machi 1975, *Kamioka Cho-shi, Shiryo-hen Chu-kan* (The History of

Kamioka Town, Historical Materials, Volume 2). Kamioka: Kamioka-machi [in Japanese].

Schlüter, C.A. (1738) *Grundlicher Unterricht von Hutte-Werken*. Braunschweig:

Shimane Prefectural Board of Education, 1999 (*Iwami-Ginzan Silver Mine No. 1: An Outline of Remains*, Matsue, Shimane Prefectural Board of Education) [in Japanese].

Song, Ying-xing (1637) *Tian-gong kai-wu*, K. Yabuuchi (trans.). Tokyo: Heibonsha, 1969.

Su, Song (1061) *Tu jing ben cao*. Fuzhou: Long yuan, 1988 [in Chinese].

Suhling, L. (1976) *Der Seigerhuttenprozess. Die Technologie des Kupferseigerns nach dem fruhen metallurgischen Schrifttum*. Stuttgart: Dr. Riederer Verlag GmbH.

Tegengren, F.R. (1924) 'Sveriges aedlare malmer och bergverk', *Sveriges Geologiska Undersoeknig* 17: 1–66.

Yamane, T. (1932) *Iwami Ginzan ni kansuru Kenkyu* (Study of the Iwami Silver Mine). Oda City: Sekito-bunka Kenkyukai (reprinted by Rinsen Shoten, Kyoto, 1974) [in Japanese].

Yasukuni, R. (1999) '"Kodo zuroku" no shoshiteki kento' ('Bibliographical study on "Kodo-zuroku"'). *Sumitomo Shiryokanpo* 30: 1–19 [in Japanese].

A technical study of silver samples from Xi'an, Shaanxi province, China, dating from the Warring States period to the Tang dynasty

Junchang Yang, Paul Jett, Lynn Brostoff and Michelle Taube

ABSTRACT Fragments of silver objects from various archaeological sites in the area of Xi'an, Shaanxi province, China, were sampled for technical study. For the initial phase of the study reported here, 22 samples were examined. Most of the objects sampled date to the Tang dynasty, but one dates to the Warring States period and seven to the Western Han dynasty. The samples are all from objects formed by hammering, with one exception – a rare example of ancient cast silver. Also, the alloys were determined to be generally high in their silver content, greater than 90%, and almost half of the samples had silver contents of 97% or more. Gilding, present on some of the samples, was determined to be mercury gilding.

Preliminary examination of the samples was performed using metallographic study and scanning electron microscopy (SEM). Analysis was performed using X-ray fluorescence (XRF) spectroscopy and laser ablation-inductively coupled plasma–mass spectroscopy (LA-ICP–MS). In addition, to better understand the metallographic structures seen in the ancient samples, test samples of silver alloys were cast, worked and studied.

Introduction

The lack of technical information on ancient Chinese silver was the impetus for the study. Published technical data on Chinese silver are very limited, particularly for material with an archaeological provenance. An investigation of the methods and materials used to make silver metalwork during the Tang dynasty was the specific focus of this study, but samples from eight important earlier works were also studied.

The role played by silver in China is modest when compared to that of other metals such as bronze or iron. Even gold, long considered to be little used in early China, seems more prominent and important than silver during the early historical periods. Dated far earlier than any other known silver object in China is a nose-ring found at Huoshaogou, Yumen city, Gansu province, in a Siba culture grave that contained material radiocarbon dated to approximately 1600 BCE. This object is unlike any other object found in China and dated to the same period; it more resembles artifacts of West Asian or Central Asian origin (Bunker 1994), and a non-Chinese origin for this piece seems more likely given the lack of other finds of silver until the late Spring and Autumn period (770–475 BCE) and Warring States period (475–221 BCE). At this time, the use of silver occurs in many forms: as inlay, sheet, or foil, as an amalgam used to decorate bronzes, and as objects in their own right, sometimes with added decorations of gilding or inlaid stones. Some of these objects reflect west Asian forms or other foreign influences (Fig. 1), and this is a continuing theme throughout the history of silver in China until the later historical periods. The many discoveries of Western silver in Chinese tombs from the Han (206 BCE–220 CE) to the Tang dynasty (618–907 CE) are well known (Qi 1999), as is the use of Western motifs in the decoration of silver, particularly during the Tang dynasty (Rawson 1982). It is also during the Tang dynasty that silverworking in China advanced greatly, both in terms of the amount produced and in diversity and originality. It is particularly interesting to note that this expansion of silverwork in China came at a time when the production of silverware in west Asia was in a steep decline and when there were severe silver shortages for decades (Meyers 2003). For example, a single hoard from the Tang dynasty – the Hejiacun hoard where nearly 200 silver vessels and boxes were found (Hansen 2003; Yang 1999) – rivals in number all the known comparable silver objects from the entire Islamic world during this period. From the Tang period onward, there is considerable production of silverware in China, notable not only for its quantity, but also for its quality and the development of many indigenous forms and motifs. For a list of silver finds in China, see Liang-Lee and Louis 1996.

In spite of its long history in China, silver has received only limited technical study, and much of that has been focused on individual objects or material removed from its archaeological or historical context. This study aimed at redressing this lack of information through the analysis and technical study of excavated material.

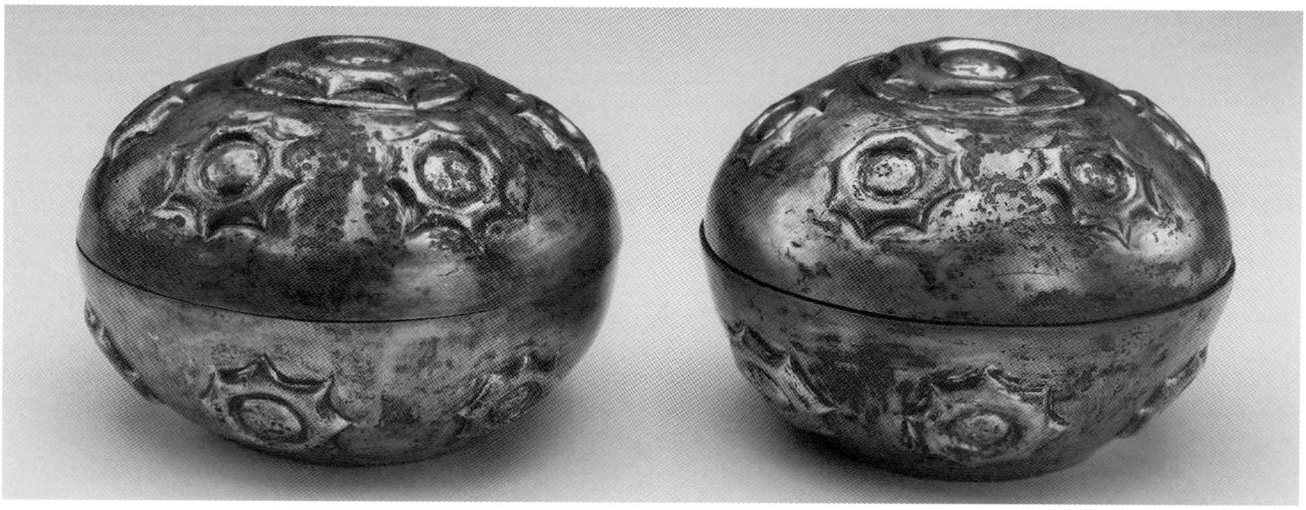

Figure 1 Pair of silver boxes, Warring States period. Accession no. RLS1997.48.515a&b. Arthur M. Sackler Gallery, Smithsonian Institution. Washington, DC.

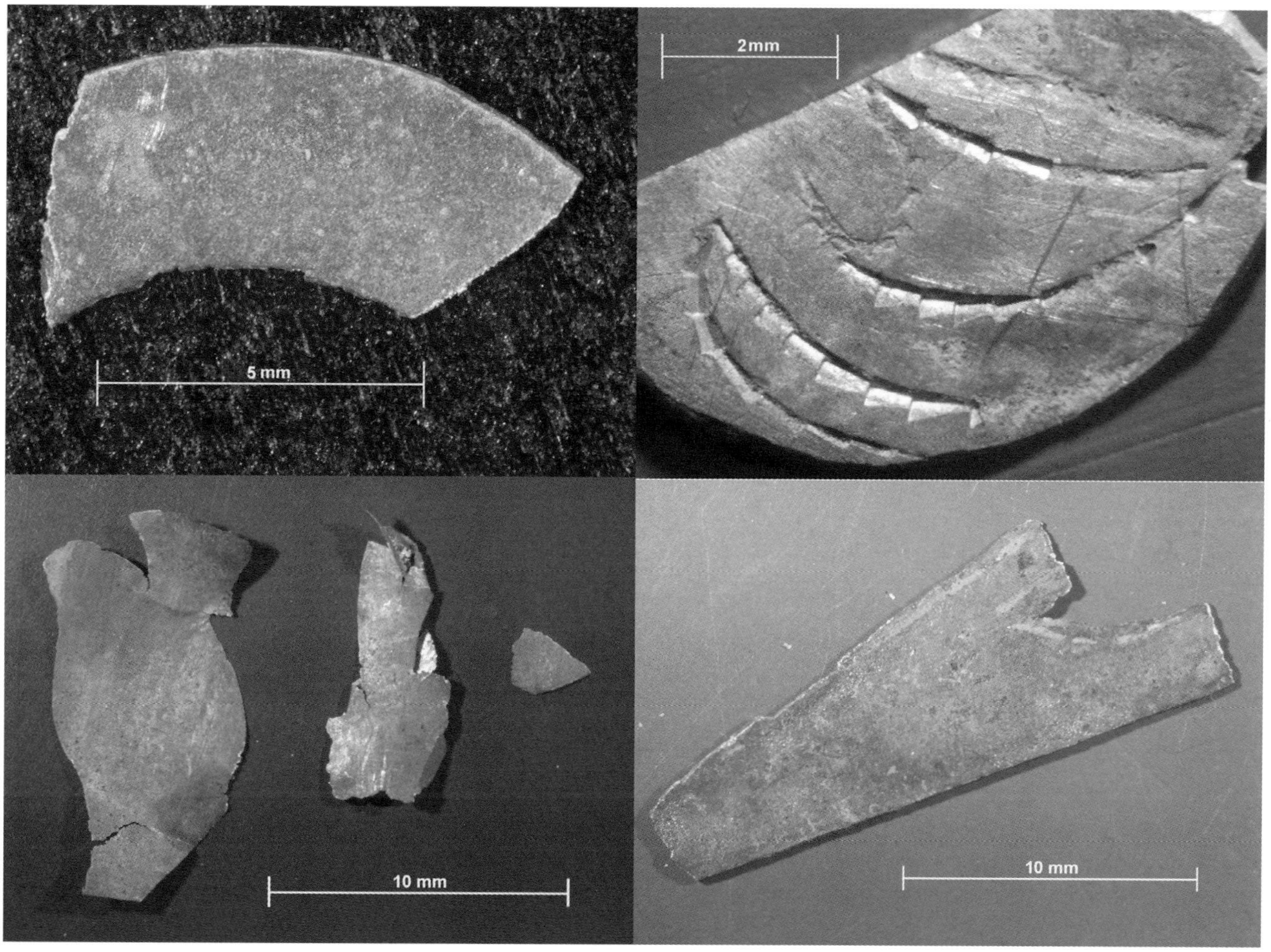

Figure 2 Some examples of the samples used for study. Clockwise from top left: #6270, #6275, #6276 & #6278.

The samples and their sources

Samples from 22 silver fragments found at 12 sites were chosen for this study (Table 1). One of the samples dates to the Warring States period, seven to the Western Han dynasty (206 BCE – 8 CE), and 14 to the Tang dynasty. Most of the silver fragments are clearly from decorative elements of other objects, such as foil for mirrors or inlay for lacquer. One heavier and thicker fragment is believed to be a portion of a bracelet or anklet (#6341), and this is the only piece that was made by casting rather than hammering. With the exception of sample #6341, which measured about 1 mm in thickness, the fragments varied between 0.2 and 0.6 mm in thickness. The fragments also varied in length and width, with the larg-

Table 1 List of samples.

Lab #	Sample type	Site	Date
6270	Silver plate	Tomb of Li Xian, Sanhecun, Pucheng County	T
6271	Silver plate	Tomb of Li Xian, Sanhecun, Pucheng County	T
6272	Silver	Lichui tomb, southern suburb of Xi'an	T
6275	Silver paper-cutter	Tomb, M22, Yuanzitou, Long County	T
6276	Fragment	Tomb, LBM2:52, Banqiaogou, Long County	H
6277	Foil	Tomb, M22, Yuanzitou, Long County	T
6278	Foil	Tomb, M22, Yuanzitou, Long County	T
6279	Fragment	Tomb, LBM2:52, Banqiaogou, Long County	H
6341	Fragment	Pit, Yangling	H
6343	Plate	Hoard, Famensi, Fufeng County	T
6352	Fragment	Yanzhi's couple's tomb, Majiagou, eastern suburb of Xi'an	T
6354	Sheet fragments	Tomb M1, Shenheyuan, southern suburb of Xi'an	WS
6359	Hairpin	Tomb, 2002ZTDM16:19, southern suburb of Xi'an	T
6360	Hairpin	Tomb, 2002ZTDM16:14, southern suburb of Xi'an	T
6361	Hairpin	Tomb, 2002ZTDM16:17, southern suburb of Xi'an	T
6486	Finger ring	Tomb, 2001ZCM63:5, northern suburb of Xi'an	H
6487	Finger ring	Tomb, 2000ZCM21:14, northern suburb of Xi'an	H
6488	Finger ring	Tomb, 1998JH:C, northern suburb of Xi'an	H
6489	Finger ring	Tomb, 1998JH:C, northern suburb of Xi'an	H
6490	Lacquer box inlay	Lichui tomb, southern suburb of Xi'an	T
6491	Mirror inlay	Tomb, southern suburb of Xi'an	T
6492	Foil from mirror	Tomb, 2006M110:6, southern suburb of Xi'an	T

WS = Warring States dynasty; H = Han dynasty; T = Tang dynasty

est being about 20 mm in length. Gilding was evident on some samples, as were chased decorations (Fig. 2).

Components of the study

After an initial examination using stereobinocular microscopy, the fragments were analyzed using X-ray fluorescence (XRF) spectroscopy. The surfaces of the samples were analyzed without preliminary preparation; this approach gave a rough estimate of the alloy compositions and indicated the range of the major elements of interest – silver, copper, gold and lead – and the appropriate range of standards to use for further analysis. XRF also indicated the presence of gilding. Based on these results, standards were prepared for further analysis using laser ablation-inductively coupled plasma–mass spectrometry (LA-ICP–MS). In the end, it was interesting to see that the XRF results, where gilding was not present, were generally in good agreement with the LA-ICP–MS results. The procedure for the LA-ICP–MS analysis is given along with the results below.

Portions of the samples were also removed for metallographic examination and study using scanning electron microscopy (SEM). Soon after beginning the study of the metallographic samples, it became clear that it would be useful to cast and work test samples of silver for comparison. The preparation and study of the test samples is described below.

Metallographic examination

As noted above, all the objects from which the samples were taken were formed by hammering, except for one piece that was cast. The cast sample (#6341) is very typical in its metallographic structure and does not display any evidence of working after casting (Fig. 3). The structures observed among the hammered samples were basically of three types. The first is a typical hammered and annealed structure that exhibits some cold-working (Fig. 4). In the particular exam-

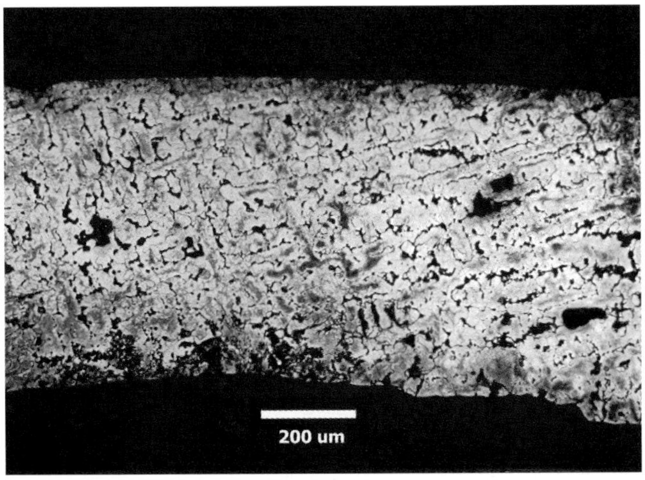

Figure 3 Sample #6341: cast structure.

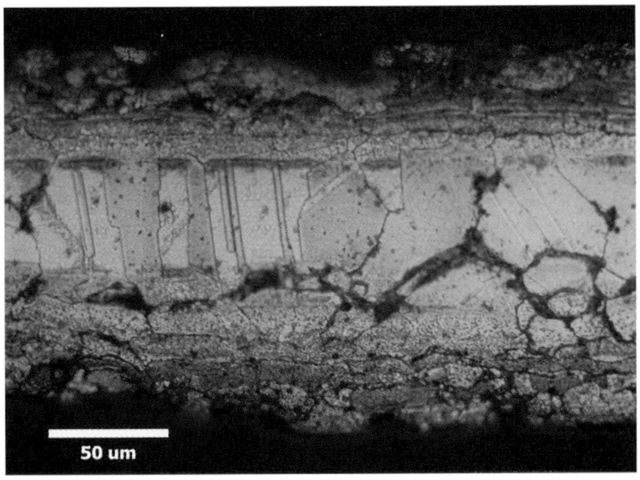

Figure 4 Sample #6271: worked structure.

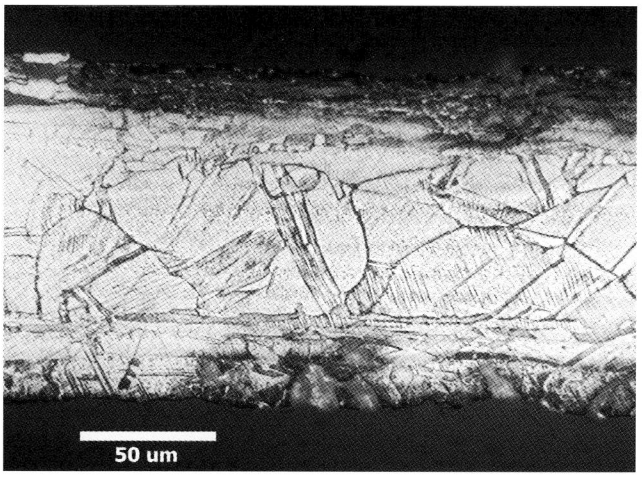

Figure 5 Sample #6270: worked structure exhibiting distorted twin lines and strain lines.

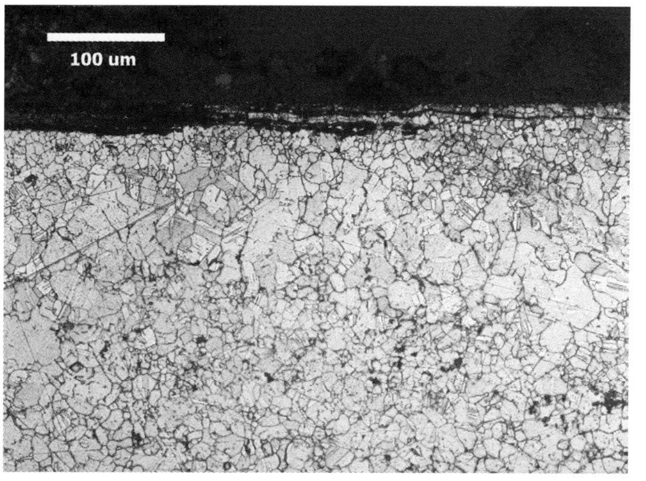

Figure 6 Sample #6352: worked structure exhibiting small grain size.

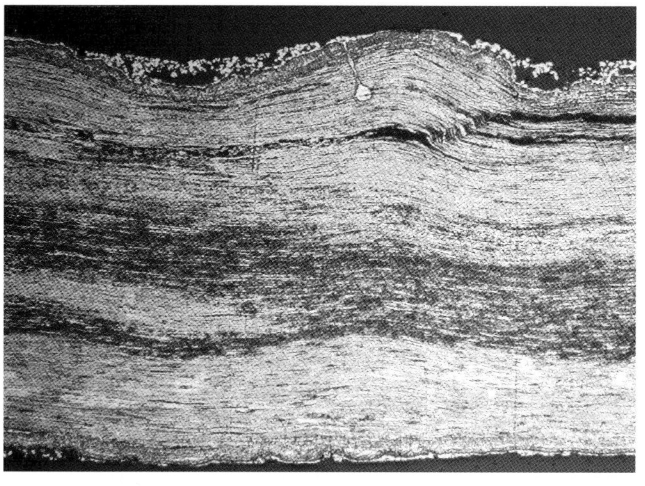

Figure 7 Sample #6361: sample showing elongated structure and a thin layer of mercury gilding.

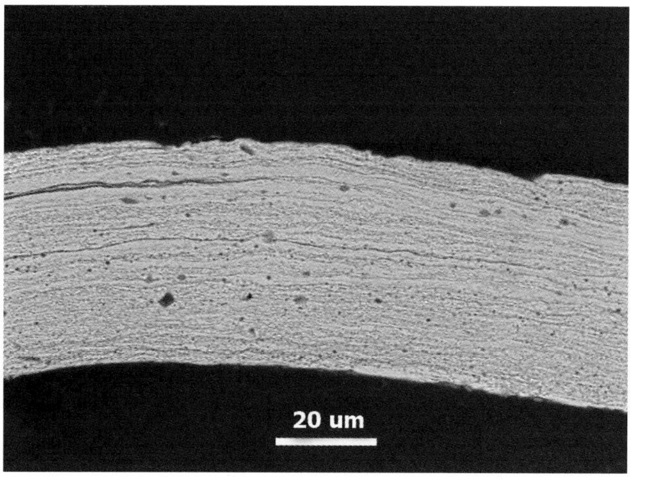

Figure 8 Modern silver sample reduced by 98% without annealing.

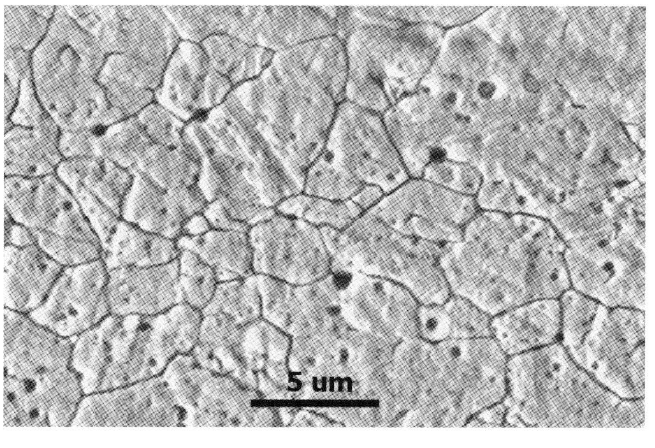

Figure 9 Modern silver sample reduced by 90%, annealed, and quickly quenched.

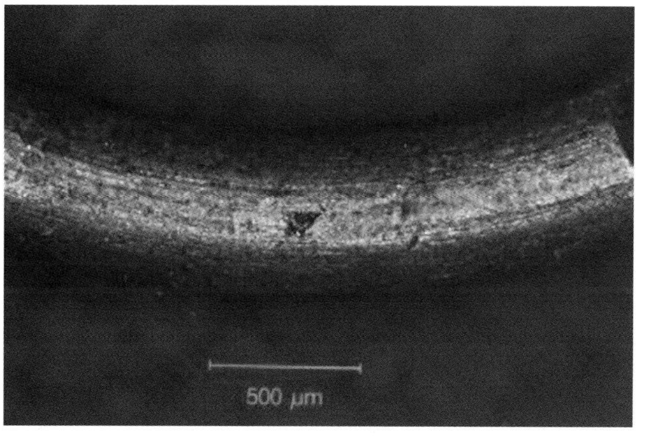

Figure 10 Wire from the hoard found at Famensi, Fufeng county. Note the parallel striations due to manufacture by drawing.

ple shown in Figure 4, there are also relatively thick layers of corrosion on the surfaces and some areas of significant intergranular corrosion. A variant of the annealed structure is shown in Figure 5, where it appears that considerable cold-working took place after annealing, which resulted in distorted twin lines and strain lines in the grains.

The second type of worked structure is shown in Figure 6 and is characterized by an extremely small grain size. Twin lines can still be seen, but it appears that little if any cold-working occurred after the final annealing. The third type of worked structure, shown in Figure 7, exhibits an extremely elongated structure that indicates little or no annealing, or at least no annealing in the latter stages of working. Also apparent in this image is mercury gilding on the surface (for more information on the working and microstructure of ancient metals, see Scott 1991: 5–10).

To understand better what can be observed in the metallographic samples of ancient silver, test samples were prepared for comparison. For this purpose, an alloy containing 98% silver, 1.3% copper and 0.7% gold was made from a mixture of pure metals. The prepared silver alloy was made by melting and casting the pure metals into an ingot mold which gave a plate of metal 3.5 mm thick. Lead was also added to the mix, but it was completely lost during the melting and casting process.

Samples cut from the ingot were then reduced in thickness using a rolling mill so that the changes in thickness could be precisely controlled. Some of the samples were annealed during the rolling process and some were not. The samples were reduced in thickness by as much as 98%.

Figure 8 shows one of the modern silver samples reduced by 98% without annealing; its structure is very similar to that seen in Figure 7. When test samples reduced to 90–98% of their original thickness were annealed and quickly quenched, extremely small grains only a few microns in length and width were produced (Fig. 9); this structure is the same as those seen in a number of the samples from the Tang dynasty, such as the one shown in Figure 6. Note also that the metallographic structure shown in Figure 9 still exhibits twinning within the grains.

It is interesting that the results of metallographic analysis and the study of prepared samples suggest that different methods of hammering and annealing were used to shape the various pieces. Leaving the silver un-annealed and work-hardened would make for a stiff, brittle sheet, but also one that was very thin and light. The annealed sheets would be more durable and flexible, and one factor in deciding how to work the sheet was probably whether or not chased decorations would be applied. Few of the samples in this study bear chased decoration but, in all cases, the samples that do have been heavily annealed. It would also make sense for very thin sheet or foil to have been well annealed for certain uses. It is possible, for example, that malleability would have been critical to successfully working with sheet or foil and applying it to substrates, particularly those substrates with irregular surfaces.

Mention should also be made of one other technical feature observed. Sample #6343, from the find of a reliquary deposit at Famensi, Fufeng county, consisted of multiple pieces, including a short fragment of wire. Microscopic study of the wire revealed that it was made by drawing (Fig. 10). This wire is a relatively early example of drawn wire from China and its identification contributes to the study of drawn wire and its history in East Asia (see for example Ogden 2003). The reliquary deposit was opened and re-interred several times during the Tang dynasty. The terminal date for the objects in the reliquary, and the wire in question, is believed to be 874 CE (Han 1999; Yang 1999).

LA-ICP–MS analysis

Results are shown in Table 2 and details regarding the methodology and instrumentation are given in the appendix. The results are summarized below. One sample, #6275, was analyzed using XRF rather than LA-ICP–MS.

Overall, little consistency was seen in the alloy compositions. The average composition of all samples in terms of the four major and minor elements was found to be about 92.9% silver, 5.7% copper, 0.7% gold and 0.5% lead, by weight percent (wt%). However, most of the objects consisted of fairly pure silver alloys, with the silver content greater than 95wt% – 10 of the samples exhibited silver contents of about 97 wt% or greater – while a few samples showed quite high amounts of copper. The composition of the cast sample, #6341, was distinctive because it had the highest lead content of all the samples, 3.52 wt%, and a relatively high copper content. Some of the other samples were notable because of their relatively low silver but high copper contents; samples #6359, 6360 and 6361 came from the same site and were similar in this regard. Sample #6361 was particularly striking in that its silver and copper contents were determined to be 54 wt% and 44 wt%, respectively. It should be noted that these values were outside of the composition range represented by the reference standards and were therefore subject to greater error. One other sample, #6491, was also relatively low in silver (77 wt%) and high in copper (21 wt%) compared to the median values.

As with the general alloy compositions, no patterns were discerned in the trace element compositions of the samples. In general, tin, zinc and bismuth were often found in relatively significant amounts, in some cases up to 0.25 wt%. In all of the samples antimony and tellurium were detected at generally low levels (< 80 ppm), while palladium and platinum were detected at very low trace levels (< 18 ppm) or were below detection limits. Results for arsenic, manganese, iron and selenium were not reliable and are not reported. The presence of these trace elements, as shown in Table 2, probably relates to the sources of silver deposits, which are commonly associated with a great variety of natural ores. Tin, zinc and bismuth may also be naturally associated with copper and lead ore components in the alloys. It is notable that similar levels of tin and zinc have been found in Sasanian silverwork dating to roughly the same period; see for example Harper and Meyers (1981).

Conclusions

The study reported here deals with the metallographic features and elemental composition of a very limited number of samples, so that few if any conclusions can be drawn about Tang silverworking. The seeming preference for worked as opposed to cast silver is not surprising and is consistent with the silver metalworking traditions of many cultures, as is the use of relatively pure silver alloys in general. However, no consistent use of alloy types was discerned in this study, in agreement with previous analyses of Tang silver performed at the Freer Gallery of Art, Smithsonian Institution (Jett nd).

Table 2 Analytical results.

Sample	6270	6271	6272	6275	6276	6277	6278	6279	6341	6343	6352	6354	6359	6360	6361	6486	6487	6488	6489	6490	6491*	6492*
Element																						
Ag (wt%)**	97.66	96.78	95.53	98.99	93.86	97.78	98.62	94.15	90.73	95.49	99.25	93.32	88.07	92.12	53.83	98.67	93.54	99.57	99.26	96.99	77.21	93.40
Cu (wt%)	1.0	2.7	2.9	1.0	4.6	1.5	1.3	4.9	5.2	2.5	0.3	2.9	10.4	6.7	44.3	0.7	3.9	0.2	0.3	2.3	20.9	5.4
variation (rsd)	4%	4%	26%		7%	40%	6%	5%	39%	10%	7%	4%	5%	4%		8%	49%	1%	235%	7%	10%	17%
ave. pp. rsd	3%	6%	13%		7%	13%	7%	3%	9%	6%	13%	2%	56%	9%	4%	7%	24%	20%	41%	2%	47%	15%
Au (wt%)	0.9	0.2	1.0	0.0	0.8	0.4	0.0	0.8	0.1	1.7	0.2	3.2	0.5	0.2	1.1	0.3	1.5	0.0	0.0	0.3	0.6	0.7
variation (rsd)	3%	3%	14%		6%	3%	11%	2%	6%	4%	22%	3%	3%	84%		30%	6%	0%	0%	6%	16%	12%
ave. pp. rsd	1%	3%	12%		5%	3%	4%	1%	7%	3%	47%	1%	2%	40%	3%	38%	4%	6%		1%	21%	12%
Pb (wt%)	0.35	0.26	0.20	0.01	0.62	0.21	0.07	0.12	3.52	0.18	0.24	0.58	0.51	0.82	0.39	0.24	0.81	0.23	0.42	0.34	1.16	0.37
variation (rsd)	9%	8%	23%		6%	7%	24%	10%	6%	7%	6%	2%	4%	3%		8%	15%	0%	93%	5%	12%	17%
ave. pp. rsd	8%	4%	11%		5%	5%	21%	13%	10%	7%	17%	3%	14%	1%	7%	11%	27%	76%	17%	1%	47%	19%
Sn (ppm)	590.77	287.56	699.31		587.90	504.64	30.20	64.46	2227.33	394.76	4.46	0.71	1768.62	452.75	2071.54	61.64	1476.43	1.19	<LOD	22.20	542.35	313.83
variation (rsd)	5%	13%	16%		36%	9%	118%	22%	5%	6%	23%	41%	8%	3%		10%	13%	1%		6%	11%	26%
ave. pp. rsd	6%	20%	11%		76%	5%	173%	29%	14%	6%	16%	22%	18%	9%	12%	9%	20%	58%		2%	67%	44%
Zn (ppm)	21.06	61.76	2546.71		0.92	87.61	0.71	<LOD	1.62	48.05	17.76	0.19	1290.53	83.41	321.82	5.24	6.12	4.61	<LOD	546.60	327.44	93.39
variation (rsd)	11%	17%	21%		51%	9%	79%		37%	23%	10%	118%	7%	3%		13%	33%	1%		6%	23%	18%
ave. pp. rsd	14%	22%	10%		70%	5%	29%		17%	28%	12%	47%	31%	4%	8%	5%	27%	88%		1%	58%	32%
Bi (ppm)	249.33	274.76	386.96		597.47	547.56	56.80	273.85	1969.95	845.78	53.41	19.57	2044.99	987.12	1239.34	867.46	985.07	6.37	159.80	148.02	379.07	823.65
variation (rsd)	19%	16%	32%		33%	6%	19%	23%	7%	8%	6%	4%	3%	3%		7%	15%	0%	164%	6%	13%	13%
ave. pp. rsd	16%	18%	17%		22%	5%	27%	26%	12%	5%	16%	1%	14%	3%	4%	7%	21%	93%	57%	1%	28%	25%
Sb (ppm)	2.21	2.56	6.94		7.54	2.41	<LOD	1.47	253.67	1.21	<LOD	<LOD	43.05	10.32	50.09	5.63	49.43	<LOD	<LOD	0.41	27.12	4.84
variation (rsd)	9%	11%	12%		12%	15%		29%	5%	30%			4%	3%		10%	6%			27%	16%	37%
ave. pp. rsd	4%	2%	7%		7%	5%		25%	14%	7%			18%	6%	4%	7%	14%			6%	56%	33%
Pd (ppm)	0.91	2.10	2.93		3.63	2.04	0.74	4.36	3.65	2.19	0.76	1.97	9.94	4.89	37.73	1.19	5.88	0.60	1.14	2.14	17.85	5.49
variation (rsd)	20%	14%	15%		11%	22%	19%	8%	13%	36%	29%	13%	8%	10%		29%	35%	1%	17%	9%	18%	43%
ave. pp. rsd	9%	4%	7%		6%	15%	1%	9%	9%	15%	17%	8%	37%	11%	5%	6%	48%	23%	9%	5%	51%	36%
Pt (ppm)	0.99	0.36	1.04		0.45	1.07	0.25	0.40	<LOD	1.41	0.52	<LOD	1.62	0.34	2.44	0.26	0.67	0.32	0.89	0.91	1.10	0.81
variation (rsd)	17%	33%	22%		19%	22%	21%	18%	1156%	23%	28%		19%	25%		46%	11%	0%	15%	14%	127%	59%
ave. pp. rsd	6%	20%	13%		9%	4%	5%	6%	143%	6%	17%		22%	16%	20%	4%	4%	20%	10%	3%	219%	28%
Te (ppm)	15.54	7.07	13.71		1.71	4.19	<LOD	0.76	2.59	18.66	0.62	3.08	40.60	17.19	51.87	<LOD	5.48	<LOD	<LOD	6.88	25.83	77.84
variation (rsd)	18%	11%	12%		10%	18%		14%	13%	7%	26%	7%	4%	4%			8%			9%	15%	17%
ave. pp. rsd	15%	8%	11%		9%	2%		8%	14%	7%	22%	4%	13%	6%	5%		5%			1%	22%	17%

* *Ag (wt%) by difference variation (rsd) = relative standard deviation (rsd) between spots ave. pp. rsd = peak to peak rsd
LOD = limits of detection
6275 - XRF results. (rsd) Ag = ± 0.04%; Cu = ± 3.6%; Au = 0%; Pb = ± 91%
* thin samples analyzed at 10 Hz

The differences found in the working of the metal are notable, however. It is logical that differing degrees of working and annealing may have been used depending on the final use or decoration of the silver. With further metallographic studies of other objects or fragments, it may be possible to discern trends in working techniques and to ultimately determine the original function of otherwise unidentifiable small or fragmentary silver finds.

The investigation of construction methods and tool marks would appear to be the most fruitful avenue to follow for future technical studies. There are no publications that give in-depth descriptions of these physical details, and it may be that trends within Tang silverwork can be determined by the study of these features.

Appendix: instrumentation and methodology of LA-ICP–MS and XRF analyses

LA-ICP–MS instrumentation consisted of a Perkin Elmer Elan 6000 ICP–MS and a Cetac LSX-200 Plus Nd:YAG 266 nm LA system. For the calibration of the silver samples, seven high purity silver reference materials were commissioned from MBH Analytical, Ltd. that contained the following elements in the ranges given: copper, 2.5–10 wt%, lead, 0.25–1.0 wt%, and gold, 0.25–2.0 wt% (three standards); copper, lead, gold, manganese, iron, zinc, arsenic, selenium, palladium, tin, antimony, tellurium, platinum and bismuth, 7–500 ppm (four standards). An attempt was made to include iridium and mercury, but the manufacturer was not successful in controlling dissolution of these elements in the standards and they occur as trace contaminants only. Small portions of the silver samples measuring about 150 × 900 μm were removed for analysis. Typically, five spots were analyzed, although in some cases more or less spots were analyzed; results are reported as the average concentration on each sample, as determined from the reference calibration curves. On each spot, the first 20–25 seconds of ablation were eliminated in order to allow the signal to stabilize and to dismiss non-representative data from the surface layers. LA-ICP–MS conditions are listed in Table 3.

XRF analysis was performed using an Omega V X-ray fluorescence spectrometer, a modified Spectrace 6000 spectrometer that allows for the direct analysis of objects without removing a sample and employs a thermoelectrically cooled Si(Li) detector, a 50 KV, 1.0 mA, Rh target X-ray tube and power supply, with primary filters on the X-ray beam in order to optimize the excitation conditions and improve the measurement of particular elements. The analyses were performed under the following conditions: 30 KV and 0.9 mA with an Al filter and a count time of 200 seconds. The results were derived from three measurements taken on different areas of an undisturbed surface on the sample. Quantification was performed using Fundamental Parameters Program Version 1.2e. Accuracy was checked by analyzing the sample along with a silver standard whose composition was previously determined and was comparable to that of the sample composition. The relative standard deviation (rsd) for the results is given in Table 2.

Table 3 LA-ICP–MS conditions.

Laser	
Laser	Nd:YAG, 266 nm
Laser pulse length	<4 ns
Laser energy	Q-switched @ 100%: ~ 4.55 mJ
Laser energy density	14 GW cm^{-2}
Spot diameter	100 μm
Pulse frequency	5–20 Hz, depending on sample thickness
Ablation cell size	52 mm diameter × 50 mm high
ICP-MS	
ICP RF power	1200.00
Mass spectrometer	Quadropole; peak resolution ~ 0.7 AMU
Analog stage voltage	–2800.00
Pulse stage voltage	1800.00
Carrier gas and flow rate	Ar: 1.23–1.27 L/min
Lens	Static, 4.9–5.2 V
Detector mode	Dual
Acquisition method	
Dwell time	10–100 ms/AMU, depending on element
Pre-ablation time	7–15 s
Total ablation time	20–50 s
Acquisition mode	Peak hopping
Acquisition timing, gold	1 or 2 sweep/reading (0.738 or 1.40 s) 1 reading/repetition Average peak processing
Acquisition timing, silver and copper alloys	1 sweep/reading (0.881 s) 3 readings/repetition Average peak processing

References

Bunker, E. (1994) 'The enigmatic role of silver in China', *Orientations* 25(11): 73–8.

Han, J. (1999) *Famen Temple.* Hong Kong: Xiangjiang Publishing Co. Ltd.

Hansen, V. (2003) 'The Hejia village hoard: a snapshot of China's silk route trade', *Orientations* 34(2): 14–19.

Harper, P.O. and Meyers, P. (1981) *Silver Vessels of the Sasanian Period.* New York: Metropolitan Museum of Art.

Jett, P. (nd) Unpublished analyses of Tang metalwork in the collections of the Freer Gallery of Art and Arthur M. Sackler Gallery, Smithsonian Institution, Washington, DC.

Liang-Lee, Y. and Louis, F. (1996) *An Index of Gold and Silver Artifacts Unearthed in the People's Republic of China.* Zürich: Museum Rietberg Zürich.

Meyers, P. (2003) 'Production of silver in antiquity: ore types identified based upon elemental compositions of ancient silver artifacts', in L. van Zelst (ed.), *Patterns and Processes: A Festschrift in honor of Dr. Edward V. Sayre*, 271–88. Suitland, MD: Smithsonian Center for Materials Research and Education.

Ogden, J. (2003) 'Connections between Islam, Europe, and the Far East in the medieval period: the evidence of jewelry technology', in P. Jett with J.C. Douglas, B. McCarthy and J. Winter (eds), *Scientific Research in the Field of Asian Art: Proceedings of the First Forbes Symposium at the Freer Gallery of Art.* London: Archetype Publications Ltd.

Qi, Dongfang (1999) *Tang dai jin yin qi yan jiu* (Research on Tang Gold and Silver). Beijing: Zhongguo she hui ke xue chu ban she.

Rawson, J. (1982) *The Ornament on Chinese Silver of the Tang Dynasty (AD 618–906).* British Museum Occasional Paper No. 40. London: British Museum.

Scott, D.A. (1991) *Metallography and Microstructure of Ancient and Historic Metals.* Marina del Rey, CA: Getty Conservation Institute.

Yang, X. (ed.) (1999) *The Golden Age of Chinese Archaeology.* Washington, DC: National Gallery of Art.

Scientific analysis of lead-silver smelting slag from two sites in China

Pengfei Xie and Thilo Rehren

ABSTRACT This paper presents data on two sets of lead-silver slag samples from northern and south-west China. At present, very few slag studies on ancient and historic slags from China are available in English, and these two sites in particular are completely unstudied. The sites and slags were selected on the assumption that they represent lead/silver production. This paper shows the technological range of slags related to the production of these two metals, and what technological information can be gained from their study. It is hoped that this paper will contribute to knowledge about ancient and historic smelting practices, by offering new and original data, and by comparing the findings from these sites to other published sites.

Introduction

Lead and silver are typically seen as very different metals. Lead is considered cheap and of little functional use apart from in architecture and as a metallurgical additive in certain bronzes. Silver in contrast is one of the precious metals, desired by mankind for its beauty, value and related functions such as jewellery and coinage. Despite this seemingly disparate nature, their ores often coexist in nature, and the smelting procedures of the two metals are closely related. Due to these natural affinities, they usually also appeared together in ancient Chinese texts, and even today their production can only be fully understood when studied together.

Silver does occur in nature as a pure or 'native' metal, and several rich silver minerals have been exploited in the past, mostly sulphides of silver together with variable quantities of other metals, such as copper, antimony or bismuth. However, the bulk of prehistoric and historic silver production is based on the smelting of silver-containing galena, lead sulphide. The silver concentrations in galena rarely exceed 0.5% by weight, and even at concentrations as low as 0.1 wt%, the value of the silver content is often higher than that of the lead metal. Smelting such mixed ore will not separate the two metals; instead, a lead-silver alloy, or bullion, will be produced. A second step is necessary to separate the two metals. During this cupellation the less noble metal (lead) is oxidised during re-melting in an open hearth, while the noble metal (silver) remains in its metallic state. Thus, at the end of the cupellation process all lead is present as lead oxide, or litharge, which can be mechanically separated from the silver metal. The litharge can then easily be re-smelted to lead metal, forming soft lead with less than *c.*100 ppm silver. Large quantities of lead metal are therefore produced (often more than 100 times the weight of the silver). Archaeological evidence of such smelting sites could include the slag from the first smelting, possibly litharge remains, and furnace fragments. The scientific study of these remains can offer much information about the technological processes, while evidence of cupellation also reveals the economic relationship between the two metals.

The slag samples studied were collected by Professor Ko and Professor Li Yanxiang of the University of Science and Technology Beijing from two Chinese sites: Tang county and Shizhu county. The Tang county of Hebei province is located in northern China, 190 km south-west of Beijing. The Shizhu county of Chongqing municipality is situated in south-west China, with the Yangtze River winding across its western border. Historically, the two sites had abundant deposits of mineral ores of lead, silver, iron and gold. They both boast a long history of mining and smelting.

Until now, no formal archaeological excavation has been carried out at these sites, and no scientific analytical study has been made of the slag and ceramic specimens from the two areas. Our research endeavoured to undertake scientific analysis of an initial selection of specimens from the two sites using optical microscopy, scanning electron microscopy with energy-dispersive spectroscopy (SEM–EDS) and X-ray fluorescence (XRF). This paper presents the results of the slag analyses.

Methodology

In the sampling process, the specimens were separated into two groups: group T1 for site no. 1 (Tang county) and group S2 for site no. 2 (Shizhu county), and then labelled

T1A, T1B, T1C, T1D, T1E, T1F and S2A, S2B, S2C, S2D, S2E respectively. After photographic documentation, polished blocks were produced for further analysis. Sample preparation for optical microscope and SEM–EDS analyses included cutting, mounting, grinding, polishing using standard metallographic procedures, and carbon coating to prevent a charge build up from the electron beam during SEM–EDS analysis. In contrast, sample preparation for XRF analysis involved cutting, crushing, milling to produce a homogenous powder, which was then mixed with wax and pressed into pellets. The equipment used was a Leica DMLM reflected light metallographic microscope equipped with a digital camera, a Hitachi S-570 SEM with an Oxford INCA EDS system and a Spectro XLab 2000 X-ray fluorescence spectrometer. The techniques were complementary, serving different purposes. SEM–EDS gives far better results for major elements in a single phase than XRF due to the good spatial resolution, while XRF is more useful for detecting minor and trace elements in the bulk material (<0.1 wt%).

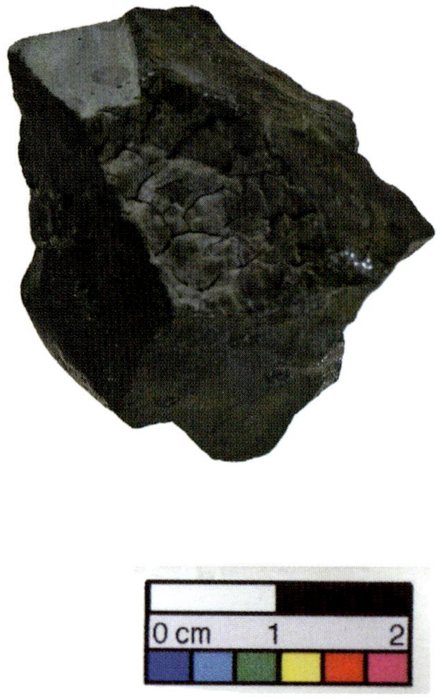

Figure 1 Slag sample T1A: typical flow structure surface (centre).

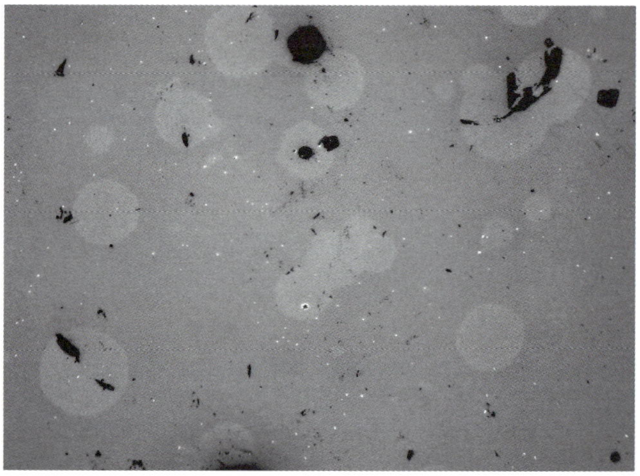

Figure 2 Slag sample T1A: devitrification spheres (light grey) in a glassy slag matrix (width of image *c*.1mm).

Results and initial interpretation

The 11 analysed specimens and the results with some initial interpretations are presented below. To avoid repetition in the presentation of analytical results, the data are presented in groups. Only representative samples are considered below. An overview of historical sources relating to silver smelting in China is given in Xie (2005). Full details of the analyses and results of our analyses are presented in Xie (2006).

Tang county

The macroscopic appearance of sample T1A is dark black with the flow texture on the surface, suggesting it might be tapping slag (Fig. 1). This sample is representative of the majority of finds studied here, including T1D and T1F, although we cannot judge how representative it is for the entire assemblage on site. The metallographic image shows that sample T1A is dominated by a glassy matrix with numerous tiny bright metallic spots, and a few crystals, bubbles and pores. The lighter grey circular features in the slag glass are noteworthy. SEM–EDS analysis did not reveal any compositional difference between the surrounding glassy matrix and the circular features; many of them contain a small inclusion or prill in their centre. We interpret these circular features as devitrification spheres around some crystal nuclei, caused by a relatively slow cooling of the slag (Fig. 2). Similar features have been noted in glassy crucible slag in early Islamic Uzbekistan (Rehren *et al.* forthcoming).

The chemical composition of these slags as determined by SEM–EDS and XRF is given in Table 1 (main oxides) and Table 2 (trace elements). The main components of these slags are silica (55–65 wt%), around 15 wt% iron oxide, 5–10 wt% lime, and lower amounts of alumina (6–8 wt%) and magnesia (2–4 wt%). Sample T1D is higher in silica than the other samples; this is due to numerous inclusions of quartz particles in the slag. These inclusions are partly absorbed in the surrounding glass, and have numerous sulphide prills in their structure, which appear to be residual ore material. SEM–EDS analyses of the melt phase between the quartz inclusions indicate a silica concentration of *c*.58 wt%, in line with the bulk composition of the other slag samples in this group. Such a composition is in agreement with the predominantly glassy character of the slag as seen in the optical microscope, and relatively typical for a high-temperature, strongly reducing furnace regime. Due to the complex chemistry, it is difficult to estimate the actual process temperature; test firings would be necessary to determine the liquidus temperature.

Only a few inclusions were found large enough to be analysed. They included lead metal prills with up to 2% silver, nearly pure silver prills, and complex sulphides rich in lead and iron, often containing up to 1% silver.

Table 1

Sample	Na_2O	MgO	Al_2O_3	SiO_2	P_2O_5	SO_3	K_2O	CaO	TiO_2	MnO	FeO	ZnO	PbO
T1A	–	4.3	6.4	55.2	0.4	0.8	1.5	11.1	–	1.5	16.9	1.9	0.4
T1C	–	4.5	6.3	55.5	0.3	0.7	1.3	10.9	0.3	1.6	16.7	1.9	0.4
T1D	–	2.0	6.8	64.7	0.2	0.1	1.6	4.7	0.3	0.7	16.7	0.4	2.2
T1F	–	3.8	7.7	58.7	0.3	0.7	2.7	9.1	0.2	2.9	12.5	1.0	0.4
T1B	1.5	2.0	14.1	61.3	–	–	2.7	12.8	0.7	0.5	4.4	–	–
T1E	0.3	22.7	7.5	49.7	–	–	1.7	14.9	0.2	0.8	2.2	–	–

(P)ED-XRF data, quantified by the Turboquant method. Data normalised to 100 wt%.

Table 2

Sample	Pb	Cu	Sr	Zn	Ag	Ba	Sb
T1A	4200	200	230	–	15	1400	–
T1C	4000	190	220	–	15	1150	–
T1D	22000	360	200	3800	220	260	80
T1F	4000	310	385	–	10	1850	–
T1B	10	15	760	25	–	750	–
T1E	5	15	350	15	–	280	–

(P)ED-XRF data, quantified by the Turboquant method.

For our investigation it is particularly important to note the presence of between 0.4 and 2 wt% each of zinc oxide and lead oxide, indicating that these are non-ferrous slags. The very low level of copper (Table 2, 200–400 ppm) suggests that this is not copper slag, but indeed lead slag. Furthermore, the presence of relatively high silver concentrations in both metal and sulphide prills suggests that the produced metal was a silver-rich bullion, and that the economic interest in this site was primarily silver production. This is further confirmed by the presence of cupellation remains at this site (to be presented in a separate publication).

The macroscopic image of sample T1C differs from the previous group in that it shows two different layers: a ceramic layer at the top and a slag layer at the bottom. The slag layer is darker than the ceramic layer, while the ceramic layer is more porous than the slag (Fig. 3). The micrograph of sample T1C shows a number of small cracks in the glassy matrix that were breaking up during polishing. We interpret this to show that the glass has a considerable amount of residual tension in it from the cooling phase. In addition, this sample shows devitrification spheres similar to the previous sample. The bright inclusions in the slag are lead sulphide and lead and silver metal prills. They are all very low in copper or iron.

The ceramic attached to this slag sample has been analysed by SEM–EDS; the average of five different area measurements was 60.2 wt% silica, 28.9 wt% alumina, 3.7 wt% iron oxide, and around 2.5 wt% each of lime and potash, plus 1.6 wt% titania, and less than 1 wt% magnesia. This is a highly refractory ceramic based on a kaolinitic clay and rich in quartz inclusions. The shape of the ceramic is tubular, although not enough was preserved to determine the diameter. This is interpreted as a fragment of a tuyère filled with some slag which was either deliberately tapped through it, or which overflowed into the tuyère when the furnace overflowed. Similar slag-filled tuyères are well known from African iron smelting sites, and seem to be a regular part of the process rather than an indication of a malfunction.

Samples T1B and T1E have a very different macroscopic appearance, microstructure and composition from the remaining samples of this site (see Tables 1 and 2). They are more porous, almost frothy (Figs 4 and 5), and have many inclusions of a hard white alloy. According to SEM–EDS analysis, these inclusions are iron-based with more than 95 wt% iron (Figs 6 and 7); the balance is most likely

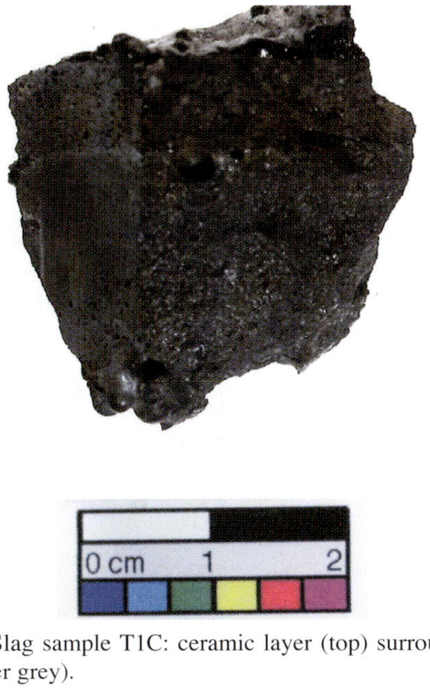

Figure 3 Slag sample T1C: ceramic layer (top) surrounding a slag core (darker grey).

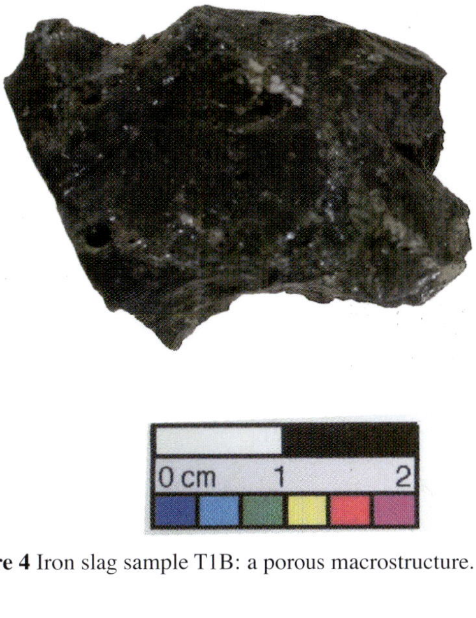

Figure 4 Iron slag sample T1B: a porous macrostructure.

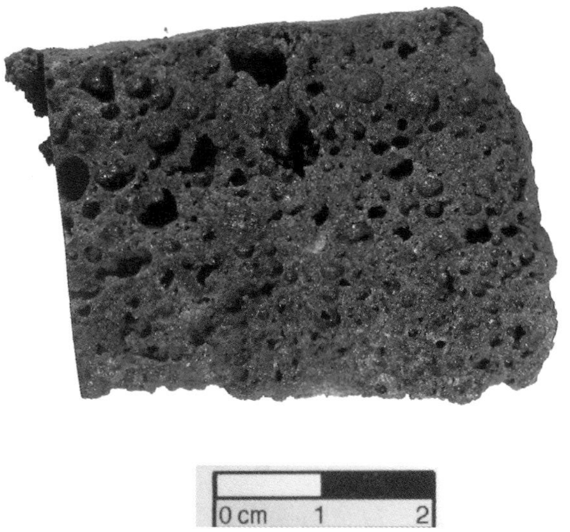

Figure 5 Iron slag sample T1E: an almost frothy macrostructure.

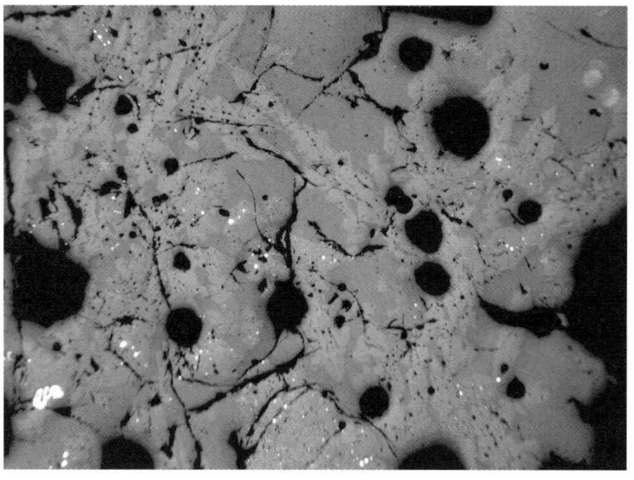

Figure 6 Micrograph of iron slag T1E: calcium silicate inclusions (grey) in a glassy matrix with numerous metallic prills (bright) and abundant porosity (black) (width of image *c*.2 mm).

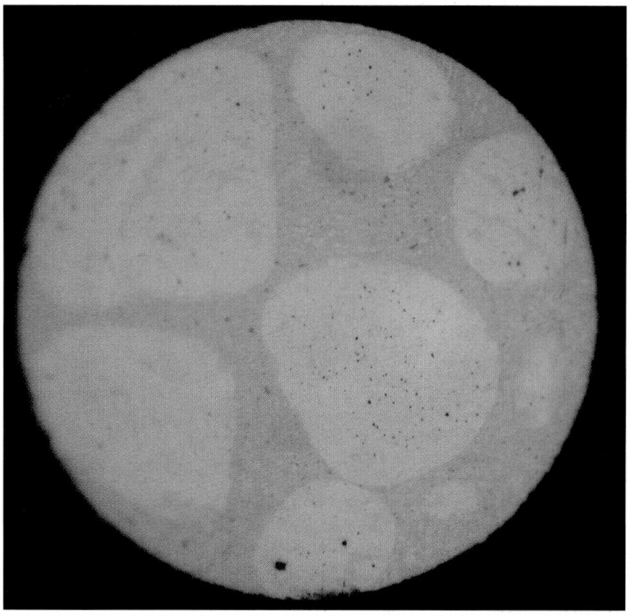

Figure 7 Iron slag sample T1E: close-up of a metal prill showing an iron-iron carbon alloy structure (width of image *c*.0.2 mm).

carbon. Chemical analysis shows that this slag has very low levels of base metals and also very low iron oxide concentrations, but high lime, magnesia and silica. This is clearly not a base metal slag; the chemical and structural composition and the numerous iron-rich prills are more characteristic of blast furnace iron slag. Intriguingly, the chemistry differs considerably between the two samples: T1E is very rich in magnesia, with more than 20 wt%, while T1B has only one-tenth of this, but a much higher alumina level. T1B is almost entirely glassy, while T1E is rich in crystals with about 45 wt% MgO, 42 wt% SiO2 and 11 wt% FeO, but very little lime and other oxides. This indicates an approximate formula of Mg_2SiO_4 with some substitution of iron oxide for magnesia, i.e. a forsterite-rich olivine.

Thus, among the few samples analysed is evidence for both lead-silver smelting and for iron smelting at this site. Both processes seem to be driven by a blast furnace technology, but are clearly based on different ores. It is impossible at present to say whether they are technically or chronologically related, or whether they are completely independent from each other and only appear together here by coincidence.

Shizhu county

The slags from site no. 2 are mostly tap slags with clear flow structures (Fig. 8). There are some green and brown corrosion features visible. No ceramic fragments are attached to these slags.

The micrograph of sample S2A shows clear evidence of tapping. Figure 9 illustrates a quenching zone near the surface where the slag quickly solidified as a glass, while in the centre, where the cooling rate would have been slower, it has formed crystals. The micrograph of sample S2B (Fig. 10) shows a magnetite skin separating two subsequent tapping layers. Such a magnetite skin forms only when slag flows into air at ambient temperature, i.e. outside the fur-

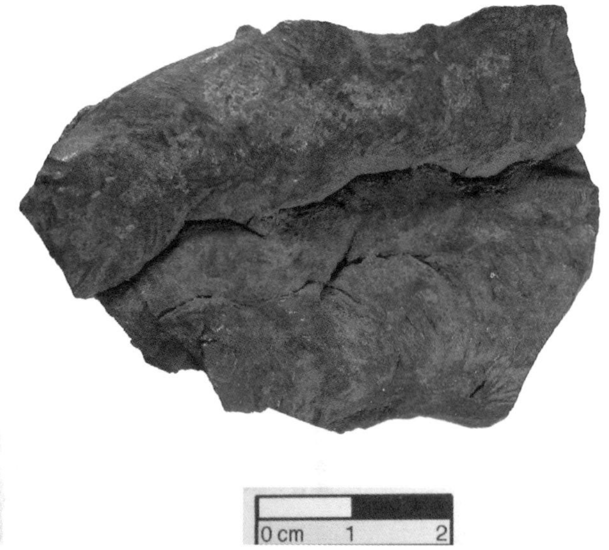

Figure 8 Slag sample S2B: clear tapping structure.

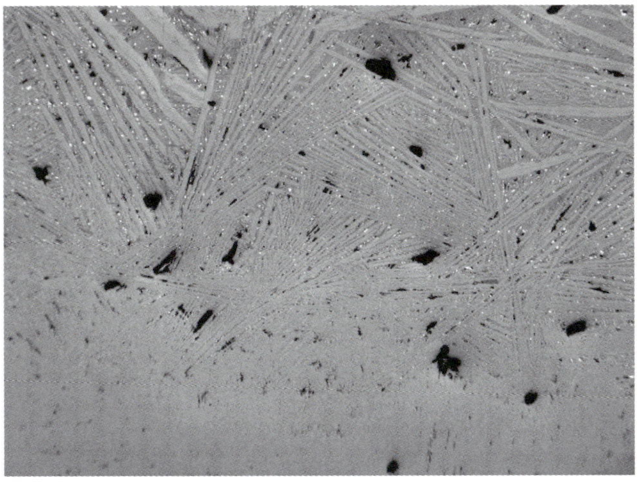

Figure 9 Micrograph of slag sample S2D: the glassy quenched outer region (bottom) and the more crystalline inner part (width of image *c.*1 mm).

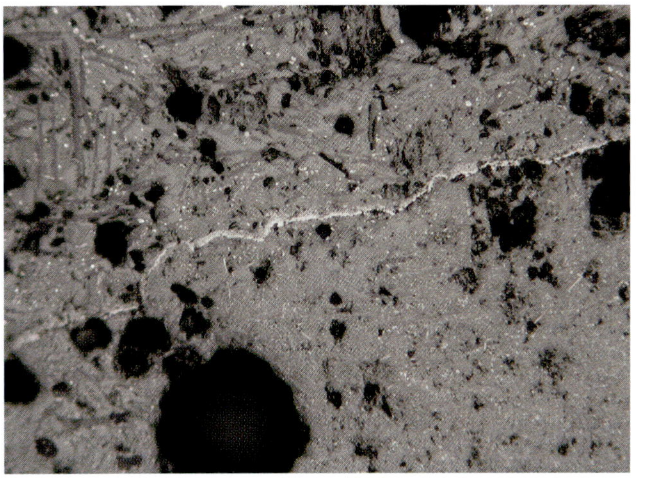

Figure 10 Micrograph of slag sample S2B: a magnetite skin (light grey, centre) separating two subsequent slag flows (width of image *c.*1 mm).

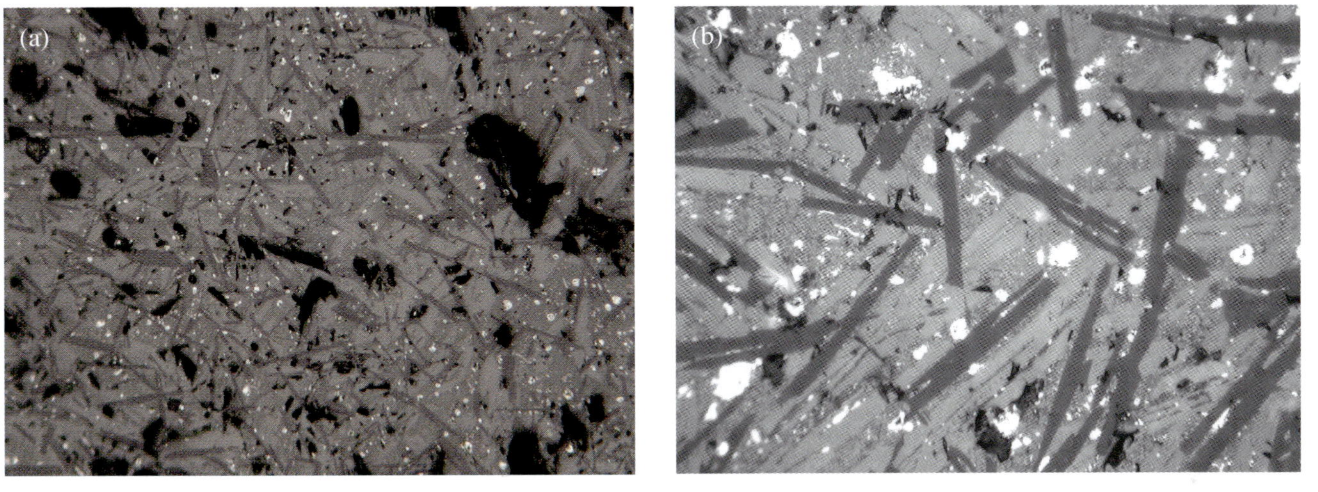

Figure 11 (a) Micrograph of slag sample S2B: typical structure rich in sulphide inclusions (bright), and various silicate crystals (different grey shades) in a glassy matrix. (b) Micrograph of slag sample S2A: several different silicate phases in a glassy matrix together with numerous sulphide inclusions (bright) (width of image *c.*0.2 mm).

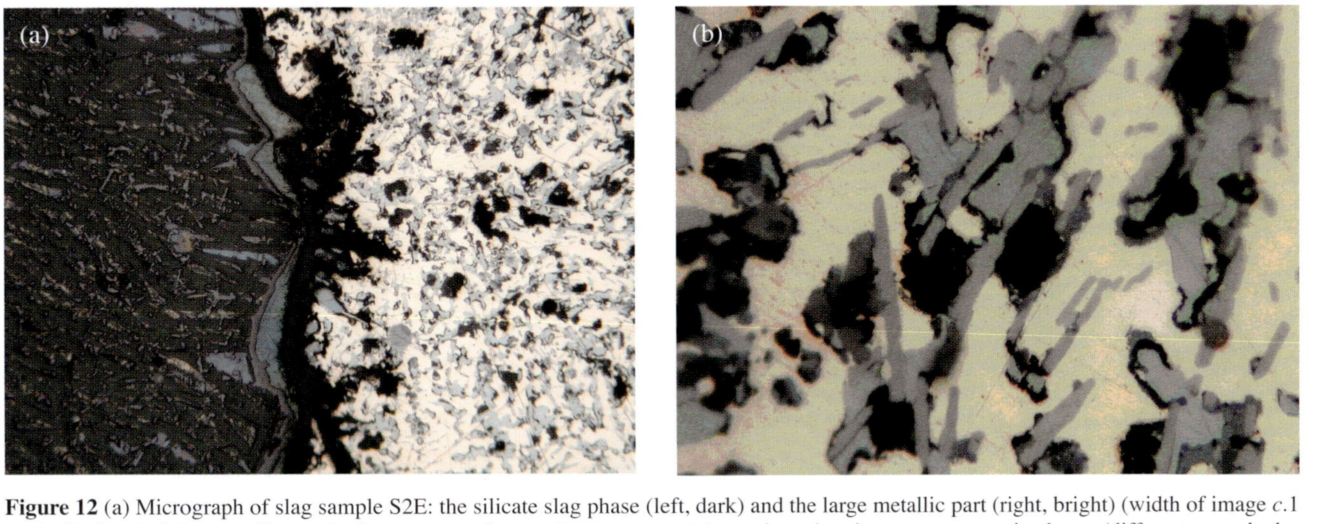

Figure 12 (a) Micrograph of slag sample S2E: the silicate slag phase (left, dark) and the large metallic part (right, bright) (width of image *c.*1 mm). (b) detail of the metallic part in the same sample showing a copper-rich matrix and various copper-arsenic phases (different grey shades and crystal shapes) (width of image *c.*0.1 mm).

Table 3

Sample	MgO	Al_2O_3	SiO_2	P_2O_5	SO_3	K_2O	CaO	TiO_2	FeO	CuO	ZnO	BaO	PbO
S2A	1.1	6.8	34.4	0.2	3.5	1.2	3.8	0.8	33.3	–	3.1	9.8	1.5
S2B	1.1	6.7	33.9	0.2	3.0	1.1	4.1	0.8	34.3	–	3.3	9.5	1.8
S2D	1.7	5.5	32.8	0.1	2.0	0.8	4.5	0.5	40.4	–	3.4	6.0	2.0
S2C	0.8	3.0	21.5	0.1	9.2	0.3	1.5	1.0	40.8	6.4	0.9	12.3	0.3
S2E	–	2.2	23.0	–	5.7	–	1.5	–	42.7	2.9	0.9	20.6	–

(P)ED-XRF data, quantified by the Turboquant method. Data normalised to 100 wt%.

nace. In contact with the oxygen of the air some of the iron in the slag near the surface is sufficiently oxidised to form magnetite crystals. This skin is then covered by the next flow of slag and is therefore preserved in the contact zone between the two flows. Often, these two flows exhibit different crystal sizes, due to different cooling speeds.

The samples have different amounts of sulphide and metal inclusions (Fig. 11). They also have various silicate crystals, mostly pyroxene and barium-rich feldspars, and are more porous than the previous group of lead slags. Their chemical composition differs significantly; they are richer in iron oxide (35–40 wt%) and have much less lime (less than 5 wt%). In addition, three have high levels of lead oxide (2 wt% PbO) and zinc oxide (3 wt% ZnO), while the fourth has only 0.3 wt% PbO and 0.9 wt% ZnO. High barium oxide levels are most certainly due to the presence of baryte ($BaSO_4$) as a gangue mineral, a typical component of many lead ores. All samples contained metal and sulphide prills; in S2A these were mostly iron arsenide (70–90 wt% Fe, 10–30 wt% As) and low amounts of copper and antimony, while the sulphides were predominantly lead sulphide. In sample S2B we only found lead sulphide inclusions, while in sample S2D a single gold prill was found (91 wt% Au, 8 wt% Fe and 1 wt% Zn), and numerous lead-iron sulphide prills. Overall, it is reasonable to assume that the three samples S2A, B and D are all lead smelting slags, although the presence of different metallic phases indicates that the ore was relatively complex, and that other metals could have been produced as well. There is less indication here for the presence of silver in the slags compared to the Tang county site; no fragments of litharge or other cupellation remains were found during fieldwork.

However, in view of the complex composition of the lead slags in this group it is not surprising to note that the other two samples from this site, S2C and E, are considerably richer in copper (3–6 wt%) and differ in other details, too. Their overall chemical characteristics are quite similar though, with a high barium content and very similar iron oxide and lime levels. Sample S2C contained numerous metal and sulphide prills: some prills had an average composition of 90 wt% Pb, 7 wt% Cu and a few percent arsenic; others were 70 wt% Cu and 30 wt% As; yet others were nearly pure iron metal with a low copper content. The sulphide inclusions were both lead- and copper-rich. Sample S2E had a huge metallic prill trapped in it; SEM–EDS analysis gave an average composition of about 70 wt% Cu, 15–20 wt% As, 5–8 wt% Fe, and minor quantities of lead, sulphur, antimony and zinc. Occasionally, individual small prills in the slag S2E were identified as being lead- or iron-dominated, but always with a high copper content. Overall, we interpret these samples as copper slag, possibly exploiting the same ore deposit but obtaining different minerals from it.

Thus, we see two different, but closely related metallurgical processes here: one smelting lead metal and the other arsenical copper. The high barium content of samples S2C and S2E matches the copper slag composition in late prehistoric Xinjing (Mei 2000: 52). They also show high copper and arsenic concentrations, and many inclusions of copper-arsenic alloy (Fig. 12). It is likely that these slags originate from the smelting of a complex ore, and that both lead and copper were produced. At this stage, it appears as if there were two physically separate operations, and not a joint lead-copper smelting process. However, more work and excavation would be necessary to clarify this technical aspect. Furthermore, it is uncertain whether the smelting also included the production of silver or whether this was predominantly a base metal metallurgy.

Table 4

Sample	Cu	Sb	Sr	As
S2A	1100	75	1300	3200
S2B	950	30	2000	1100
S2D	830	25	1100	1350
S2C	70000	–	1150	3700

(P)ED-XRF data, quantified by the Turboquant method.

From the above analyses, we summarise that sample group S2 consists of three lead (silver?) slag samples, and two arsenical copper (lead?) slag samples. The three lead slag samples (S2A, S2B and S2D) are mostly glassy and have only small metal or sulphide inclusions, while the two copper slags (S2C and S2E) appear less glassy and have many more metal and sulphide inclusions. So far, no firm evidence for silver production has been found among the slags from this site.

Conclusions

Through scientific analysis of 11 slag samples from two sites, we find that seven are lead and possibly lead/silver smelting slags, two are iron slags, and the remaining two are copper slags from a mixed (arsenic) copper-lead ore.

If we compare the composition of the lead-silver slags from the two sites (see Tables 1 and 3), we find that in considering the major elements, group S2 slags have higher concentrations of FeO, BaO, ZnO, SO_3, TiO_2 and As_2O_3 than group T1, while group T1 slags have higher levels of SiO_2, MgO, CaO and MnO. Among the trace elements, group S2 samples have much higher concentrations of copper, arsenic and strontium, while group T1 samples show higher levels of silver. These differences were clearly caused by different ore deposits and specifically adopted smelting processes, as would be expected from their different geological origin.

From the Tang county site there is evidence for the use of tuyères to blow air into the furnaces. They are made from a high-quality refractory clay and would have lasted a long time even at high temperatures. It is assumed that they filled with slag when the blowing stopped at the end of the process and the slag level had increased in the furnace. More fieldwork is necessary to investigate this aspect further, by looking at fragments of tuyères without slag filling and possible other furnace fragments that would provide important information about the furnace type. Further archaeological work is also necessary to date the metallurgical activities at both sites.

Lead smelting in China began very early and the metal was mass-produced in the Chinese Bronze Age to meet the enormous demand for lead-containing bronze objects (Sun and Li 2003). Archaeological excavations have demonstrated that as early as the Shang dynasty (16th–11th century BC), the Chinese people were able to make almost pure lead metal (Li 1984). However, it is noteworthy that all the slags studied here are from a type typical of advanced blast furnace technology, as is known from ancient text and illustrations (see also Mei and Rehren 2005).

Early lead slags are known from numerous sites in ancient Europe; from the Bronze Age through the Roman period and the Middle Ages into the early modern period. This paper cannot give a comprehensive overview of their chemistry and mineralogy but suffice it to say that lead slags are far more variable in their appearance and composition than typical copper or iron slags, reflecting the wider chemical variability of lead ores and the associated gangue minerals. The closest parallels in terms of bulk composition, glassy appearance and strongly reducing conditions are from various medieval to early modern sites in the Schwarzwald, south-west Germany (Goldenberg 1994), thought to be based on a water-powered blast furnace technology.

Of particular interest is the evidence for multiple metallurgical operations at both sites. In Tang county, the lead-silver slag is mixed with typical iron blast furnace slag; the macroscopic and microscopic appearance of both slags is so similar that it can be assumed that very similar furnaces were employed in their production, even though their chemical composition is significantly different, as befits the smelting of two such different metals. In the Shizhu county site, two related metals were produced: lead and copper. We can speculate that they came from the same ore deposit, but may have been smelted separately. Clearly, these sites have a huge potential for further archaeological and archaeometallurgical work.

Acknowledgements

We would like to express our gratitude to the late Professor P. Ucko of the UCL Institute of Archaeology, who as Executive Director of the International Centre for Chinese Heritage and Archaeology facilitated Pengfei Xie's study at UCL; the Institute of Historical Metallurgy and Materials of USTB, and in particular Professor Ko and Professor Li Yanxiang, for permission to study the material; and Dr Marcos Martinón-Torres, Simon Groom and Kevin Reeves of the UCL Institute of Archaeology for their generous help in various aspects of this research.

References

Goldenberg, G. (1994) *Archäometallurgische Untersuchungen zur Entwicklung des Metallhüttenwesens im Südschwarzwald* (doctoral thesis, University of Freiburg, Germany).

Li, Minsheng (1984) 'Xianqin yongqian de lishi gaikuang', *Wenwu* 10: 85–8 [in Chinese].

Mei, Jianjun (2000) *Copper and Bronze Metallurgy in Late Prehistoric Xinjiang: Its Cultural Context and Relationship with Neighboring Regions*. BAR International Series 865. Oxford: Archaeopress.

Mei, Jianjun and Rehren, Th. (2005) 'Copper smelting from Xinjiang, northwest China. Part I: Kangcun village, Kuche county, c.18th century AD', *Historical Metallurgy* 39: 96–105.

Rehren, Th., Groom, S. and Anarbaev, A. (forthcoming) 'Steel making at Akhsiket: an analysis of Uzbek crucible steel slags'.

Sun, Shuyun and Li, Yanxiang (2003) *Zhongguo Gudai Yejinjishu Zhuanlun*. Beijing: Chinese Science and Culture Press [in Chinese].

Xie, Pengfei (2005) *Documents of Ancient Chinese Silver Smelting Technology* (unpublished master's thesis, Institute of Historical Metallurgy and Materials, University of Science and Technology Beijing).

Xie, Pengfei (2006) *Scientific Analysis of Slag Samples from Two Sites in China* (unpublished MSc dissertation UCL Institute of Archaeology, London).